VIRAL INFECTIONS AND CELL BIOLOGY

VIRAL INFECTIONS AND CELL BIOLOGY

VOLUME 1

Simin Liu

Kruger Brentt
Publishers
2025

Kruger Brentt Publishers UK. LTD.
Company Number 9728962

Regd. Office: 68 St Margarets Road, Edgware, Middlesex HA8 9UU

ISBN: 978-1-78715-330-1 (Vol. 1)

For information on all our publications visit our website at http://krugerbrentt.com/

PREFACE

"Viral Infections and Cell Biology" offers an in-depth exploration into the intricate interactions between viruses and host cells, providing insights into the molecular mechanisms underlying viral infections and their impact on cellular processes. This book serves as an essential resource for researchers, clinicians, students, and educators interested in understanding the complex interplay between viruses and the host cellular machinery.

Viral infections pose significant threats to human health, agriculture, and ecosystems, causing a wide range of diseases, including respiratory infections, gastroenteritis, hepatitis, and cancer. Understanding the molecular mechanisms of viral infection is crucial for developing effective antiviral therapies, vaccines, and diagnostic tools.

As editors of this volume, we have endeavored to compile a diverse array of perspectives, methodologies, and case studies from leading experts and practitioners in the field of viral infections and cell biology. Through a blend of theoretical foundations, experimental approaches, and clinical insights, this book covers a wide range of topics, including viral entry and replication mechanisms, host-virus interactions, immune responses to viral infections, and the development of antiviral strategies.

Our overarching goal in presenting this book is to provide readers with a comprehensive understanding of the molecular and cellular basis of viral infections. By offering insights into the latest advancements, best practices, and emerging trends in viral pathogenesis and host response, we aim to facilitate the development of innovative approaches for the prevention, diagnosis, and treatment of viral diseases.

We extend our sincere gratitude to all the contributors who have generously shared their expertise, experiences, and insights in the field of viral infections and cell biology. It is our hope that this book will serve as a valuable reference and guide for researchers, clinicians, students, and educators as they work towards combating viral diseases and improving global health outcomes.

Simin Liu

CONTENTS

01 CELLULAR MECHANISMS INVOLVED IN VIRAL INFECTIONS

1. INTRODUCTION: AN OVERVIEW

Viruses consist of nucleic acids and proteins, and occasionally, lipids and glycans. Isolates of viral particles exhibit significant variations in their structural, molecular, and genetic organization. Viruses vary in size, ranging from a few tens of nanometers to micrometers. Additionally, viral genomes can contain anything from a few genes to several hundreds. Viruses, being obligate intracellular parasites, utilize host cells to duplicate, increase in number, and then propagate from one cell to another and from one host to another. Up to this point, a multitude of viruses have been sequenced, but only a small number are recognized as causing lethal diseases. Nevertheless, viral infections exert a tremendous influence on public health, agricultural productivity, and the economy. International trade, urbanization, deforestation, climate change, and human activity are all contributing causes to the emergence, re-emergence, and global spread of viruses. Notable examples include the Zika pandemic and the ongoing global issue caused by the infamous coronavirus SARS-CoV^{-2}. Gaining comprehension of the infection process at the molecular and cellular level is an essential requirement for the advancement of diagnostic tools, preventive methods, and therapeutic tactics against viruses.

The scope of research on the cellular biology of viral infections is now confined to a very small group of viruses. This limitation presents a significant obstacle in our ability to effectively prepare for future outbreaks of new viral infectious illnesses. Numerous advancements in the subject have taken place in recent decades, providing insight into the cellular life cycles of significant pathogens including human immunodeficiency (HIV) and hepatitis viruses. Conversely, some of these inquiries have resulted in significant progress in the field of molecular and cellular biology. Some notable discoveries include the identification of protein G in the vesicular stomatitis virus, which helped to define the exocytotic machinery, the use of various viruses as functional cargo to study endocytic pathways, the study of viral fusion proteins and their role in developmental biology, and the application of retroviral reverse transcriptase in different biotechnology applications. The list is not restricted solely to cell biology and has the potential to encompass a wide range of topics such as immunology, neurology, and molecular medicine. Viruses serve as intriguing molecular and cellular connections among various scientific areas. The primary objective of this special issue is to showcase unique findings and discussions on many facets of viral infection cell biology and their relevance to other areas of biology.Infection starts when viruses attach to the host cell surface, which invariably implies binding to one or more cellular receptors, including proteins, carbohydrates, and lipids. The expression of these molecules is often tissue-specific. Although other factors and processes contribute to the infection program, the identification of receptors and coreceptors helps to define the virus tropism, and sometimes, to explain the virus-induced pathogenesis. Wielgat and colleagues discuss a possible link among coronaviruses, sialic acids as attachment factors, immune escape, and disease severity [1]. Ideally, preventing infection requires approaches targeting early virus–host cell interactions, before the release of the viral genome into the cytosol. In this sense, Luteijn et al. identified a peptide with broad-spectrum antiviral activity that targets phosphatidylserine in the viral particle envelope and prevents infection by poxviruses, HIV, hepatitis B virus, and Rift Valley fever virus [2]. This finding is consistent with the fact that viruses often use phosphatidylserine and apoptotic mimicry as a mechanism of entry into host cells [3].

After attachment to the cell surface and prior to viral replication, viruses must gain access to the cytosol, and most viruses are sorted into the endocytic machinery [4]. To release their genomic material into the cytosol, enveloped viruses fuse their membrane with that of endosomal vesicles, and nonenveloped viruses usually inflict damage in the endosomal membrane. Daussy and colleagues reviewed the different strategies by which nonenveloped viruses enter the cytosol and provided the latest knowledge on how these viruses control and prevent cellular stress responses following membrane injuries such as inflammation and autophagy [5]. After delivery into the cytosol, the virus journey is not necessarily over, as they very often must reach a specific subcellular location to initiate viral replication, a point illustrated by Grikscheit et al. [6]. By combining live cell microscopy and computer-based image analysis, the researchers provided evidence that the Ebola virus (EBOV) finely regulates actin polymerization to ensure the directed long-distance transport of its nucleocapsid.

After penetration into the cytosol, viral replication usually takes place in viral factories wherein nucleic acids and specific viral and cellular proteins accumulate. Two studies by the Hoenen laboratory reported that EBOV recruits carbamoyl-phosphate synthetase 2, aspartate transcarbamylase, dihydroorotase (CAD) and the nuclear RNA export factor NXF1 within inclusion bodies in infected cells [7,8]. CAD facilitates the replication and transcription of the EBOV genome, while NXF1 contributes to the expression of EBOV proteins. Van Huizen and McInerney reviewed how Old World alphaviruses manipulate the phosphatidylinositol-3-kinase (PI3K)-AKT signaling pathway and discuss the benefit for viral replication [9]. Of note, viral factories can be either devoid of any lipidic layer, such as Negri bodies, or delimited by cellular membranes [10]. In the case of hantaviruses, Davies et al. established that the Tula virus induces a dramatic reorganization of the Golgi to replicate [11]. Once viral replication begins, cells are infected.

The fight between host cell defenses and viruses is permanent at each stage of the viral life cycle. The work by the Grandvaux team points out the existence of a noncanonical pathway leading to an antiviral, immunoregulatory response [12]. The data support the view that the two factors STAT2 and IRF9 regulate distinct pools of genes with functions related to antiviral and immunoregulatory responses when the levels of both interferon β and tumor necrosis factor are elevated. Another example is that bovine herpesvirus 1 infection results in the proteasomal degradation of the transporter associated with antigen processing (TAP), a key player in major histocompatibility complex class I (MHC-I)-restricted antigen presentation. Wachalska et al. found that the process depends on the host cell factor p97, which is involved in the complex ER-associated degradation pathway [13]. Saulle and colleagues observed that ERAP2/Iso3 expression is induced by infection with different viruses, including SARS-CoV-2 [14]. ERAP2/Iso3 is an alternatively spliced isoform of rs2248374-G ERAP2 that, unlike the full transcript, is incapable of trimming peptides to be loaded onto MHC-I molecules. These results suggest that viruses trigger ERAP2/Iso3 expression to escape the adaptive immune response, which could contribute to the severity of virus-induced disease. In summary, cells have acquired many molecular weapons to control viral infections, but viruses have developed equally as many strategies to evade cell defenses.

To infect and replicate in host cells, viruses display not only the capacity to escape immune responses, but also to control cellular stress responses such as degradative pathways, autophagy, and cell death. Vector-borne viruses typically persist in their vector, but not in other hosts [15,16]. In this regard, much more work is needed to improve our understanding of the vector-borne virus life cycle and transmission to humans and other vertebrates.

Viruses often assemble within viral factories, but they sometimes depend on the endoplasmic reticulum (ER), Golgi, and subsequent exocytic pathways for the maturation and folding of their own surface and fusion proteins. This is the case, for instance, for viruses that bud at the plasma membrane such as influenza virus, HIV, and coronaviruses. The henipavirus fusion protein also requires endocytic uptake for cleavage by endosomal proteases, followed by recycling to the cell surface for incorporation into budding viral particles. In this regard, Fischer and colleagues highlight the interdependence of the exocytic and endocytic machinery to produce active viral fusion proteins and infectious Cedar particles, a middle pathogenic henipavirus [17].

The pathways by which viruses leave host cells are complex and difficult to investigate, but it is obvious that they are just as diverse as viral entry pathways, as various exocytic host factors and processes have been observed. Regarding this issue, the laboratory of Urtzi Garaigorta identified Erlin-1, a cholesterol-binding protein in the ER, as a proviral host cell factor participating in the production of infectious hepatitis C virus particles [18]. Chen and colleagues exemplified the central role of ion flow in the trafficking of intracellular vesicles, addressing the importance of calcium channels and pumps in virus penetration, assembly, and exit [19]. The review by Oechslin et al. nicely illustrates all these molecular and cellular processes in the context of hepatitis E virus (HEV), presenting the most recent information on the HEV life cycle, from virus entry and replication to assembly and release of new viral progenies [20].

Viral particles are engaged to transfer the viral genome from an infected cell to a noninfected cell. However, an increasing number of studies on cell-associated viral transmission have indicated the existence of alternative routes for the spread of viral material. Sid Ahmed et al. analyzed how the cellular environment may impact HIV spread in vivo [21]. This review developed from a recent paper by the same group showing that a 3D collagen environment restricts cell-free HIV infection, but promotes cell-associated viral propagation [22]. Other viruses interfere with cell and tissue differentiation as they spread. In this respect, Bilz and colleagues established that rubella virus hampers the formation of embryoid bodies and other 3D cell aggregates derived from human-induced pluripotent stem cells [23]. Whether this blockage is due to viral replication or required for viral spread remains to be clarified. This study suggests that the clinical signs of congenital rubella syndrome are due to an impairment of human development.

In this special issue, I intend to illustrate the possible viral life cycle stages, addressing virology issues from the perspective of cell biology, and from other angles. I strongly encouraged researchers to engage in discussions on the latest knowledge, technical issues, and technological challenges as well as prospects in the fields of viral infections, cell biology, and beyond. Based on 6 reviews, 1 opinion, 12 original research articles, and 1 communication reporting on a wide variety of viruses, this special issue can be considered a success. One of these reviews also demonstrates the interest in using viruses to develop new therapeutic approaches against noninfectious diseases. Nguyen et al. examined the potential of oncolytic viruses genetically modified to encode antitumor interleukine-12 in cancer therapy, with a special focus on herpes simplex virus [24]. Two research articles documenting new methods to study and analyze viral infections include the one by Tekes and colleagues reporting a protocol for culturing organoids derived from the ileum and colon and showing their usefulness in studying feline coronavirus infection [25], and the one by Potratz et al. describing a novel quantitative approach, based on computer analysis of microscopic images to analyze the cell tropism of neurotropic rabies virus [26].

However, I wish to clarify that this special issue does not include a comprehensive overview of the cell biology of viral infections. Certain talks may be absent due to a scarcity of specialists in certain instances, but these exclusions are not intentional. Topics of significance include virus receptors, endocytic pathways, and subsequent processing of genetic material. The intriguing aspects of exosomes in the spread of viruses and the involvement of protein aggregation in virus-induced illnesses are captivating [27,28]. I would have preferred to include a stronger focus on technology, for as by incorporating literature on nanobodies, electron microscopy techniques, and OMICS. I anticipate that this particular edition will offer guidance for future research directions in the field of viral infection cell biology and encourage investigation in areas that have received less attention and are still in their early stages of development. This special issue would not have been possible without the invaluable contributions of authors and the invaluable assistance of Dolly Wang and other editorial assistants from Cells. Hence, I would like to express my gratitude to all these exceptional individuals.

Lastly, amidst the ongoing pandemic, it is challenging to edit a special issue on the cell biology of viral infections without acknowledging the current viral zoonotic pandemic and the novel coronavirus. The World Health Organization (WHO) has compiled a list of pathogens in a blueprint, highlighting the need for immediate development of diagnostics, medicines, and research for these infections. The list comprises coronaviruses and several additional zoonotic viruses, among others. The consecutive outbreaks and global spread of SARS-CoV, Chikungunya virus, Middle East respiratory syndrome coronavirus, EBOV, and Zika virus in the last two decades serve as clear indications that should have raised our awareness. However, the ongoing interest in emerging viruses is limited to times of epidemics and pandemics, which has not resulted in consistent funding and consequently, the promotion of long-term research on developing zoonotic viruses. Multiple factors, too numerous to elaborate on here, are accountable for this circumstance. It is reasonable to consider the potential time saved in combating SARS-CoV version 2.0 if a portion of the financial losses incurred during the worldwide lockdown had been allocated towards research on emerging viruses, particularly coronaviruses. The viruses included on the WHO blueprint are a minuscule fraction of the virology research cited by PubMed.gov, amounting to less than 5% of the total (Figure 1). While not fully comprehensive, this approach offers a valuable insight into the combined endeavor that has been undertaken thus far, highlighting the significant research that still needs to be conducted in the ongoing fight against these viruses and potential future pandemics.

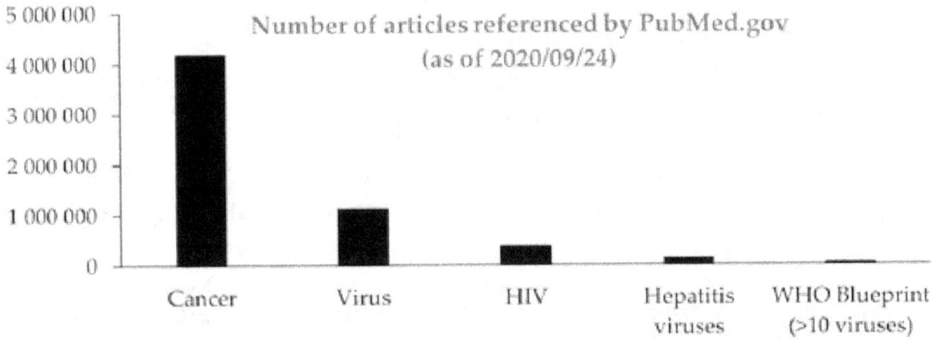

Figure 1. Number of articles on emerging viruses referenced by PuMed. gov. The keywords cancer, virus, HIV, and hepatitis viruses (A to E) were inputted into the search engine. Similarly, the WHO blueprint list represents the sum of results obtained for the next 10 keywords: Crimean–Congo hemorrhagic fever, Ebola, Marburg, Lassa fever, Middle East and severe acute respiratory syndrome coronavirus, Nipah, henipa, Rift Valley fever, and Zika.

2. CONCISE SUMMARY

Viruses demonstrate a refined simplicity, being both fundamental and deeply unsettling. While only a small number of diseases pose a direct threat to life, they have significant consequences for both human health and the economy, as demonstrated by the current coronavirus pandemic. Viruses are little pathogens that can infect a wide range of organisms, including animals, plants, prokaryotes, and archaebacteria. These organisms are obligate intracellular parasites, meaning they rely on living within host cells and manipulate many molecular and cellular processes to reproduce, increase in number, and then spread to other cells. This special issue focuses on the cellular mechanisms involved in viral infections. It includes a compilation of original research articles, communications, opinions, and reviews that explore different aspects of the interactions between viruses and host cells. Collectively, these publications offer insights into the most recent studies on the cellular mechanisms of viral infections, as well as innovative advancements in technology.

REFERENCES

1. Wielgat, P.; Rogowski, K.; Godlewska, K.; Car, H. Coronaviruses: Is Sialic Acid a Gate to the Eye of Cytokine Storm? From the Entry to the Effects. Cells 2020, 9, 1963.
2. Luteijn, R.D.; Praest, P.; Thiele, F.; Sadasivam, S.M.; Singethan, K.; Drijfhout, J.W.; Bach, C.; de Boer, S.M.; Lebbink, R.J.; Tao, S.; et al. A Broad-Spectrum Antiviral Peptide Blocks Infection of Viruses by Binding to Phosphatidylserine in the Viral Envelope. Cells 2020, 9, 1989.
3. Amara, A.; Mercer, J. Viral apoptotic mimicry. Nat. Rev. Microbiol. 2015, 13, 461–469.
4. Boulant, S.; Stanifer, M.; Lozach, P.Y. Dynamics of virus-receptor interactions in virus binding, signaling, and endocytosis. Viruses 2015, 7, 2794–2815.

5. Daussy, C.F.; Wodrich, H. "Repair Me if You Can": Membrane Damage, Response, and Control from the Viral Perspective. Cells 2020, 9, 2042.

6. Grikscheit, K.; Dolnik, O.; Takamatsu, Y.; Pereira, A.R.; Becker, S. Ebola Virus Nucleocapsid-Like Structures Utilize Arp2/3 Signaling for Intracellular Long-Distance Transport. Cells 2020, 9, 1728.

7. Brandt, J.; Wendt, L.; Bodmer, B.S.; Mettenleiter, T.C.; Hoenen, T. The Cellular Protein CAD is Recruited into Ebola Virus Inclusion Bodies by the Nucleoprotein NP to Facilitate Genome Replication and Transcription. Cells 2020, 9, 1126.

8. Wendt, L.; Brandt, J.; Bodmer, B.S.; Reiche, S.; Schmidt, M.L.; Traeger, S.; Hoenen, T. The Ebola Virus Nucleoprotein Recruits the Nuclear RNA Export Factor NXF1 into Inclusion Bodies to Facilitate Viral Protein Expression. Cells 2020, 9, 187.

9. Van Huizen, E.; McInerney, G.M. Activation of the PI3K-AKT Pathway by Old World Alphaviruses. Cells 2020, 9, 970.

10. Nevers, Q.; Albertini, A.A.; Lagaudriere-Gesbert, C.; Gaudin, Y. Negri bodies and other virus membrane-less replication compartments. Biochim. Biophys. Acta Mol. Cell Res. 2020, 1867, 118831.

11. Davies, K.A.; Chadwick, B.; Hewson, R.; Fontana, J.; Mankouri, J.; Barr, J.N. The RNA Replication Site of Tula Orthohantavirus Resides within a Remodelled Golgi Network. Cells 2020, 9, 1569.

12. Mariani, M.K.; Dasmeh, P.; Fortin, A.; Caron, E.; Kalamujic, M.; Harrison, A.N.; Hotea, D.I.; Kasumba, D.M.; Cervantes-Ortiz, S.L.; Mukawera, E.; et al. The Combination of IFN β and TNF Induces an Antiviral and Immunoregulatory Program via Non-Canonical Pathways Involving STAT2 and IRF9. Cells 2019, 8, 919.

13. Wąchalska, M.; Graul, M.; Praest, P.; Luteijn, R.D.; Babnis, A.W.; Wiertz, E.J.H.J.; Bieńkowska-Szewczyk, K.; Lipińska, A.D. Fluorescent TAP as a Platform for Virus-Induced Degradation of the Antigenic Peptide Transporter. Cells 2019, 8, 1590.

14. Saulle, I.; Vanetti, C.; Goglia, S.; Vicentini, C.; Tombetti, E.; Garziano, M.; Clerici, M.; Biasin, M. A New ERAP2/Iso3 Isoform Expression Is Triggered by Different Microbial Stimuli in Human Cells. Could It Play a Role in the Modulation of SARS-CoV-2 Infection? Cells 2020, 9, 1951.

15. Léger, P.; Lozach, P.-Y. Bunyaviruses: From transmission by arthropods to virus entry into the mammalian host first-target cells. Future Virol. 2015, 10, 859–881.

16. Mazelier, M.; Rouxel, R.N.; Zumstein, M.; Mancini, R.; Bell-Sakyi, L.; Lozach, P.Y. Uukuniemi Virus as a Tick-Borne Virus Model. J. Virol. 2016, 90, 6784–6798.

17. Fischer, K.; Groschup, M.H.; Diederich, S. Importance of Endocytosis for the Biological Activity of Cedar Virus Fusion Protein. Cells 2020, 9, 2054.

18. Whitten-Bauer, C.; Chung, J.; Gómez-Moreno, A.; Gomollón-Zueco, P.; Huber, M.D.; Gerace, L.; Garaigorta, U. The Host Factor Erlin-1 is Required for Efficient Hepatitis C Virus Infection. Cells 2019, 8, 1555.

19. Chen, X.; Cao, R.; Zhong, W. Host Calcium Channels and Pumps in Viral Infections. Cells 2020, 9, 94.

20. Oechslin, N.; Moradpour, D.; Gouttenoire, J. On the Host Side of the Hepatitis E Virus Life Cycle. Cells 2020, 9, 1294.

21. Sid Ahmed, S.; Bundgaard, N.; Graw, F.; Fackler, O.T. Environmental Restrictions: A New Concept Governing HIV-1 Spread Emerging from Integrated Experimental-Computational Analysis of Tissue-Like 3D Cultures. Cells 2020, 9, 1112.

22. Imle, A.; Kumberger, P.; Schnellbacher, N.D.; Fehr, J.; Carrillo-Bustamante, P.; Ales, J.; Schmidt, P.; Ritter, C.; Godinez, W.J.; Muller, B.; et al. Experimental and computational analyses reveal that environmental restrictions shape HIV-1 spread in 3D cultures. Nat. Commun. 2019, 10, 2144.

23. Bilz, N.C.; Willscher, E.; Binder, H.; Böhnke, J.; Stanifer, M.L.; Hübner, D.; Boulant, S.; Liebert, U.G.; Claus, C. Teratogenic Rubella Virus Alters the Endodermal Differentiation Capacity of Human Induced Pluripotent Stem Cells. Cells 2019, 8, 870.

24. Nguyen, H.-M.; Guz-Montgomery, K.; Saha, D. Oncolytic Virus Encoding a Master Pro-Inflammatory Cytokine Interleukin 12 in Cancer Immunotherapy. Cells 2020, 9, 400.

25. Tekes, G.; Ehmann, R.; Boulant, S.; Stanifer, M.L. Development of Feline Ileum- and Colon-Derived Organoids and Their Potential Use to Support Feline Coronavirus Infection. Cells 2020, 9, 2085.

26. Potratz, M.; Zaeck, L.; Christen, M.; te Kamp, V.; Klein, A.; Nolden, T.; Freuling, C.M.; Müller, T.; Finke, S. Astrocyte Infection during Rabies Encephalitis Depends on the Virus Strain and Infection Route as Demonstrated by Novel Quantitative 3D Analysis of Cell Tropism. Cells 2020, 9, 412.

27. Leger, P.; Nachman, E.; Richter, K.; Tamietti, C.; Koch, J.; Burk, R.; Kummer, S.; Xin, Q.; Stanifer, M.; Bouloy, M.; et al. NSs amyloid formation is associated with the virulence of Rift Valley fever virus in mice. Nat. Commun. 2020, 11, 3281.

28. Ezzat, K.; Pernemalm, M.; Palsson, S.; Roberts, T.C.; Jarver, P.; Dondalska, A.; Bestas, B.; Sobkowiak, M.J.; Levanen, B.; Skold, M.; et al. The viral protein corona directs viral pathogenesis and amyloid aggregation. Nat. Commun. 2019, 10, 2331.

02 CORONAVIRUSES: DOES SIALIC ACID SERVE AS AN ENTRY POINT FOR THE CYTOKINE STORM IN THE EYE? FROM THE INTRODUCTION TO THE CONSEQUENCES

1. INTRODUCTION: AN OVERVIEW

As per the International Committee on Taxonomy of Viruses, the subfamily Coronavirinae consists of four genera of coronaviruses (CoV), which are known for their extensive biodiversity and ability to cause infections. [1]. In recent decades, epidemiological studies have mostly examined the spread of CoVs, including α-, β-, γ-, and δ-CoVs. These viruses have been found to cause respiratory and intestinal illnesses in mammals and birds. While most of these infections are mild or show no symptoms, there have also been cases of severe and deadly outcomes. [2]. In the last years, the extensive studies on the highly pathogenic human coronaviruses, especially on SARS- and MERS-CoVs helped to understand their biology but has also driven the discovery of new therapeutic strategies in case of epidemics [3,4]. However, a novel CoV named SARS-CoV-2 was described as the causative agent of the severe acute respiratory syndrome with high invasiveness and mortality, and disease it causes, called COVID-19, was officially declared a pandemic by the World Health Organization (WHO) [5,6,7]. Given the extensive invasion of SARS-CoV-2, it became evident that our understanding of CoVs was inadequate in terms of their structure, infection processes, serology, epidemiology, and ultimately, effective therapy. The CoVs are a type of enclosed viruses that have a single-stranded big RNA genome, which is approximately 30,000 nucleotides in size. [8]. The viral particle contains several classes of proteins that play structural and non-structural functions. The non-structural proteins belong to the family of replicases that are processed by proteases into 16 proteins functionally involved in genome transcription and replication [9]. Among four structural proteins, nucleocapsid (N) protein form complexes with RNA, whereas spike (S), membrane (M) and envelope (E) proteins participate in viral envelope formation, mediates attachment to the host cell surface receptors and may suppress the host immune response [10,11]. Given the invasive potential, the S protein play pivotal role in the fusion between the viral and host cell membranes, and in some CoVs interferes with interaction between infected and surrounding uninfected cells. The mechanisms underlying the attachment of virus via S protein recruit molecular systems which are specific feature of each CoV genera and a potential candidates in targeting therapy that may prevent a virus from being able to enter a cell. The N-terminal domain (NTD) and C-domain, called receptor binding domain (RBD), of S protein recognize and bind host membrane proteins [12]. As shown previously, the infection with morbid and deadly human CoVs, MERS- and SARS-CoV, is mediated by the high-affinity interaction between viral RBD and respectively, dipeptidyl peptidase 4 (DPP4) and angiotensin-converting enzyme 2 (ACE2) in host cells [13]. However, there is evidence that the main viral entry mechanisms may be supported by sialic acid-based machinery resulting in high diffusion speed [14,15,16]. Considering that sialoglycans are a prevalent element of cellular glycocalyx in different tissues, their potential regulatory role in CoVs can provide new insights into predictive diagnoses and future therapy. In the context of SARS-CoV-2 infections, it is crucial to address the excessive and unforeseen immune response that leads to severe acute respiratory syndrome. Additionally, the available medication therapy options are limited and not yet fully established. This research provides a concise examination of how sialic acids contribute to the tropism and infectivity of CoVs, as well as their involvement in regulating the excessive response of the body's immune system.

2. THE ROLE OF SIALIC ACIDS IN CORONAVIRUSES INFECTIONS

2.1. Sialic Acids–Structural and Functional Players in the Epidemiological Events

Sialic acids are a type of keto-sugar with nine carbon atoms that are chemically bonded to the outermost end of glycoconjugates found on the surface of cells, as well as in secreted glycoproteins and glycolipids. The sialoglycans possess biological and physical characteristics, such as the presence of negative charge and high hydrophilicity in cell membranes, which govern the interactions between cells and between cells and the extracellular matrix [17,18,19,20]. Due to sialic acid glycotopes ubiquity it is biologically important in biological recognition and immune responses [21,22,23,24,25,26,27,28]. According to Läubli and Varki sialoglycans expressed in the mammalian glycocalyx can be considered as self-associated molecular patterns (SAMPs) that can interact with the individual immune receptors presented on the same cell membranes and orchestrate inflammatory reactions within damaged tissues [29]. In contrast to SAMPs, the pathogen-associated molecular patterns (PAMPs) express the ability to recognize several immune receptors that underlie the mechanisms of chronic inflammation and neurodegeneration, as well as impaired immune surveillance in pathogen infections. The recruitment of the interplay between sialic acids and sialic acid-binding immunoglobulin-like lectins (Siglecs) underlie the proinflammatory signaling in response to some pathogens eg. Haemophilus influenzae and Escherichia coli [30,31,32]. In the field of virology, sialic acids were the first identified virus receptors that facilitate post-attachment events in the entry process [14]. The invasion by recognizing and binding to cell surface sialic acids, and in consequence fusion with the cellular membrane, is one of the best-established mechanisms of viral entry in host cells, as demonstrated with highly pathogenic influenza A, -B, and -C strains in vivo and in vitro studies [33,34,35]. Depending on the serotype of envelope glycoproteins that interact selectively with sialoglycans of the host, viruses can recognize and bind α2-3-, α2-6- and/or α2-8-linked sialic acid moieties, however, these preferences are additionally limited by natural host origin. For example, the receptor-binding specificity of human influenza viruses is focused preferably on α2-3- and α2-6-linked sialic acids which predominates on epithelial cells in the nasal mucosa, pharynx, larynx, trachea and ocular cells and strongly correlate with clinical symptoms [36,37,38,39]. The binding of α2-3- and/or α2-6-linked sialic acids as cellular receptors recruited in the host cell entry processes was described in several viral families isolated from humans, including Adenoviridae, Picornaviridae, Reoviridae, Paramyxoviridae and Polyomaviridae [40,41,42,43,44]. In the context of the virus-host interaction, some viruses genera employ the sialic acid- and Siglec-based mechanisms that avoid or deactivates host defence after pathogen invasion and underlie in the pathogenesis of acquired immune deficiency syndrome (AIDS) [45,46,47,48].

2.2. Corona virinae-Obviousnesses and Surprises

Individual corona viruses use specific cell receptors, which was used in differentiation and taxonomic classification to one of four genera. The ability to recognize host cell receptors by specific types of viruses is not synonymous with determining their pathogenic properties. It has been shown that NL63-CoV, one of seven known CoVs to infect humans, exerts lower affinity for human ACE2 (hACE2) than SARS-CoVs. These differences are related to the primary and secondary structure of S protein and determinate the clinical course of infection [49]. The infection of the upper respiratory tract with the low pathogenic CoVs including 229E-CoV, OC43-CoV, NL63-CoV, and HKU-CoV is manifested by the mild and cold-like symptoms. In contrast, highly pathogenic MERS-CoV, SARS-CoV and SARS-CoV-2 cause atypical pneumonia with the high rate of mortality [2]. The pathological changes in lungs following SARS- and MERS-CoV infection are accompanied by an excessive immune response and described as "cytokine storm" [50,51,52]. The clinical and experimental observation showed that individuals with severe or lethal SARS present very low serum level of anti-inflammatory IL-10 but high levels of proinflammatory cytokines and chemokines, e.g., IL-1, IL-6, IL-12, CCL2, CXCL10 and CXCL9, in comparison to patients with mild disease [53]. These changes were correlated with increased populations of neutrophils and monocytes in peripheral blood and lungs of infected individuals. In line, the in vitro studies with cultured monocytic THP-1 cell line and human peripheral blood monocytes infected with MERS-CoV confirmed the up-regulated production of pro-inflammatory factors. In animal models employed to study SARS, young mice developed lethal disease featured by pulmonary oedema, diffuse alveolar damage and excessive influx of monocytes/macrophages accompanied by delayed interferone-α/β (IFN-α/β) response, whereas targeting IFN signaling showed protective effects and better prognositic outcomes [4].

One of the greatest difficulties in the management of SARS-Cov-2 infection that underlie COVID-19 pandemic is the not fully understood hypersensitivity or relative resistance in some individuals in the context of the immune response [54]. These differences are genetically dependent and closely related to the structure and function of particular immune receptors. It has been shown that the polymorphisms of Toll-like receptor 1 (TLR1) and TLR4, linked to single nucleotide polymorphism, predispose individuals to "cytokine storm" in response to bacterial products [55,56]. The level of cytokine and chemokine expression results from the interplay between PAMPs and the pathogen recognition receptors (PRRs) of the host. The PAMPs-PRRs axis is featured by high specificity of recognition and interaction capacity between a pathogenic molecule (microbial, protozoan or viral) and engaging immune receptor. In result, the enhanced immune signalling pathways lead to a stage of hyperresponsiveness to PRRs agonists [57]. It has been shown that respiratory syncytial virus (RSV)-infected airway epithelial cells become hypersensitive to endotoxin by enhancing TLR4 expression with production of IL-8 and tumor necrosis factor α (TNFα) as well as increased mitogen-activated protein (MAP) kinase activity [58].

As previously mentioned, the recognition of patterns based on sialic acid by Siglec receptors can be seen as the essential element in the identification of self and nonself molecules by the innate immune system. The role of the sialic acid-Siglec immune checkpoint in bacterial, viral, and malignant invasions has been characterized as part of the molecular mimicry system. However, its involvement in CoVs infection and related diseases is not completely known. [28]. Extensive study on the COVID-19 pandemic has provided significant insights into the biology of coronaviruses (CoVs), namely about their infection mechanisms and their significance in prognosis and therapy.

In the following we focus on the involvement of sialoglycans in the human CoVs biology, including the highly invasive species of the genus Betacoronavirus, in the field of mechanisms of host cell infection and the control of the immune response.

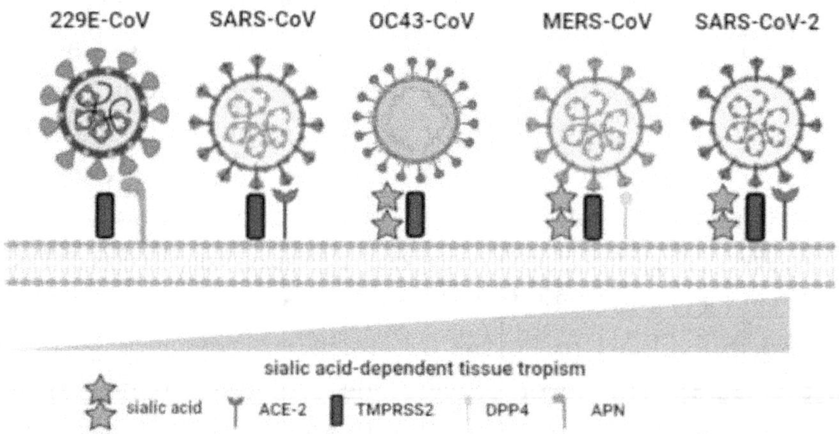

Figure 1. Receptor recognition pattern during the CoVs infections. The CoVs invade host cells through the attachment, binding and entry mechanism based on the sialic acid and protein receptors. After sialic acid-mediated virus attachment and its spike protein activation by transmembrane serine protease 2 (TMPRSS2), the entry event is associated with the binding of specific protein receptor: ACE-2 (angiotensin converting enzyme type 2); DPP4 (dipeptidyl peptidase 4); APN (aminopeptidase N). The ability to recognize sialoglycans determines virus tissue tropism and clinical manifestations among infected organs.

2.3. MERS-CoV versus SARS-CoV—One Genus and Two Scenarios

Genomic and phylogenetic analyses have established several similarities between MERS- and SARS-CoV. However, these two viruses differ in various structural aspects that play a vital role in their ability to cause infection [64]. As highly pathogenic human β-CoVs, they cause a wide spectrum of clinical manifestations in respiratory tracts, however, the rate of clinical changes seem to be unfavorable in MERS-CoV infections [4]. While the clinical course of SARS resulted in acute respiratory illness presents in three distinct phases, the MERS clinical manifestations in respiratory tracts rapidly progress to pneumonia with a 35% overall mortality rate [65]. Comparative studies from human autopsies and animal models of SARS and MERS strongly suggest a pathogenic role for "cytokine storm" derived from inflammatory monocyte-macrophages (IMMs) and neutrophils. In line with this, enhanced serum cytokine and chemokine levels during MERS-CoV infection strongly correlate with increased neutrophil and monocyte numbers in the lungs and the peripheral blood, indicating a possible role for these cells in described pathology [4,66].

The attachment and fusion with the host cell membrane are crucial steps in viral infection, as previously stated. Following the interaction with the target cell, the S protein undergoes cleavage into S1 and S2 subunits, which are responsible for receptor binding and membrane fusion, respectively. In most coronaviruses (CoVs), the C-terminal domain of the S1 subunit identifies particular host receptors, while the N-terminal domain, depending on the type of virus, identifies specific sugar molecules. This enables the first attachment of viral particles to the cell surface. The S protein in SARS-CoV undergoes cleavage by the host transmembrane protease serine 2 (TMPRSS2), resulting in the formation of S1 and S2 subunits [67,68,69]. To enter host cells, virus utilizes human ACE2 as the functional receptor recognized and linked via viral RBD of S1 subunit. This mechanism was confirmed in multiple in vitro studies with naturally ACE-2-expressing colonic epithelial LoVo cells and alveolar epithelial A549 cells [70,71]. Whereas the structure of RBD is systematically studied, the NTD of SARS-CoV is not fully characterized. However, the current state of structural analyses suggests that SARS-Cov, unlike other β-Covs, does not recognize glycans, including sialic acids, via NTD in S1 subunit [72]. For MERS-CoV, human lung epithelial cells susceptibility correlates with DPP4 expression in these cells. Analysis of MERS-CoV S1 subunit structure by cryoelectron microscopy showed four individually folded domains (S1A-S1D) implicated in receptor binding. The engagement of binding domain closely depends on virus species [73]. Among β-CoVs, low invasive bovine CoV (BCoV) and human OC43-CoV recruit S1A domain for binding of sialic acid-based entry receptors [74]. Binding of MERS-CoV to the cell surface entry receptor DPP4 occurs via S1B whereas sialic acid-binding activity was assigned to S subdomain S1A. In addition, the analysis with 60-metric lumazine synthase nanoparticle presenting S1A domains revealed high preference for α2.3-linked over α2.6-linked sialic acids. The importance of sialic acid in the mechanism of MERS-CoV infection was presented by Li et al. [73]. They showed that lower sialylated Vero cells exert lower susceptibility to MERS-CoV infection compared to highly sialic acid expressing human airway Calu-3 epithelial cells. Moreover, enzymatic depletion of sialic acids by neuraminidase from the cell surface or their modification by 9-O-acetylation or 5-N-glycolylation hindered or dampened these interactions and thereby viral entry [73]. As demonstrated in influenza viruses infections, the sialic acid-binding preference and distribution of sialoglycans is a critical regulator of viral tropism. Therefore, site of MERS-CoV replication colocalize with α2,3-linked sialic acid in vivo. According to Li et al., the differential distribution of α2,3-linked sialic acid in humans and camels correlates with predominant sites of MERS-CoV replication in the lower and upper respiratory tracts, respectively [73]. Finally, the identified MERS-CoV interaction with the host sialosides in the dual-receptors binding mechanism, has been characterized by fast kinetics and was required for viral entry [75].

2.4. SARS-CoV-2 versus SARS-CoV—Are Sialic Acids Really Important?

The comparing analysis of genetic sequences revealed that SARS-CoV-2 shared about 79% of its genome with SARS-CoV [76,77]. In addition, the overall sequence similarity for the whole S protein was 78% and 76% for RBD [78]. The close identity between S1 subdomains indicate that SARS-Cov-2 also utilizes ACE2 as the receptor. Despite similarities, SARS-CoV-2 show several structural differences that may influence infection kinetics and clinical symptoms in individuals. First, quantification of the kinetics of ligand-receptor interaction by surface plasmon resonance showed that SARS-CoV-2 binding to ACE2 is about 20-fold higher compared to SARS-CoV [79]. Second, clinical course of COVID-19 in some patients range from mild symptoms to deadly "cytokine storm", however, asymptomatic individuals, similar to those in MERS, may pose a significant public health problem [80]. To better understand the mechanisms underlying these phenomenon, the total genomic sequence of human CoVs was compared with the library of viruses. Studies on the complementarity of SARS-CoV-2 S protein identified another possible binding region responsible for the interaction with the host receptors. Despite the SARS-CoV-2 is less genetically similar to MERS-CoV (about 50%) than SARS-CoV (about 79%), it has been hypothesized that infection with both SARS-CoV-2 and MERS-CoV depends on similar mechanism of the attachment to the host cells that facilitates virus entry and mediates the immune response [81,82,83]. The structural properties of several SARS-CoV-2 and MERS-CoV strains were analysed using CROSSalign, a tool for prediction of the structural similarities of two RNA profiles of different length. In the preliminary studies, group of Vandelli have found that the structural region between nucleotides 23,000 and 24,000 of S region that corresponds to amino acids 330–500 of S protein domain, is highly conserved across SARS-CoV-2 strains and responsible for ACE2 binding. However, the structural profile in SARS-CoV-2 at nucleotides 22,500–23,000 corresponds to amino acids 243–302 that share identity in MERS-CoV, but not in SARS-CoV, and bind to sialic acids during virus attachment to the host cells [82]. This observation was confirmed by Milanetti and coworkers in the preliminary analysis using quantitative assessment of the geometrical shape complementarity between interacting proteins. They used experimental structures of SARS-CoV-2, MERS-CoV and SARS-CoV S protein and sialic acid complexes, extracted their binding regions and described through Zernicke descriptors. The selected region in the analyzed viruses expressed the highest similarity when SARS-CoV-2 and MERS-CoV were compared. In addition, the MERS-CoV S protein region interacting with sialic acids and its analogue in SARS-CoV-2 showed electrostatic similarity whereas the compared region in SARS-CoV present not compatible charge that disturb the interaction with sialylated receptors. Milanetti suggest that the sialic acid binding capacity in SARS-CoV-2 seem to be a part of complex cell entry mechanism that explain highly effective infection of the upper and its progession in low airways [83]. This hypothesis seem to be confirmed by the findings of Fantini and coworkers [81]. They identified the specific area of NTD that interacts with oligosaccharide part of ganglioside GM1. Molecular and structural studies revealed that amnio acid residues 100–175 of the NTD, especially Phe-135, Asn-137 and Arg-158 are critical for the interaction with the host GM1 molecule.

Beside the upper respiratory tract, the eye exposure to infectious fluids may be associated with increased risk for SARS-CoV-2 transmission. According to goup of Craeger the conjunctival and corneal epithelia and the nasolacrimal system contain both α2-3-linked and α2-6-linked sialic acid receptors preferentially interacting with numerous human viruses [84]. The eye contamination with CoVs was confirmed by Loon and coworkers using PCR analysis of tears samples collected from probably positive patients [85]. Recently, the clinical observation by Colavita et al. have revealed that SARS-CoV-2 RNA was detected in ocular swab whereas it was undetectable in nasal swab form patient diagnosed for COVID-19 [86].

As previously stated, alterations in sialylation pattern play a crucial role in glycan-mediated biological recognition and contribute to immune dysregulation and associated diseases. According to research from the World Health Organization (WHO), individuals with cardiovascular and chronic respiratory problems are more likely to experience severe symptoms and have a worse prognosis when infected with COVID-19, compared to those without any pre-existing chronic medical conditions [87]. The clinical-related advances in glycobiology are focused on the prognostic value of sialylated epitopes as markers of pathology. For example, it has been shown that serum sialic acid forecast both coronary heart disease and stroke fatality and reflects the progress of an atherosclerotic process [88,89,90]. The clinical observation and experimental models described an enhanced level of total serum sialic acid and sialoglycans on the human vein endothelial cells as a marker of pathology and significant risk factor of coronary artery disease (CAD) and angiopathy [91,92]. Moreover, the glycome analysis in asthma and COPD showed that the alterations in sialoglycans correspond to the progress of disease [93,94,95]. According to Guo, the lung pathology inducing factors, e.g., tobacco, can be considered as the crucial players in the promotion of SARS-CoV-2 infection and associated with the risk of severe COVID-19 [95]. Given the harmful effect of cigarette smoke (CS) on respiratory and circulatory parameters, heavy smokers are at higher risk of developing severe COVID-19 symptoms [96,97]. Beside the clinical consequences, CS induces multiple molecular changes in the lung cells. Our previous studies revealed that human alveolar epithelial A549 cells present enhanced α2-3- and α-2-8-, but not α2-6-sialylation pattern when exposed to CS [98]. Additionally, Leung et al. found that patients with chronic obstructive pulmonary disease (COPD) and individuals who are still smoking presents higher levels of ACE-2 in their airways [99]. Thus, it is logical to propose that a dual mechanism involving ACE2 and sialic acid could potentially increase their susceptibility to severe COVID-19 infections.

2.5. Does Siglec-Sialic Acid Axis Participates in CoVs Infections?

The extent of immune system activity is the primary factor in safeguarding the body's systems against infections. Various investigations have shown that when multiple individuals are exposed to the same infection, their responses can vary significantly, ranging from having no symptoms to experiencing a severe immune reaction known as a "cytokine storm" that can lead to death. The most severe prognosis connected to coronaviruses is attributed to the immune system's hyperresponsiveness, rather than direct damage caused by the virus itself. The influence of genetics on immunological profiles plays a significant role in determining how different populations respond to infections. However, the specific immunity mechanisms involved in CoVs infection, particularly SARS-CoV-2, are not yet completely understood. [100,101].

In the context of participation of sialic acid in the mechanisms of CoVs entry, we discussed above scenario showed cellular sialoglycans as receptors for recognizing proteins expressed in the viral envelope. On the other hand, several pathogens express sialoglycan epitopes that act as a ligand recognized by cell membrane receptors [102] (Figure 2).

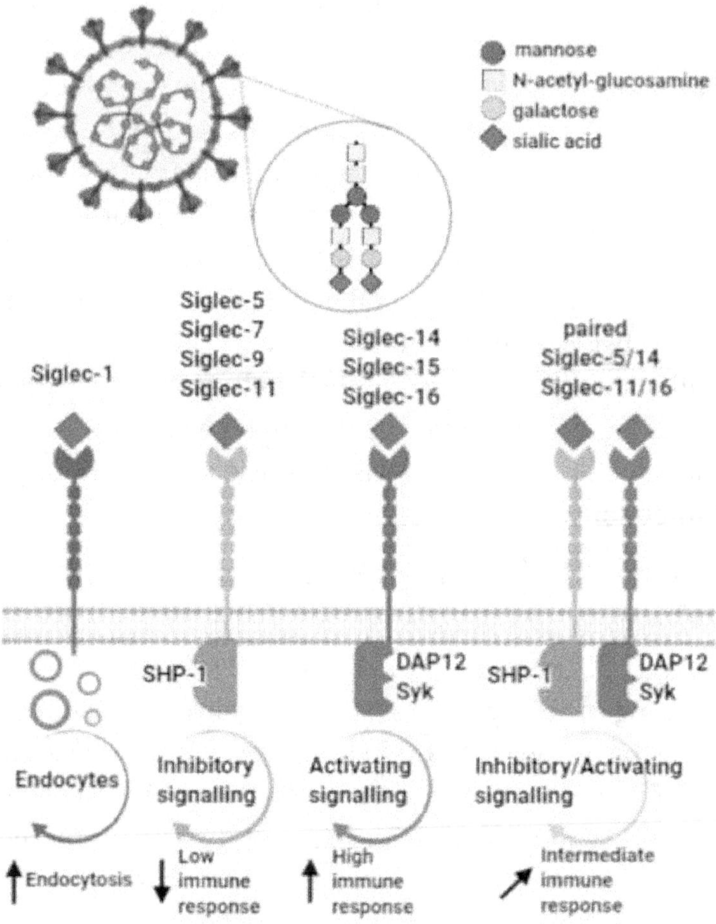

Figure 2. The possible effects of the CoVs targeting to Siglecs. The viral envelope glycans interact with sialic acid-binding lectins expressed in the host cell-dependent manner. Depending on the individual Siglec expression profile, the interactions with the viral sialic acid ligands exerts the immune response of different intensity. Abbreviations: DAP12-DNAX activation protein of 12 kDa; Syk-spleen tyrosine kinase; SHP-1-Src homology region 2 domain-containing phosphatase-1; SHP-2-Src homology region 2 domain-containing phosphatase-2.

The enzymes linked with the virus can influence the cell membrane glycocalyx and reveal Siglecs by disrupting their cis ligands. Evidence has demonstrated that the unmasking of Siglec-1 (also known as sialoadhesin or Sn) enhances the absorption of CoV into macrophages [103,104,105]. Another mechanism of potential importance includes interactions between sialic acids and Siglec receptors resulting in divergent immune system regulation. The role of CoV-originated sialoglycans in immune modulation was not extensively studied and the poor data have rather speculative value. Is it reasonable to consider the interplay between Siglecs and sialic acids as a controlling mechanism in CoVs invasion machinery? Varki and Angata hypothesize that expression of sialic acids by the envelope of CoV can affect Siglec receptors biology in the hosts and thereby regulate the reactivity of innate immune cells [102]. The inhibitory receptors, such as Siglec-5, -7 and -9, are implicated in molecular mimicry mechanism that allow several pathogens to avoid immunesurveillance as shown in the hepatitis B virus (HBV) infections [106,107]. In the context of Siglec-mediated immune regulation, the clinical differences in the CoVs infections can be interpreted by the variety of human phenotypes. Among the Siglec family members, the paired receptors express extremely similar amino acid sequence of extracellular domain and comparable tissue distribution resulting in the same ligand recognition capacity whereas the diverse intracellular signalling pathways trigger the opposite effects. The paired Siglec-5 and Siglec-14 proteins, gene fusion products, are widely expressed on myeloid-derived cells residing and infiltrating human airways, however their distribution in individuals is closely related to the frequencies of wild-type and SIGLEC-14 null alleles in various populations [108,109]. Therefore, the expression of activating Siglec-14 receptor predominates in the European population whereas inhibitory Siglec-5 is widely expressed in Asians. Clinically, the expression of Siglec-5 is closely linked to reduced bactericidal and virucidal abilities during infections with Streptococcus, Neisseria, Pseudomonas, Campylobacter and HIV [110]. Moreover, expression of Siglec-14 predispose to the potentiated inflammatory response in bacterial (Haemophilus influenzae) and viral infections (influenza virus), which are the cause of chronic respiratory diseases. It was confirmed by elevated expression of IL-1β in primary macrophages from SIGLEC14+/+, but not SIGLEC14-/- individuals [111]. Interestingly, the expression and function of paired Siglec-5/14 receptors undergo multiple modification in response to several pathology promoting factors. In the experimental model of exposure to cigarette smoke (CS), an increase in the expression of paired Siglec-5/14 receptors in THP-1 cells was observed [98]. It was accompanied by simultaneous alterations in immune activity with an enhanced intracellular interleukin IL-1β and interleukin 10 (IL-10) expression, and reduced phagocytic capacity. Given the CS-induced functional alterations in the immune response, it is reasonable to speculate that sialic acid-Siglec axis can be an important mechanism of exacerbations in respiratory diseases and the immune system performance in viral infections in smokers.

Additionally, previous studies have revealed that conventional therapies may influence Siglec-mediated immune regulation. During the SARS epidemic in 2003, the early anti-inflammatory treatment of corticosteroids enhanced plasma viral titre leading to exacerbated disease [112]. The strong suppressive action of corticosteroids may explain these effect, however, Siglec-based mechanism can be also involved. The cytometric analysis of cell population isolated from patients with chronic obstructive pulmonary disease (COPD) has demonstrated that inhaled corticosteroids increase the expression of Siglec-5/14 in CD14+ cells [113]. Similarly, the enhancement of Siglec-9 in neutrophils was observed after management with dexamethasone (Dex) [114]. Therefore, the clinical assessment of Siglecs profiles can be considered before the therapeutic intervention with corticosteroids in individuals with respiratory comorbidities in CoVs infections.

Notably, ongoing clinical trials have demonstrated that chloroquine (CLQ) and hydroxychloroquine (CLQ-OH), which are 4-aminoquinoline derivatives, can effectively block the binding of the COVID-19 virus to the ACE2 receptor [115]. Based on the in vitro studies, it is believed that chloroquine inhibits terminal glycosylation and thereby prevent SARS-CoV-2 from binding with gangliosides [116]. As structural and molecular modelling approaches have shown, this effect seems to be linked with direct interaction of chloroquine (CLQ) and hydroxychloroquine (CLQ-OH) with structural motif including amino acid residues 111–158 of the NTD that prevent of binding the sialylated GM1 with the high affinity [81]. From a molecular perspective, the therapeutic potential of CLQ and CLQ-OH is indicated. However, recent randomized studies of CLQ-OH as postexposure prophylaxis for COVID-19 could not conclusively validate this effect. Therefore, further clinical and molecular observations are required.

3. CONCISE SUMMARY

Although the biological importance of sialic acid is well-documented, its role in the development of CoVs infections remains incompletely understood. Recent findings in the field of virus attachment and penetration into host cells indicate that the ability to identify certain sialoglycan epitopes is a distinguishing feature among coronaviruses. Sialic acid binding is a vulnerable element of the dual mechanism through which viruses penetrate, making it a possible focus for future therapeutic investigation. Conducting investigations on the relationship between sialic acid and Siglec in both living organisms (in vivo) and laboratory conditions (in vitro) can offer valuable insights into the immune-related causes of viral infections and their effects on various clinical outcomes in the population, including the prevalence of asymptomatic individuals.

The current state of knowledge in this domain is characterized by significant advancements. However, there are still numerous unresolved concerns within the discipline of glycobiology: (i) Can sialic acid function as a host cellular marker to control the likelihood of CoVs attachment? (ii) Do many coronaviruses express sialic acid epitopes to modulate Siglec function? Is the interaction between sialic acid and Siglecs in CoV infections significant for cell function and activation? (iv) Can conventional medications modulate the cellular sialome in the attachment process of CoVs? (v) Are comorbidities and pathological conditions that are linked to an elevated sialylation pattern a risk factor for SARS-CoV-2 infections and/or their severe progression? The primary problem in current virology is the avoidance of viral attachment and penetration. Given the ongoing SARS-CoV-2 pandemic, it is necessary to develop new immunotherapeutic strategies for COVID-19 that align with our current understanding of molecular mechanisms. According to the information provided in this review, targeting the sialome-modulating machinery of patients can potentially inhibit viral invasion through various mechanisms. These mechanisms include modifying the host receptors, regulating endocytosis, preventing interaction between the virus and host cells, and modulating the immune response.

Coronaviruses (CoVs) are a wide-ranging group of enveloped viruses that infect both humans and animals, and are known to cause respiratory and intestinal diseases. The significant pathogenicity of human Coronaviruses (CoVs), such as SARS-CoV, MERS-CoV, and SARS-CoV-2, is strongly linked to the invasion mechanisms involved in the attachment and entry of viral particles into the host cells. There is growing evidence that sialylated substances found in the cellular glycocalyx play a significant role in the mechanism of CoVs infection. Furthermore, the interaction between sialic acid and the immunological lectins of the host is known to elicit an immune response of varying intensity in specific clinical phases. In this article, we will discuss the latest discoveries in the field of glycobiology about the function of sialic acid in tissue tropism, viral entry kinetics, and immune modulation in CoV infections.

REFERENCES

1. International Committee on Taxonomy of Viruses Executive Committee. The new scope of virus taxonomy: Partitioning the virosphere into 15 hierarchical ranks. Nat. Microbiol. 2020, 5, 668–674.

2. Pillaiyar, T.; Meenakshisundaram, S.; Manickam, M. Recent discovery and development of inhibitors targeting coronaviruses. Drug. Discov. Today 2020, 25, 668–688.

3. Tao, X.; Hill, T.E.; Morimoto, C.; Peters, C.J.; Ksiazek, T.G.; Tseng, C.T. Bilateral entry and release of Middle East respiratory syndrome coronavirus induces profound apoptosis of human bronchial epithelial cells. J. Virol. 2013, 87, 9953–9958.

4. Channappanavar, R.; Perlman, S. Pathogenic human coronavirus infections: Causes and consequences of cytokine storm and immunopathology. Semin. Immunopathol. 2017, 39, 529–539.

5. Huang, C.; Wang, Y.; Li, X.; Ren, L.; Zhao, J.; Hu, Y.; Zhang, L.; Fan, G.; Xu, J.; Gu, X.; et al. Clinical features of patients infected with 2019 novel coronavirus in Wuhan, China. Lancet 2020, 395, 497–506.

6. Zhu, N.; Zhang, D.; Wang, W.; Li, X.; Yang, B.; Song, J.; Zhao, X.; Huang, B.; Shi, W.; Lu, R.; et al. A novel coronavirus from patients with pneumonia in China, 2019. N. Engl. J. Med. 2020, 382, 727–733.

7. Walls, A.C.; Park, Y.J.; Tortorici, M.A.; Wall, A.; McGuire, A.T.; Veesler, D. Structure, function, and antigenicity of the SARS-CoV-2 spike glycoprotein. Cell 2020, 181, 281–292.

8. Schwegmann-Wessels, C.; Herrler, G. Sialic acids as receptor determinants for coronaviruses. Glycoconj. J. 2006, 23, 51–58.

9. Harcourt, B.H.; Jukneliene, D.; Kanjanahaluethai, A.; Bechill, J.; Severson, K.M.; Smith, C.M.; Rota, P.A.; Baker, S.C. Identification of severe acute respiratory syndrome coronavirus replicase products and characterization of papain-like protease activity. J. Virol. 2004, 78, 13600–13612.

10. McBride, R.; van Zyl, M.; Fielding, B.C. The coronavirus nucleocapsid is a multifunctional protein. Viruses 2014, 6, 2991–3018.

11. Arndt, A.L.; Larson, B.J.; Hogue, B.G. A conserved domain in the coronavirus membrane protein tail is important for virus assembly. J. Virol. 2010, 84, 11418–11428.

12. Zeng, F.; Hon, C.C.; Yip, C.W.; Law, K.M.; Yeung, Y.S.; Chan, K.H.; Peiris, J.S.M.; Leung, F.C.C. Quantitative comparison of the efficiency of antibodies against S1 and S2 subunit of SARS coronavirus spike protein in virus neutralization and blocking of receptor binding:

Implications for the functional roles of S2 subunit. FEBS Lett. 2006, 580, 5612–5620.

13. Wan, Y.; Shang, J.; Sun, S.; Tai, W.; Chen, J.; Geng, Q.; He, L.; Chen, Y.; Wu, J.; Shi, Z.; et al. Molecular mechanism for antibody-dependent enhancement of coronavirus entry. J. Virol. 2020, 94, e02015–e02019.
14. Matrosovich, M.; Herrler, G.; Klenk, H.D. Sialic acid receptors of viruses. Top. Curr. Chem. 2015, 367, 1–28.
15. Tortorici, M.A.; Walls, A.C.; Lang, Y.; Wang, C.; Li, Z.; Koerhuis, D.; Boons, G.-J.; Bosch, B.-J.; Rey, F.A.; de Groot, R.J.; et al. Structural basis for human coronavirus attachment to sialic acid receptors. Nat. Struct. Mol. Biol. 2019, 26, 481–489.
16. Qing, E.; Hantak, M.; Perlman, S.; Gallagher, T. Distinct roles for sialoside and protein receptors in coronavirus infection. mBio 2020, 11, e02764-19.
17. Schnaar, R.L. The biology of gangliosides. Adv. Carbohydr. Chem. Biochem. 2019, 76, 113–148.
18. Trinchera, M.; Aronica, A.; Dall'Olio, F. Selectin ligands sialyl-Lewis a and sialyl-Lewis x in gastrointestinal cancers. Biology 2017, 6, 16.
19. Bork, K.; Horstkorte, R.; Weidemann, W. Increasing the sialylation of therapeutic glycoproteins: The potential of the sialic acid biosynthetic pathway. J. Pharm. Sci. 2009, 98, 3499–3508.
20. Bas, M.; Terrier, A.; Jacque, E.; Dehenne, A.; Pochet-Béghin, V.; Beghin, C.; Dezetter, A.-S.; Dupont, G.; Engrand, A.; Beaufils, B.; et al. Fc sialylation prolongs serum half-life of therapeutic antibodies. J. Immunol. 2019, 202, 1582–1594.
21. Varki, A. Sialic acids in human health and disease. Trends Mol. Med. 2008, 14, 351–360.
22. Miyagi, T.; Takahashi, K.; Hata, K.; Shiozaki, K.; Yamaguchi, K. Sialidase significance for cancer progression. Glycoconj. J. 2012, 29, 567–577.

23. Vajaria, B.N.; Patel, K.R.; Begum, R.; Patel, P.S. Sialylation: An avenue to target cancer cells. Pathol. Oncol. Res. 2016, 22, 443–447.

24. Pearce, O.M.; Läubli, H. Sialic acids in cancer biology and immunity. Glycobiology 2016, 26, 111–128.

25. Chiodelli, P.; Urbinati, C.; Paiardi, G.; Monti, E.; Rusnati, M. Sialic acid as a target for the development of novel antiangiogenic strategies. Future Med. Chem. 2018, 10, 2835–2854.

26. McMillan, S.J.; Crocker, P.R. CD33-related sialic-acid-binding immunoglobulin-like lectins in health and disease. Carbohydr. Res. 2008, 343, 2050–2056.

27. Cagnoni, A.J.; Pérez Sáez, J.M.; Rabinovich, G.A.; Mariño, K.V. Turning-off signaling by siglecs, selectins, and galectins: Chemical inhibition of glycan-dependent interactions in cancer. Front. Oncol. 2016, 6, 109.

28. Bärenwaldt, A.; Läubli, H. The sialoglycan-Siglec glyco-immune checkpoint—A target for improving innate and adaptive anti-cancer immunity. Expert Opin. Targets 2019, 23, 839–853.

29. Läubli, H.; Varki, A. Sialic acid-binding immunoglobulin-like lectins (Siglecs) detect self-associated molecular patterns to regulate immune responses. Cell. Mol. Life Sci. 2020, 77, 593–605.

30. Sarkar, M.; Reneer, D.V.; Carlyon, J.A. Sialyl-Lewis x-independent infection of human myeloid cells by Anaplasma phagocytophilum strains HZ and HGE1. Infect. Immun. 2007, 75, 5720–5725.

31. Seidman, D.; Hebert, K.S.; Truchan, H.K.; Miller, D.P.; Tegels, B.K.; Marconi, R.T.; Carlyon, J.A. Essential domains of Anaplasma phagocytophilum invasins utilized to infect mammalian host cells. PLoS Pathog. 2015, 11, e1004669.

32. Maginnis, M.S. Virus-receptor interactions: The key to cellular invasion. J. Mol. Biol. 2018, 430, 2590–2611.

33. Kumlin, U.; Olofsson, S.; Dimock, K.; Arnberg, N. Sialic acid tissue distribution and influenza virus tropism. Influenza Other Respir. Viruses 2008, 2, 147–154.

34. Connor, R.J.; Kawaoka, Y.; Webster, R.G.; Paulson, J.C. Receptor specificity in human, avian, and equine H2 and H3 influenza virus isolates. Virology 1994, 205, 17–23.

35. Matrosovich, M.; Tuzikov, A.; Bovin, N.; Gambaryan, A.; Klimov, A.; Castrucci, M.R.; Donatelli, I.; Kawaoka, Y. Early alterations of the receptor-binding properties of H1, H2, and H3 avian influenza virus hemagglutinins after their introduction into mammals. J. Virol. 2000, 74, 8502–8512.

36. Shinya, K.; Ebina, M.; Yamada, S.; Ono, M.; Kasai, N.; Kawaoka, Y. Avian flu: Influenza virus receptors in the human airway. Nature 2006, 440, 435–436.

37. Ibricevic, A.; Pekosz, A.; Walter, M.J.; Newby, C.; Battaile, J.T.; Brown, E.G.; Holtzman, M.J.; Brody, S.L. Influenza virus receptor specificity and cell tropism in mouse and human airway epithelial cells. J. Virol. 2006, 80, 7469–7480.

38. Nicholls, J.M.; Chan, M.C.; Chan, W.Y.; Wong, H.K.; Cheung, C.Y.; Kwong, D.L.W.; Wong, M.P.; Chui, W.H.; Poon, L.L.M.; Tsao, S.W.; et al. Tropism of avian influenza A (H5N1) in the upper and lower respiratory tract. Nat. Med. 2007, 13, 147–149.

39. França, M.; Stallknecht, D.E.; Howerth, E.W. Expression and distribution of sialic acid influenza virus receptors in wild birds. Avian Pathol. 2013, 42, 60–71.

40. Connolly, J.L.; Barton, E.S.; Dermody, T.S. Reovirus binding to cell surface sialic acid potentiates virus-induced apoptosis. J. Virol. 2001, 75, 4029–4039.

41. Villar, E.; Barroso, I.M. Role of sialic acid-containing molecules in paramyxovirus entry into the host cell: A minireview. Glycoconj. J. 2006, 23, 5–17.

42. Gee, G.V.; Dugan, A.S.; Tsomaia, N.; Mierke, D.F.; Atwood, W.J. The role of sialic acid in human polyomavirus infections. Glycoconj. J. 2006, 23, 19–26.

43. Zocher, G.; Mistry, N.; Frank, M.; Hähnlein-Schick, I.; Ekström, J.-O.; Arnberg, N.; Stehle, T. A sialic acid binding site in a human picornavirus. PLoS Pathog. 2014, 10, e1004401.

44. Chandra, N.; Liu, Y.; Liu, J.X.; Frängsmyr, L.; Wu, N.; Silva, L.M.; Lindström, M.; Chai, W.; Domellöf, F.P.; Feizi, T.; et al. Sulfated glycosaminoglycans as viral decoy receptors for human adenovirus type 37. Viruses 2019, 11, 247.

45. Xie, J.; Christiaens, I.; Yang, B.; Trus, I.; Devriendt, B.; Cui, T.; Wei, R.; Nauwynck, H.J. Preferential use of Siglec-1 or Siglec-10 by type 1 and type 2 PRRSV strains to infect PK15S1-CD163 and PK15S10-CD163 cells. Vet. Res. 2018, 49, 67.

46. Perez-Zsolt, D.; Martinez-Picado, J.; Izquierdo-Useros, N. When dendritic cells go viral: The role of Siglec-1 in host defense and dissemination of enveloped viruses. Viruses 2019, 12, 8.

47. Mikulak, J.; Di Vito, C.; Zaghi, E.; Mavilio, D. Host immune responses in HIV-1 infection: The emerging pathogenic role of Siglecs and their clinical correlates. Front. Immunol. 2017, 8, 314.

48. Arokiasamy, S.; Balderstone, M.J.M.; De Rossi, G.; Whiteford, J.R. Syndecan-3 in inflammation and angiogenesis. Front. Immunol. 2020, 10, 3031.

49. Mathewson, A.C.; Bishop, A.; Yao, Y.; Kemp, F.; Ren, J.; Chen, H.; Xu, X.; Berkhout, B.; van der Hoek, L.; Jones, I.M.; et al. Interaction of severe acute respiratory syndrome-coronavirus and NL63 coronavirus spike proteins with angiotensin converting enzyme-2. J. Gen. Virol. 2008, 89, 2741–2745.

50. Lau, S.K.P.; Lau, C.C.Y.; Chan, K.H.; Clara, P.Y.L.; Chen, H.; Jin, D.-Y.; Chan, J.F.W.; Woo, P.C.Y.; Yuen, K.-Y. Delayed induction of proinflammatory cytokines and suppression of innate antiviral response by the novel Middle East respiratory syndrome coronavirus: Implications for pathogenesis and treatment. J. Gen. Virol. 2013, 94, 2679–2690.

51. Li, Y.; Chen, M.; Cao, H.; Zhu, Y.; Zheng, J.; Zhou, H. Extraordinary GU-rich single-strand RNA identified from SARS coronavirus contributes an excessive innate immune response. Microbes Infect. 2013, 15, 88–95.

52. Sun, X.; Wang, T.; Cai, D.; Hu, Z.; Chen, J.; Liao, H.; Zhi, L.; Wei, H.; Zhang, Z.; Qiu, Y.; et al. Cytokine storm intervention in the early stages of COVID-19 pneumonia. Cytokine Growth Factor. Rev. 2020, 53, 38–42.

53. Tisoncik, J.R.; Korth, M.; Simmons, C.P.; Farrar, J.; Martin, T.R.; Katze, M.G. Into the eye of the cytokine storm. Microbiol. Mol. Biol. Rev. 2012, 76, 16–32.

54. Li, X.; Geng, M.; Peng, Y.; Meng, L.; Lu, S. Molecular immune pathogenesis and diagnosis of COVID-19. J. Pharm. Anal. 2020, 10, 102–108.

55. Molteni, M.; Gemma, S.; Rossetti, C. The role of Toll-like receptor 4 in infectious and noninfectious inflammation. Mediat. Inflamm. 2016, 2016, 6978936.

56. Perrin-Cocon, L.; Aublin-Gex, A.; Sestito, S.E.; Shirey, K.A.; Patel, M.C.; André, P.; Blanco, J.C.; Vogel, S.N.; Peri, F.; Lotteau, V. TLR4 antagonist FP7 inhibits LPS-induced cytokine production and glycolytic reprogramming in dendritic cells, and protects mice from lethal influenza infection. Sci. Rep. 2017, 7, 40791.

57. Thompson, M.R.; Kaminski, J.J.; Kurt-Jones, E.A.; Fitzgerald, K.A. Pattern recognition receptors and the innate immune response to viral infection. Viruses 2011, 3, 920–940.

58. Hu, T.; Yu, H.; Lu, M.; Yuan, X. TLR4 and nucleolin influence cell injury, apoptosis and inflammatory factor expression in respiratory syncytial virus-infected N2a neuronal cells. J. Cell. Biochem. 2019, 120, 16206–16218.

59. Zeng, Q.; Langereis, M.A.; van Vliet, A.L.; Huizinga, E.G.; de Groot, R.J. Structure of coronavirus hemagglutinin-esterase offers insight into corona and influenza virus evolution. Proc. Natl. Acad. Sci. USA 2008, 105, 9065–9069.

60. Hulswit, R.J.G.; Lang, Y.; Bakkers, M.J.G.; Li, W. Human coronaviruses OC43 and HKU1 bind to 9-O-acetylated sialic acids via a conserved receptor-binding site in spike protein domain A. Proc. Natl. Acad. Sci. USA 2019, 116, 2681–2690.
61. Schultze, B.; Krempl, C.; Ballesteros, M.L.; Shaw, L.; Schauer, R.; Enjuanes, L.; Herrler, G. Transmissible gastroenteritis coronavirus, but not the related porcine respiratory coronavirus, has a sialic acid (N-glycolylneuraminic acid) binding activity. J. Virol. 1996, 70, 5634–5637.
62. Cui, T.; Theuns, S.; Xie, J.; Van den Broeck, W.; Nauwynck, H.J. Role of porcine aminopeptidase N and sialic acids in porcine coronavirus infections in primary porcine enterocytes. Viruses 2020, 12, 402.
63. Wong, A.H.M.; Tomlinson, A.C.A.; Zhou, D.; Satkunarajah, M.; Chen, K.; Sharon, C.; Desforges, M.; Talbot, P.J.; Rini, J.M. Receptor-binding loops in alphacoronavirus adaptation and evolution. Nat. Commun. 2017, 8, 1735.
64. Eckerle, I.; Müller, M.A.; Kallies, S.; Gotthardt, D.N.; Drosten, C. In-vitro renal epithelial cell infection reveals a viral kidney tropism as a potential mechanism for acute renal failure during Middle East Respiratory Syndrome (MERS) Coronavirus infection. Virol. J. 2013, 10, 359.
65. Alsolamy, S.; Arabi, Y.M. Infection with Middle East respiratory syndrome coronavirus. Can. J. Respir. 2015, 51, 102.
66. Gu, J.; Gong, E.; Zhang, B.; Zheng, J.; Gao, Z.; Zhong, Y.; Zou, W.; Zhan, J.; Wang, S.; Xie, Z.; et al. Multiple organ infection and the pathogenesis of SARS. J. Exp. Med. 2005, 202, 415–424.
67. Belouzard, S.; Chu, V.C.; Whittaker, G.R. Activation of the SARS coronavirus spike protein via sequential proteolytic cleavage at two distinct sites. Proc. Natl. Acad. Sci. USA 2009, 106, 5871–5876.
68. Shirato, K.; Kawase, M.; Matsuyama, S. Middle East respiratory syndrome coronavirus infection mediated by the transmembrane serine protease TMPRSS2. J. Virol. 2013, 87, 12552–12561.

69. Hoffmann, M.; Kleine-Weber, H.; Pöhlmann, S. A multibasic cleavage site in the spike protein of SARS-CoV-2 is essential for infection of human lung cell. Mol. Cell. 2020, 78, 779–784.

70. Chan, P.K.; To, K.F.; Lo, A.W.; Cheung, J.L.K.; Chu, I.; Au, F.W.L.; Tong, J.H.M.; Tam, J.S.; Sung, J.J.Y.; Ng, H.-K. Persistent infection of SARS coronavirus in colonic cells in vitro. J. Med. Virol. 2004, 74, 1–7.

71. Ma, D.; Chen, C.B.; Jhanji, V.; Yuan, X.-L.; Liang, J.-J.; Huang, Y.; Cen, L.-P.; Ng, T.K. Expression of SARS-CoV-2 receptor ACE2 and TMPRSS2 in human primary conjunctival and pterygium cell lines and in mouse cornea. Eye 2020, 34, 1–8.

72. Ou, X.; Liu, Y.; Lei, X.; Li, P.; Mi, D.; Ren, L.; Guo, L.; Guo, R.; Chen, T.; Hu, J.; et al. Characterization of spike glycoprotein of SARS-CoV-2 on virus entry and its immune cross-reactivity with SARS-CoV. Nat. Commun. 2020, 11, 1620.

73. Li, W.; Hulswit, R.J.G.; Widjaja, I.; Stalin Raj, V.; McBride, R.; Peng, W.; Widagdo, W.; Alejandra Tortorici, M.; van Dieren, B.; Lang, Y.; et al. Identification of sialic acid-binding function for the Middle East respiratory syndrome coronavirus spike glycoprotein. Proc. Natl. Acad. Sci. USA 2017, 114, E8508–E8517.

74. Huang, X.; Dong, W.; Milewska, A.; Golda, A.; Qi, Y.; Zhu, Q.K.; Marasco, W.A.; Baric, R.S.; Sims, A.C.; Pyrc, K.; et al. Human coronavirus HKU1 spike protein uses O-acetylated sialic acid as an attachment receptor determinant and employs hemagglutinin-esterase protein as a receptor-destroying enzyme. J. Virol. 2015, 89, 7202–7213.

75. Petrosillo, N.; Viceconte, G.; Ergonul, O.; Ippolito, G.; Petersen, E. COVID-19, SARS and MERS: Are they closely related? Clin. Microbiol. Infect. 2020, 26, 729–734.

76. Ceccarelli, M.; Berretta, M.; Venanzi Rullo, E.; Nunnari, G.; Cacopardo, B. Differences and similarities between Severe Acute Respiratory Syndrome (SARS)-CoronaVirus (CoV) and SARS-

CoV-2. Would a rose by another name smell as sweet? Eur. Rev. Med. Pharm. Sci. 2020, 24, 2781–2783.

77. Wang, H.; Li, X.; Li, T.; Zhang, S.; Wang, L.; Wu, X.; Liu, J. The genetic sequence, origin, and diagnosis of SARS-CoV-2. Eur. J. Clin. Microbiol. Infect. Dis. 2020, 39, 1629–1635.

78. Wan, Y.; Shang, J.; Graham, R.; Baric, R.S.; Li, F. Receptor recognition by the novel coronavirus from Wuhan: An analysis based on decade-long structural studies of SARS coronavirus. J. Virol. 2020, 94, e00127-20.

79. Xia, S.; Liu, M.; Wang, C.; Xu, W.; Lan, Q.; Feng, S.; Qi, F.; Bao, L.; Du, L.; Liu, S.; et al. Inhibition of SARS-CoV-2 (previously 2019-nCoV) infection by a highly potent pan-coronavirus fusion inhibitor targeting its spike protein that harbors a high capacity to mediate membrane fusion. Cell. Res. 2020, 30, 343–355.

80. Al-Tawfiq, J.A. Asymptomatic coronavirus infection: MERS-CoV and SARS-CoV-2 (COVID-19). Travel. Med. Infect. Dis. 2020, 35, 101608.

81. Fantini, J.; Di Scala, C.; Chahinian, H.; Yahi, N. Structural and molecular modelling studies reveal a new mechanism of action of chloroquine and hydroxychloroquine against SARS-CoV-2 infection. Int. J. Antimicrob. Agents 2020, 55, 105960.

82. Vandelli, A.; Monti, M.; Milanetti, E.; Rupert, J.; Zacco, E.; Bechara, E.; Ponti, R.D.; Tartaglia, G.G. Structural analysis of SARS-CoV-2 and predictions of the human interactome. bioRxiv 2020.

83. Milanetti, E.; Miotto, M.; Di Rienzo, L.; Monti, M.; Gosti, G.; Ruocco, G. In-Silico evidence for two receptors based strategy of SARS-CoV-2. bioRxiv 2020.

84. Creager, H.M.; Kumar, A.; Zeng, H.; Maines, T.R.; Tumpey, T.M.; Belser, J.A. Infection and replication of influenza virus at the ocular surface. J. Virol. 2018, 92, e02192-17.

85. Loon, S.C.; Teoh, S.C.; Oon, L.L.; Se-Thoe, S.-Y.; Ling, A.-E.; Leo, Y.-S.; Leong, H.-N. The severe acute respiratory syndrome coronavirus in tears. Br. J. Ophthalmol. 2004, 88, 861–863.

86. Colavita, F.; Lapa, D.; Carletti, F.; Lalle, E.; Bordi, L.; Marsella, P.; Nicastri, E.; Bevilacqua, N.; Giancola, M.L.; Corpolongo, A.; et al. SARS-CoV-2 isolation from ocular secretions of a patient with COVID-19 in Italy with prolonged viral RNA detection. Ann. Intern. Med. 2020, 173, 242–243.

87. World Health Organization. Report of the WHO-China Joint Mission on Coronavirus Disease 2019 (COVID-19); 14–20 February 2020; WHO: Geneva, Switzerland, 2020.

88. Chappey, B.; Beyssen, B.; Foos, E.; Ledru, F.; Guermonprez, J.L.; Gaux, J.C.; Myara, I. Sialic acid content of LDL in coronary artery disease: No evidence of desialylation in subjects with coronary stenosis and increased levels in subjects with extensive atherosclerosis and acute myocardial infarction: Relation between desialylation and in vitro peroxidation. Arter. Thromb. Vasc. Biol. 1998, 18, 876–883.

89. Nigam, P.K.; Narain, V.S.; Kumar, A. Sialic acid in cardiovascular diseases. Indian J. Clin. Biochem. 2006, 21, 54–61.

90. Gopaul, K.P.; Crook, M.A. Sialic acid: A novel marker of cardiovascular disease? Clin. Biochem. 2006, 39, 667–681.

91. Zhang, C.; Chen, J.; Liu, Y.; Xu, D. Sialic acid metabolism as a potential therapeutic target of atherosclerosis. Lipids Health Dis. 2019, 18, 173.

92. Zhang, Y.; Zheng, Y.; Li, J.; Nie, L.; Hu, Y.; Wang, F.; Liu, H.; Fernandes, S.M.; Zhong, Q.; Li, X.; et al. Immunoregulatory Siglec ligands are abundant in human and mouse aorta and are up-regulated by high glucose. Life Sci. 2019, 216, 189–199.

93. Kirkeby, S.; Jensen, N.E.; Mandel, U.; Poulsen, S.S. Asthma induction in mice leads to appearance of alpha2-3- and alpha2-6-linked sialic acid residues in respiratory goblet-like cells. Virchows Arch. 2008, 453, 283–290.

94. Rathod, S.; Shori, T.; Sarda, T.S.; Raj, A.; Jadhav, P. Comparative analysis of salivary sialic acid levels in patients with chronic obstructive pulmonary disease and chronic periodontitis patients: A biochemical study. Indian J. Dent. Res. 2018, 29, 22–25.

95. Guo, F.R. Active smoking is associated with severity of coronavirus disease 2019 (COVID-19): An update of a meta-analysis. Tob. Induc. Dis. 2020, 18, 37.

96. Vardavas, C.I.; Nikitara, K. COVID-19 and smoking: A systematic review of the evidence. Tob. Induc. Dis. 2020, 18, 20.

97. Liu, W.; Tao, Z.W.; Wang, L.; Yuan, M.-L.; Liu, K.; Zhou, L.; Wei, S.; Deng, Y.; Liu, J.; Liu, H.-G.; et al. Analysis of factors associated with disease outcomes in hospitalized patients with 2019 novel coronavirus disease. Chin. Med. J. 2020, 133, 1032–1038.

98. Wielgat, P.; Trofimiuk, E.; Czarnomysy, R.; Holownia, A.; Braszko, J.J. Sialylation pattern in lung epithelial cell line and Siglecs expression in monocytic THP-1 cells as cellular indicators of cigarette smoke-induced pathology in vitro. Exp. Lung Res. 2018, 44, 167–177.

99. Leung, J.M.; Yang, C.X.; Tam, A.; Shaipanich, T.; Hackett, T.-L.; Singhera, G.K.; Dorscheid, D.R.; Sin, D.D. ACE-2 expression in the small airway epithelia of smokers and COPD patients: Implications for COVID-19. Eur. Respir. J. 2020, 55, 2000688.

100. Reghunathan, R.; Jayapal, M.; Hsu, L.Y.; Chng, H.-H.; Tai, D.; Leung, B.P.; Melendez, A.J. Expression profile of immune response genes in patients with Severe Acute Respiratory Syndrome. BMC Immunol. 2005, 6, 2.

101. Frieman, M.; Heise, M.; Baric, R. SARS coronavirus and innate immunity. Virus Res. 2008, 133, 101–112.

102. Varki, A.; Angata, T. Siglecs—The major subfamily of I-type lectins. Glycobiology 2006, 16, 1R–27R.

103. Teuton, J.R.; Brandt, C.R. Sialic acid on herpes simplex virus type 1 envelope glycoproteins is required for efficient infection of cells. J. Virol. 2007, 81, 3731–3739.

104. Vanderheijden, N.; Delputte, P.L.; Favoreel, H.W.; Vandekerckhove, J.; Van Damme, J.; van Woensel, P.A.; Nauwynck, H.J. Involvement of sialoadhesin in entry of porcine reproductive and respiratory

syndrome virus into porcine alveolar macrophages. J. Virol. 2003, 77, 8207–8215.

105. Delputte, P.L.; Nauwynck, H.J. Porcine arterivirus infection of alveolar macrophages is mediated by sialic acid on the virus. J. Virol. 2004, 78, 8094–8101.

106. Zheng, Y.; Ma, X.; Su, D.; Zhang, Y.; Yu, L.; Jiang, F.; Zhou, X.; Feng, Y.; Ma, F. The roles of Siglec7 and Siglec9 on natural killer cells in virus infection and tumour progression. J. Immunol. Res. 2020, 2020, 6243819.

107. Zhao, D.; Jiang, X.; Xu, Y.; Yang, H.; Gao, D.; Li, X.; Gao, L.; Ma, C.; Liang, X. Decreased Siglec-9 expression on natural killer cell subset associated with persistent HBV replication. Front. Immunol. 2018, 9, 1124.

108. Yamanaka, M.; Kato, Y.; Angata, T.; Narimatsu, H. Deletion polymorphism of SIGLEC14 and its functional implications. Glycobiology 2009, 19, 841–846.

109. Tsai, C.M.; Riestra, A.M.; Ali, S.R.; Fong, J.J.; Liu, J.Z.; Hughes, G.; Varki, A.; Nizet, V. Siglec-14 enhances NLRP3-inflammasome activation in macrophages. J. Innate. Immun. 2019, 1–11.

110. Chang, Y.C.; Nizet, V. The interplay between Siglecs and sialylated pathogens. Glycobiology 2014, 24, 818–825.

111. Ali, S.R.; Fong, J.J.; Carlin, A.F.; Busch, T.D.; Linden, R.; Angata, T.; Areschoung, T.; Parast, M.; Varki, N.M.; Murray, J.C.; et al. Siglec-5 and Siglec-14 are polymorphic paired receptors that modulate neutrophil and amnion signaling responses to group B Streptococcus. J. Exp. Med. 2014, 211, 1231–1242.

112. Auyeung, T.W.; Lee, J.S.; Lai, W.K.; Choi, C.H.; Lee, H.K.; Lee, J.S.; Li, P.C.; Lok, K.H.; Ng, Y.Y.; Ming, W.W.; et al. The use of corticosteroid as treatment in SARS was associated with adverse outcomes: A retrospective cohort study. J. Infect. 2005, 51, 98–102.

113. Wielgat, P.; Mroz, R.M.; Stasiak-Barmuta, A.; Szepiel, P.; Chyczewska, E.; Braszko, J.J.; Holownia, A. Inhaled corticosteroids increase

siglec-5/14 expression in sputum cells of COPD patients. Adv. Exp. Med. Biol. 2015, 839, 1–5.

114. Zeng, Z.; Li, M.; Wang, M.; Wu, X.; Li, Q.; Ning, Q.; Zhao, J.; Xu, Y.; Xie, J. Increased expression of Siglec-9 in chronic obstructive pulmonary disease. Sci. Rep. 2017, 7, 10116.

115. Singh, A.K.; Singh, A.; Shaikh, A.; Singh, R.; Misra, A. Chloroquine and hydroxychloroquine in the treatment of COVID-19 with or without diabetes: A systematic search and a narrative review with a special reference to India and other developing countries. Diabetes Metab. Syndr. 2020, 14, 241–246.

116. Liu, J.; Cao, R.; Xu, M.; Wang, X.; Zhang, H.; Hu, H.; Li, Y.; Hu, Z.; Zhong, W.; Wang, M. Hydroxychloroquine, a less toxic derivative of chloroquine, is effective in inhibiting SARS-CoV-2 infection in vitro. Cell. Discov. 2020, 6, 16.

03 AN ANTIVIRAL PEPTIDE WITH A WIDE RANGE OF EFFECTIVENESS INHIBITS VIRAL INFECTION BY ATTACHING TO PHOSPHATIDYLSERINE IN THE VIRAL ENVELOPE

1. INTRODUCTION: AN OVERVIEW

Given the persistent danger posed by conventional viruses and the constant appearance of new virus species, it is imperative to continuously seek out novel antiviral agents. Promising treatment candidates are broad-spectrum antiviral inhibitors[1]. These compounds act on common elements shared by many groups of viruses, including the synthesis of viral RNA and DNA, viral proteases, and glycosylation of viral proteins (reviewed by [2]). An important class of broad-spectrum antivirals targets the lipid envelope of viruses [3,4,5,6,7]. These include not only antibodies and chemical compounds, but also cationic peptides present as defense peptides in the human host and other organisms. In essence, the applicability of such peptides as antiviral compounds targeting the envelope of viruses will depend on their ability to discriminate viral from human membranes.

Enveloped viruses, in contrast to non-enveloped ones, have their nucleocapsid surrounded by a lipid membrane. Depending on the virus, this lipid bilayer is derived from the plasma membrane or intracellular organelles. Although the membrane is host cell-derived, it may differ functionally and structurally from the membrane of origin in several aspects, including the lipid composition. Compared to host cell membranes, the envelope of many viruses is enriched for the phospholipid phosphatidylserine (PS). PS is the most abundant anionic lipid in eukaryotic cells. In resting cells, PS is mainly contained within the cell, with only very limited exposure to the extracellular environment. PS is enriched in some intracellular organelles, but the majority of PS is asymmetrically distributed on the inner leaflet of the plasma membrane [8]. This arrangement is achieved through the continuous translocation of phospholipids from the outer membrane leaflet to the inner membrane leaflet by ATP-dependent flippases. Consequently, the outer layer of the cellular plasma membrane does not contain any phosphatidylserine (PS). The loss of cellular organization in apoptotic cells is caused by the inactivation of flippases and the activation of scramblases. The PS exposure on apoptotic cells and cell debris leads to their absorption by surrounding cells via PS receptors. Numerous viruses use this absorption mechanism by exposing phosphatidylserine (PS) on their envelope, thus promoting the entry of the virus. The elevation of phosphatidylserine (PS) levels on the viral membrane might be attributed to the absence of a functional flippase enzyme. Alternatively, certain viruses have the ability to gather phosphatidylserine (PS) at specific locations where the virus is released, or they release from subcellular microdomains that have a high concentration of PS [9].

Another crucial characteristic of the host membrane is its ability to regenerate itself upon harm. This rapid procedure comprises multiple actions, including the identification of the injured spot, release of cytosolic vesicles in close proximity, restructuring of the cytoskeleton, uptake of the damaged membrane, and restoration of membrane stability [10,11]. The lipid membrane loses its self-renewal capacity once surrounding the viral nucleocapsid, thereby becoming particularly prone to membrane damage [5]. The structural and functional differences between virus and host cell membranes make viral membranes ideal targets for antiviral therapy.

Here, we identify a peptide with broad-spectrum antiviral activity, whose sequence is derived from the cowpox virus protein CPXV012. Within virus-infected cells, this protein helps to evade the immune system by inhibiting the transporter associated with antigen processing

(TAP), thereby interfering with MHC I-dependent antigen presentation [12,13,14,15,16]. We have identified the specific part of the CPXV012 protein that inhibits TAP. This region is a peptide that consists of the C-terminal domain of CPXV012. We currently have no evidence to suggest that this peptide is synthesized during CPXV infection. Remarkably, during functional studies including this peptide, we observed a substantial drop in the proportion of infected cells following infection. This discovery served as the foundation for the present investigation. In this study, we demonstrate that this peptide effectively hinders the infection of numerous enveloped viruses by directly interacting with the virus particles. Analysis of several forms of the CPXV012 peptide demonstrated that the presence of basic amino acids inside the peptide is crucial for its inhibitory impact. In addition, CPXV012 exhibits a considerable affinity for lipid monolayers and membranes that have a high concentration of the anionic phospholipid PS. These findings indicate that the CPXV012 peptide disrupts the structural stability of viral membranes that contain a high concentration of phosphatidylserine (PS). Therefore, it could potentially be investigated as an antiviral treatment for a wide variety of enveloped viruses.

2. MATERIALS AND METHODS

2.1. Cells and Viruses

MelJuSo (MJS) and BHK21 cells were cultivated with RPMI 1640 containing 10% FCS, 100 U/mL penicillin, 100 µg/mL streptomycin, and 2 mM L-glutamine (complete medium). HEK293T, HeLa, Huh7.5, and Vero cells were grown in DMEM supplemented with 10% FCS (5% for Vero cells), 100 U/mL penicillin, 100 µg/mL streptomycin, 1% nonessential amino acids, 2 mM L-glutamine, and 1% sodium pyruvate (complete DMEM). HepRG cells were differentiated in Williams E medium supplemented with 10% FCS (Fetalclone II from Hyclone, Thermo Scientific; Breda, The Netherlands), 100 U/mL penicillin, 100 µg/mL streptomycin, glutaMax, 0.023 IE/mL human insulin (Sanofi Aventis, Amsterdam, Netherlands), 4.7 µg/mL hydrocortisone (Pfizer, Capelle a/d IJssel, The Netherlands), 80 µg/mL Gentamicin, and 1.8% DMSO. HepG2.2.15 cells were cultivated in Williams E medium with 10% FCS, 100 U/mL penicillin, 100 μg/mL streptomycin, and 1% nonessential amino acids. LC5-RIC cells (EASY-HIT assay) were maintained as described previously [17]. Recombinant modified vaccinia virus Ankara (MVA) expressing eGFP under the natural early/late promoter p7.5 was used in this study (MVA-eGFP) [18]. MVA-eGFP was propagated and titrated in chicken embryonic fibroblasts (CEFs) according to standard methodology [19]. Vaccinia virus expressing eGFP (VACV-eGFP) was a generous gift from Dr. Jon Yewdell (NIH, Bethesda, USA). VACV-eGFP was propagated and titrated on CV-1 cells. The Cowpox virus strain Brighton Red (CPXV-BR) producing RFP/eGFP virus was initially acquired from Dr. Karsten Tischer at FU Berlin, Germany. Dr. Peter O'Hare from Imperial College London, UK, generously contributed HSV-1 that expresses the VP16-eGFP. Stocks were produced and titrated using Vero cells.

HBV particles were isolated from the supernatant of HepG2.2.15 cells using the method outlined in a previous study [20]. HIV-1 stocks that can be transmitted to others were generated using the previously described method. [17]. Measles virus expressing eGFP (MV-eGFP) was generated as previously described [21]. Vesicular stomatitis virus expressing Luciferase (VSV-deltaG (Luc)) was a kind gift from Dr. Gert Zimmer (Institute of Virology and Immunology, Mittelhäusern, Switzerland). Recombinant adenovirus expressing GFP and ovalbumin (AdGOva) was kindly provided by Dr. Percy Knolle (Institutes of Molecular Medicine and Experimental Immunology, Bonn, Germany) and virus stocks were prepared as previously described [22]. Non-spreading Rift Valley fever virus (RVFV) replicon particles were produced and titrated as described previously [23]. The production and titration of the Coxsackie B3 virus expressing Renilla Luciferase (RLuc-CVB3) was carried out following the previously published method. [24].

2.2. Peptides

The peptide employed in this study is composed of amino acids 36 to 69 of the CPXV012 protein (QEGISRFKICPYHWYKQHMSLLFRRYYHKLDSII). The control peptide is a fragment of the UL49.5 protein, which is a TAP inhibitor in the pseudorabies virus. It consists of amino acid residues 28 to 59 of this protein, with the following sequence: STEGPLPLLREESRINFWNAAUAARGVPVDQP. The "U" in the amino acid sequence refers to alpha-amino-n-butyric acid, as seen below.The synthesis of synthetic peptides was carried out utilizing standard Fmoc chemistry. Preloaded Tentagel resins were utilized, along with PyBop/N-methylmorpholine for in situ activation and a solution of 20% piperidine in N-methylpyrrolidinone for Fmoc removal [25]. The process of coupling was carried out for a duration of 75 minutes. Following the removal of Fmoc, the peptides were cleaved using a mixture of trifluoroacetic acid and water in a ratio of 19:1 by volume. Additional scavengers were included in the cleavage solution when the peptide sequence contained C or W residues. Peptides were extracted by the process of ether/pentane precipitation. The purity of the peptides was assessed using reversed-phase HPLC–mass spectrometry, while their integrity was evaluated using MALDI-TOF mass spectrometry, which revealed the predicted molecular masses. Every peptide was synthesized with a biotin group at the beginning and an amide group at the end. The cysteines were substituted with the isosteric alpha-amino-n-butyric acid. The lyophilized peptides were reconstituted in dimethyl sulfoxide (DMSO). The NanoDrop spectrophotometer (Thermo Scientific, Amsterdam, The Netherlands) was used to verify the amounts of peptides by measuring the absorbance of Trp at 280 nm. The peptides were divided into smaller portions and placed in Eppendorf LoBind microcentrifuge tubes (Sigma-Aldrich, Zwijndrecht, The Netherlands). These tubes were then stored at a temperature of -20 °C for a brief period of time.

2.3. Cell Viability Assays

The cells were exposed to CPXV012 peptide, control peptide, or DMSO at the specified concentrations and incubated at 37 °C for a duration of 20 to 24 hours. Subsequently, the vitality of the cells was assessed utilizing several assays. Flow cytometry was used to differentiate between viable and non-viable cells by performing DAPI labeling. The Cell Titer-Blue Cell Viability Assay (Promega, Leiden, The Netherlands) was conducted following the guidelines provided by the manufacturer. Neutral Red uptake was conducted using the Applichem brand from Darmstadt,

Germany, using the previously reported method [26]. The WST-1 salt conversion assay (Roche, Woerden, The Netherlands) was conducted following the manufacturer's instructions, either 7 or 20 hours after exposure to the peptides. There was no discernible distinction between the measurements taken 7 hours and 20 hours after exposure.

2.4. Virus Inhibition Assays

2.4.1. MVA, VACV, and Cowpox

MJS, BHK21, HeLa, and HEK-293T cells (1 × 105) were seeded in 24-well plates in a total volume of 400 μL and incubated overnight at 37 °C. Cells were infected with MVA-eGFP, VACV-eGFP, or CPXV-RFP/eGFP at 37 °C for 18 to 20 h in medium containing 0.1% DMSO and the indicated concentrations of either CPXV012 peptide or control peptide UL49.5 or 0.1% DMSO only as vehicle control. Flow cytometric analysis was performed to quantify the percentage of infected cells as indicated by eGFP or RFP expression.

Where mentioned, MVA-eGFP was pretreated with 50, 100, or 150 μg/mL CPXV012 peptide, control peptide, or DMSO vehicle at 37 °C for 1 h. Thereafter, the virus-CPXV012 peptide mixture was diluted 1:10 by volume and used to infect MJS cells (corresponding to an MOI of 10) resulting in a final concentration of 5, 10, or 15 μg/mL peptide in the culture medium. As a control, MJS cells were incubated with the same concentrations of CPXV012 peptide (5, 10, and 15 μg/mL) during infection with MVA-eGFP that was not pretreated with peptides. After 20 h, the cytometric analysis was performed to quantify the percentage of infected cells indicated by eGFP-expression.

For qPCR analysis (Figure 1B), MJS cells were infected with MVA-eGFP (MOI 10) in the presence of 100 μg/mL of either CPXV012 peptide or control peptide in medium with 0.1% DMSO or in 0.1% DMSO only as vehicle control at 37 °C for 20 h. RNA was isolated using the RNeasy Mini Kit (Qiagen, Venlo, The Netherlands). A total of 3 μg of RNA was digested with 10 U of DNase I (Roche, Woerden, The Netherlands) and cDNA synthesis was performed by using 200 U Superscript III RNase H reverse transcriptase (Invitrogen, Amsterdam, The Netherlands), 7.5

pmol oligo(dT)12–18 (Invitrogen), 20 U of RNasin (Promega), and 10 mM each deoxynucleoside triphosphate (Qiagen). For semiquantitative analysis of viral mRNAs, cDNA was used as the template for a PCR with the indicated primers: B8R (fwd ATCCGCATTTCCAAAGAATG, rev ACATGTCACCGCGTTTGTAA), H3L (fwd GTCTTGAAGGCAATGCATGA, rev TCCCGATGATAGACCTCCAG) and G8R (fwd ATC GAT AAA CTG CGC CAA AT), and rev CTC CGC GGT AGA ACA CTG AT). Quantitative RT-PCR analysis was performed by using the LightCycler DNA Master SYBR green I kit (Roche) and LightCycler 1.5 (Roche). Gene expression was determined using the 2-ΔΔCT method. The results are quantified relative to the housekeeping gene GAPDH [27,28].

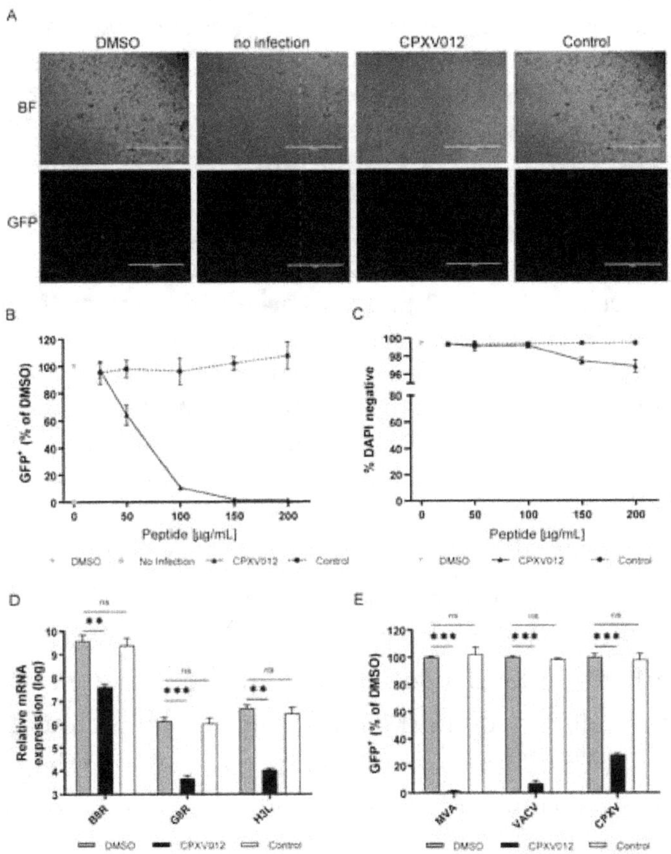

Figure 1. CPXV012 peptide inhibits poxvirus infection. (A) MJS cells were infected with MVA-eGFP (MOI 10) under different conditions (0.1% DMSO control, CPXV012 and control peptide UL49.5 in 0.1% DMSO). At 20 h.p.i., infection was assessed by fluorescence microscopy (bar size 1 mm). Results

are representative of at least three independent experiments. BF: Bright field. GFP: Green fluorescent protein (MVA infection). (B) MJS cells were infected with MVA-eGFP (MOI 10) in the presence of 0.1% DMSO vehicle control, or CPXV012 peptide, or control peptide, both in 0.1% DMSO. At 18 to 20 h post infection, cells were harvested and analyzed for viral gene expression. The number of eGFP-positive cells was quantified using flow cytometry. Mean ± S.E.M. of three independent experiments is shown. (C) MJS cells were treated for 20 h with the indicated concentrations of CPXV012 or control peptide. Subsequently, cells were harvested and stained with DAPI for flow cytometric analysis. S.E.M. of three independent experiments is shown. (D) MJS cells infected with MVA-eGFP (MOI 10) in the presence of 100 µg/mL CPXV012 or UL49.5 (control) peptide in 0.1% DMSO or 0.1% DMSO only were lysed 20 h post infection. RNA was isolated and qPCR was performed for expression of viral genes B8R, H3L, and G8R. Mean ± S.E.M. of three independent experiments is shown. Log-transformed data were analyzed with one-way ANOVA followed by multiple comparisons Dunnett's test (the mean of each column was compared to that of the DMSO control). (E) CPXV012 peptide inhibits infection with the poxviruses MVA-eGFP, VACV-eGFP, and CPXV-RFP/eGFP. MJS cells were infected with VACV-eGFP (MOI 10) or CPXV-RFP/eGFP (MOI 10) in the presence of 150 µg/mL CPXV012 or UL49.5 control peptide in 0.1% DMSO or 0.1% DMSO only. After 18 to 20 h, infection was quantified by cytometric analysis of fluorescent cells. Mean ± S.E.M. of three independent experiments is shown using different batches of peptide. Data were analyzed with one-way ANOVA followed by multiple comparisons Dunnett's test (the mean of each column was compared to that of the DMSO control). ($p < 0.01$; *** $p < 0.001$).**

2.4.2. HSV-1

MJS cells (105/well) were seeded in a 24-well plate and cultured overnight at 37 °C. HSV-1-eGFP was mixed with CPXV012 peptide, control peptide UL49.5, or DMSO only in complete medium. DMSO concentration was 0.1% with or without peptide. The medium of cells was replaced with virus/peptide inoculum (MOI 0.1) and cells were incubated for 16 h at 37 °C. Cells were harvested, fixed using 1% formaldehyde, and the number of eGFP-positive cells was determined using flow cytometry.

2.4.3. HBV

HepRG cells were cultivated on 48-well plates and differentiated for a duration of two weeks. The media was eliminated and the cells were not treated, or the CPXV012 peptide was introduced to the medium containing 0.1% DMSO at the specified concentrations. A solution with a concentration of 0.1% dimethyl sulfoxide (DMSO) was employed as a control for comparison. The cells were infected with Hepatitis B virus (HBV) at a multiplicity of infection (MOI) of 200 and incubated for 16 hours. The medium was eliminated, and the cells were rinsed three times with PBS. Subsequently, fresh medium was introduced, and the cultures were cultured for an additional 12 days with a medium replacement every three days. Afterwards, the liquid above the sediment was gathered following the process of spinning at a high speed to eliminate any solid particles from the cells (5 minutes, 350 times the force of gravity). The concentration of HBeAg was determined using commercially available immunoassays, as previously explained [20]. The HBV DNA was isolated from the liquid portion using the BioSprint 96 One-For-All Vet Kit (Qiagen) and its amount was measured using qPCR.

2.4.4. HIV-1

The EASY-HIT assay was performed as described previously [17]. Briefly, 1 × 104 LC5-RIC cells were plated in 96-well plates and incubated at 37 °C for 24 h. Thereafter, cells were treated with medium containing 0.1% DMSO and the indicated concentrations of CPXV012 peptide or 0.1% DMSO only as vehicle control for 1 h prior to infection with HIV inoculum. After 48 h, cultures were assayed for cellular reporter gene expression by quantification of total fluorescence intensity of each culture using a fluorescence microplate reader (step 1). To assess titers of infectious virus in the culture supernatant (virion production in primary infected cells), 20 μL of medium was added to fresh LC5-RIC cells and incubated for another 72 h before reporter gene expression was quantified (step 2).

2.4.5. Measles and VSV

Vero cells (1 × 105 cells/well) were plated in 12-well plates and left untreated or treated with medium containing 0.1% DMSO and the indicated concentrations of CPXV012 peptide or 0.1% DMSO only as vehicle control. Cells were infected with MV-eGFP (MOI 0.1). Once maximum giant cell formation was observed at 48 h post infection, microscopic fluorescence images were taken by using an inverted microscope CKX41 (Olympus) with an LCachN/10×/0.40 Phc/1/FN22 UIS objective. Thereafter, the medium was removed, and infection was measured based on the expression of eGFP. Fluorescence was detected using an Infinite 200 PRO Tecan reader (fluorescence bottom reading for cell-based assays). For VSV infections, Huh7.5 cells (2 × 105 cells/well) were seeded in 12-well plates, and peptides were added at the concentrations indicated. DMSO was used as vehicle control. Cells were infected with VSVdeltaG (Luc) (MOI 0.6). At 18 h post infection, luciferase was measured using the luciferase assay system (Promega) and normalized to the protein content of the individual sample as determined by a Bradford assay.

2.4.6. Adenovirus

HEK-293 cells (4 × 105 cells/well) were seeded in 12-well plates in a total volume of 1 mL and incubated overnight at 37 °C. Cells were treated with medium containing 0.1% DMSO and the indicated concentrations of CPXV012 peptide or 0.1% DMSO only as vehicle control at 37 °C for 1 h. Afterward, cells were infected with an MOI of 10 with AdGOva for 24 h. Photometric analysis was performed with the Infinite 200 PRO Tecan to quantify the percentage of infected cells based on GFP-expression.

2.4.7. Rift Valley Fever Virus

MJS cells (2 × 104 cells/well) were seeded in a 96-well plate and incubated for two days at 37 °C. Cells were treated with medium containing 0.1% DMSO and the indicated concentrations of either CPXV012 peptide or control peptide UL49.5 or 0.1% DMSO only as vehicle control. After 15 min, RVFV-eGFP was added to the cells. Twenty-four hours after infection, cells were harvested and the percentage of eGFP-positive cells was determined using flow cytometry.

2.4.8. Coxsackievirus B3

MJS cells (104 cells/well) were seeded in a 96-well plate. After overnight incubation at 37 °C, cells were infected with RLuc-CVB3 in the presence of medium containing 0.1% DMSO and the indicated concentrations of CPXV012 peptide or 0.1% DMSO only as vehicle control. After 7 h of infection, cells were lysed and renilla luciferase expression levels were quantified using the Renilla Luciferase Assay System kit (Promega).

2.5. Preparation of Large Unilamellar Vesicles (LUVs)

Calcein-encapsulated LUVs were prepared using 1,2-dioleoyl-sn-glycero-3-phosphocholine (DOPC) or a mixture of DOPC and 1,2-dioleoyl-sn-glycero-3-phospho-L-serine (DOPS) (Avanti Polar Lipids) in a 7:3 molar ratio. A stock solution of DOPC and DOPS in chloroform (10 mM) were mixed in a glass tube. The solvent was evaporated with dry nitrogen gas yielding a lipid film that was subsequently kept in a vacuum desiccator for 20 min. Lipid films were hydrated for 30 min in buffer containing 10 mM Tris, 50 mM NaCl at pH 7.4 resulting in a total lipid concentration of 10 mM. For calcein-encapsulated LUVs, 50 mM of calcein was added during hydration. The lipid suspension was freeze-thawed ten cycles, at temperatures of -80 and +40 °C, respectively, and eventually extruded 10 times through 0.2 μM-pore size filters (Anotop 10, Whatman, UK). For the preparation of calcein-encapsulated LUVs, free calcein was separated from calcein-filled LUVs using size exclusion column chromatography (Sephadex G-50 fine) and eluted with 10 mM Tris-HCl, 150 mM NaCl buffer at pH 7.4. Finally, the phospholipid content of lipid stock solutions and vesicle preparations was determined as inorganic phosphate according to Rouser [29].

2.6. Circular Dichroism

Circular dichroism (CD) experiments were performed as described previously (15). Briefly, the CD spectra of 625 μM LUVs and/or 100 μg/mL peptides diluted in 10 mM MES buffer (pH 6.2) were recorded on a Jasco 810 spectropolarimeter (Jasco, Easton, MD) over a wavelength range of 200 to 250 nm. Each reported spectrum is the average of five independent scans recorded every 1 nm at a scan rate of 20 nm/min at room temperature in cuvettes with a path length of 1.0 mm.

2.7. Langmuir Monolayers

Peptide-induced changes in the surface pressure of a monomolecular layer (monolayer) of phospholipids at a constant surface area were measured using a Langmuir Microtrough XL device (Kibron, Helsinki, Finland). A Teflon trough was filled with 16 mL PBS (pH 7.4) and lipid monolayers of DOPC or DOPC/DOPS (ratio 7:3) were spread from a 0.5 mM stock solution in chloroform at the air–buffer interface. The buffer below the lipid monolayer (subphase) was continuously stirred using a magnetic stirrer. Upon stabilization of the initial surface pressure to 25 mN/m, a freshly prepared stock of peptide in DMSO was injected into the subphase, resulting in a final peptide concentration of 0.25 μM.

2.8. Membrane Permeability Assay

Membrane permeability was measured in standard 96-well transparent microtiter plates in a plate reader (Spectrafluor, Tecan, Salzburg, Austria). Peptides (5 μL of 0.2 mM in DMSO) were added to calcein-containing LUVs (195 μL of lipid vesicles (50 μM) in 10 mM Tris–HCl, 100 mM NaCl buffer (pH 7.4)). As positive control human islet amyloid polypeptide (hIAPP; Bachem) 5 μL of a 0.2 mM in DMSO) was added to calcein-containing LUVs. As a negative control, murine IAPP ((mIAPP; Bachem) 5 μL of 0.2 mM in DMSO) was added to calcein-containing LUVs. For blank only, 5 μL DMSO was added to calcein-containing LUVs. Directly after the addition of all components, the microtiter plate was shaken for 10 s. Fluorescence was measured from the top, every 5 min, using a 485 nm excitation filter and a 535 nm emission filter at 25 °C. The maximum leakage at the end of each measurement was determined by adding 1 μL of 10% Triton X-100 to a final concentration of 0.05% (v/v). The release of fluorescent dye was calculated according to Equation (1):

$$L(t) = (Ft - F0)/(F100 - F0) \quad (1)$$

L(t) is the fraction of dye released (normalized to membrane leakage), Ft is the measured fluorescence intensity, and F0 and F100 are the fluorescence intensities at times t = 0 and after addition of Triton X-100, respectively. All membrane leakage assays were performed two times, each in duplicate, on different days, using different IAPP stock solutions.

2.9. Statistical Analysis

Statistical significance was analyzed by one-way ANOVA testing, followed by Dunnett's multiple comparisons test (the mean of each column was compared to that of the DMSO control), using GraphPad Prism 8.0.1. In Figure 1D, statistical analysis was performed on the plotted data transformed as follows: Y = Log(Y). p-values of <0.05 were considered significant (* $p < 0.05$; ** $p < 0.01$, *** $p < 0.001$).

3. RESULTS

3.1. CPXV012 Peptide Prevents Infection by Different Viruses in Cell Culture

We tested a number of peptides for their ability to inhibit TAP and subsequent MHC-I antigen presentation. These peptide sequences were based on the active domains of viral TAP-inhibiting proteins CPXV012 (from cowpox virus) and UL49.5 (from pseudorabies virus) [30]. While testing their ability to block MHC I antigen presentation during virus infection, we found that the CPXV012 peptide was uniquely capable of inhibiting virus reporter expression. To assess the effect of CPXV012 peptide on viral infection, different infection inhibition assays were performed using a variety of viruses and cell types. Cells were treated with CPXV012 peptide (34 amino acids, QEGISRFKICPYHWYKQHMSLLFRRYYHKLDSII) or a similar-length control peptide derived from the pseudorabies virus protein UL49.5. The viral inoculum was mixed with the peptide at different concentrations and immediately added to cells. After incubation for the indicated times, infection was determined based on viral gene expression, or viral DNA or mRNA synthesis.

Infection of the human melanoma cell line MelJuSo (MJS) with recombinant modified vaccinia virus Ankara (MVA) expressing eGFP (MVA-eGFP) was confirmed via microscopy (Figure 1A). The cell monolayer showed the typical cytopathic effect upon infection (in the presence of DMSO, peptide vehicle control; bright field) and eGFP fluorescence could be detected. A similar pattern was verified in the presence of the control peptide. In contrast, upon infection and incubation with the CPXV012 peptide (100 µg/mL), the monolayer was microscopically indiscernible from that of the uninfected control. MVA-eGFP infection was inhibited in a concentration-dependent manner (Figure 1B). Inhibition starts at 50 µg/mL and is stronger upon increasing peptide concentrations, reaching 98.3% (± 0.3) at 150 µg/mL CPXV012 peptide (Figure 1B). To test potential cytotoxicity of the peptides, the amount of DAPI-negative (live, i.e., cells whose membrane was not compromised/permeable) cells was quantified. We did not observe any decrease in live cells up to the highest concentrations (200 µg/mL) (Figure 1C). To further test any effect on cell viability, other assays were performed. The WST-1 assay measures the net metabolic activity of the cells (it is based on the enzymatic conversion of the WST-1 salt into the colored dye formazan in viable cells), and the Neutral Red uptake assay relies on the staining of lysosomes in viable cells upon active transport of the cationic dye. MJS cell viability was not severely impaired upon addition of up to 200 µg/mL CPXV012 peptide using the WST-1 assay, and the Neutral Red assay. Cell titer blue assays were performed to test the viability of other cell lines (Vero and Huh7.5 cells) used in this study in the setting of other viral infections. Again, no significant effect was observed at the highest peptide concentrations.

To confirm the inhibitory effect of CPXV012 peptide on infection of MJS with MVA-eGFP, qPCR analysis for expression of viral genes was performed (Figure 1D). CPXV012 peptide decreased the expression of the early-expressed gene B8R up to two logs, relative to its expression upon infection in the presence of DMSO vehicle only. The decrease in expression of the intermediate-expressed gene G8R and late expressed gene H3L was more pronounced (more than two logs). No effect was seen if cells were treated with the control peptide compared to the DMSO sample (Figure 1D). To assess whether the inhibitory effect on MVA-eGFP infection also occurs in other cell types, HEK-293T, BHK21, and HeLa cells were used. In all cell lines, a strong decrease in infection (87.4 to 94.2%) was observed.

To investigate whether other poxviruses besides MVA were inhibited by CPXV012 peptide, we tested the effect on Vaccinia virus (VACV) strain WR and cowpox virus strain BR (CPXV). Again, infection of MJS with both viruses was inhibited (93 ± 1.7% for VACV and 72 ± 1.1% for CPXV) by CPXV012 peptide (150 μg/mL) while no effect was seen with the control peptide in comparison to the DMSO sample (Figure 1E). Thus, CPXV012 peptide inhibits infection of different members of the poxvirus family.

To further characterize the inhibitory effect of CPXV012, in vitro infection inhibition assays were performed with a collection of viruses that differ in genome content (RNA or DNA) and structural composition (enveloped or non-enveloped). These viruses include herpes simplex virus-1 (HSV-1), Hepatitis B virus (HBV), HIV-1, adenovirus, measles virus, vesicular stomatitis virus (VSV), coxsackievirus B3 (CVB3), and Rift Valley fever virus (RVFV) (Figure 2 and Figure 3).

The inhibitory effect of CPXV012 peptide on HSV-1-eGFP, a large dsDNA virus from the herpesvirus family, was evaluated upon infection of MJS cells. The cells were infected with HSV-1-eGFP in the presence of peptide or DMSO vehicle control. After 16 h, the amount of eGFP-positive cells was determined using flow cytometry (Figure 2A). As observed for the poxviruses, CPXV012 peptide showed a dose-dependent inhibition of HSV-1 infection (Figure 2A; 75.9 ± 5.7% when using 150 μg/mL peptide).

For HBV, an enveloped dsDNA virus of the hepadnavirus family, infectivity was monitored by quantifying the amount of viral envelope protein (HBeAg) and viral DNA in the supernatant of HepRG-infected cells (Figure 2B). Both HBeAg and viral DNA were decreased in the supernatant of cells treated with CPXV012 peptide in a concentration-dependent manner. At a peptide concentration of 100 μg/mL, a decrease of 84.0 ± 3.0% of HBeAg and 73.6 ± 2.3% of viral DNA was observed.

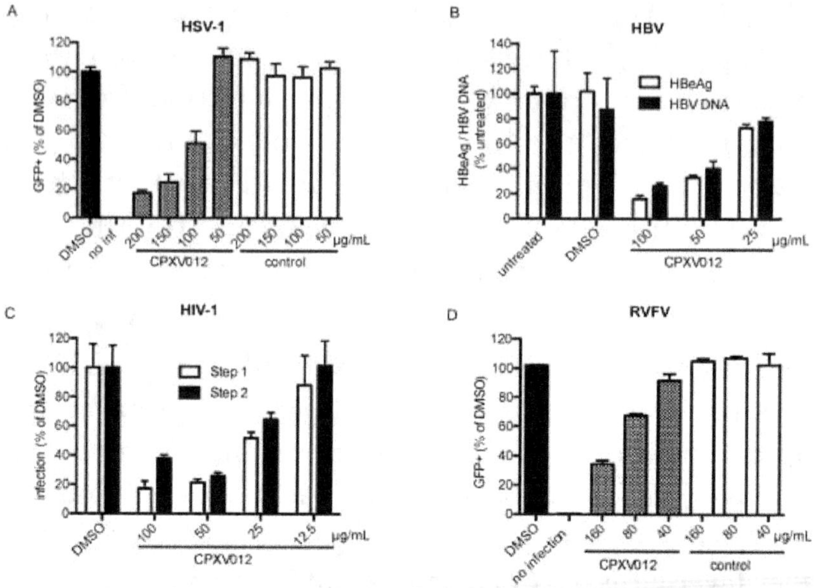

Figure 2. CPXV012 peptide inhibits infection with HSV-1, HBV, HIV, and RVFV. (A) CPXV012 peptide inhibits HSV-1 infection. MJS cells were left uninfected or infected with HSV-1 (MOI 0.1) in the presence of CPXV012 or UL49.5 as control at the indicated peptide concentrations in 0.1% DMSO or 0.1% DMSO only. At 16 h after infection, cells were harvested, and the amount of eGFP-positive cells was quantified using flow cytometry. Mean ± S.E.M. of three independent experiments is shown. (B) Effect of CPXV012 peptide on HBV infection. Differentiated HepRG cells were left untreated or CPXV012 peptide in 0.1% DMSO was added using the concentrations indicated or 0.1% DMSO only as control. Cells were infected with HBV (MOI 200) for 16 h. Subsequently, cells were washed and fresh medium was added. After 12 days, the supernatant was analyzed for HBeAg and HBV DNA. Mean ± S.E.M. of three independent experiments is shown. (C) CPXV012 peptide inhibits HIV infection. LC5-RIC cells were treated with the indicated concentrations of CPXV012 peptide in 0.1% DMSO or 0.1% DMSO only as control for 1 h and infected with HIV for 48 h according to the EASY-HIT assay system (17). Cellular reporter expression was quantified using a fluorescence microplate reader to assess the ability of HIV to infect LC5-RIC cells (Step 1). Next, 20 µL of culture supernatant was added to fresh LC5-RIC cells and incubated for another 72 h before fluorescence detection was performed to assess virion production from the first round of infection (Step 2). Data are shown as the percentage of infection referred to the respective DMSO control which was set to 100%. Mean ± S.E.M. of three independent experiments is shown. (D) CPXV012 peptide inhibits infection with RVFV. MJS cells were left uninfected or

infected with RVFV-eGFP in the presence of CPXV012 or control peptide (UL49.5) at the indicated peptide concentrations in 0.1% DMSO or 0.1% DMSO only. After 24 h, cells were harvested and the amount of eGFP-positive cells was quantified using flow cytometry. Mean ± S.E.M. of three independent experiments is shown.

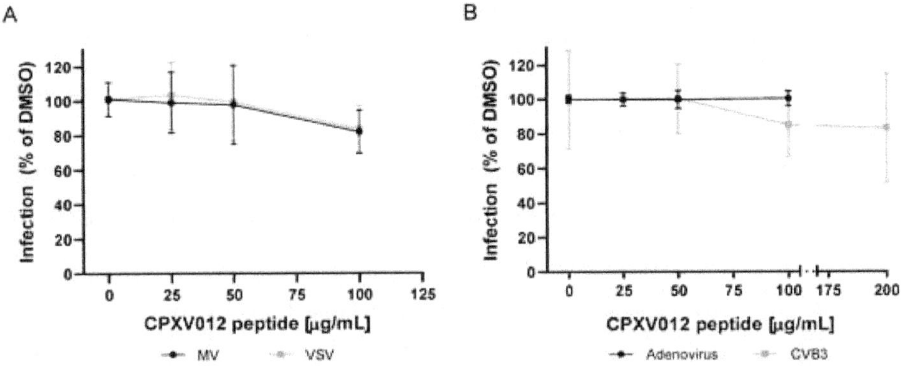

Figure 3. CPXV012 peptide does not inhibit infection with Measles virus, VSV, Adenovirus, or Coxsackie B3 virus and no difference was observed with the control peptide UL49.5 (data not shown). (A). Measles virus infection: Vero cells were treated with the indicated concentrations of CPXV012 peptide or DMSO as control, and infected with eGFP-expressing measles virus (MV-eGFP) (MOI 0.1). Once maximum giant cell formation was observed (approximately 48 h post infection), infection was quantified by detection of eGFP using a microplate reader. VSV infection: Huh7.5 cells were treated with the indicated concentrations of CPXV012 peptide in 0.1% DMSO or 0.1% DMSO only, and infected with luciferase-expressing VSV-deltaG (Luc) (MOI 0.6). At 18 h post infection, luciferase activity was measured to assess VSV infection of the culture. Data are shown as percentage of infection referred to the respective DMSO control which was set to 100%. Mean ± S.E.M. of each three independent experiments is shown. (B) Adenovirus infection: HEK-293 cells were treated with the indicated concentrations of CPXV012 peptide or DMSO for 1 h and infected with eGFP-expressing adenovirus (AdGOva) (MOI 10) for 24 h. Infection was quantified by cytometric analysis of eGFP expression. Coxsackievirus B3 infection: MJS cells were infected with RLuc-CVB3 in the presence of the indicated concentration of peptide or DMSO as vehicle control. After 7 h, cells were lysed and Renilla Luciferase expression was quantified. Data are shown as percentage of infection referred to the respective DMSO control which was set to 100%. Mean ± S.E.M. of each three independent experiments are shown.

The infectivity of HIV-1, an enveloped ssRNA virus, was monitored using the EASY-HIT assay on LC5-RIC reporter cells. In this essay, both early and late phases of HIV replication are assessed: in step 1, the levels of a fluorescent reporter protein induced during the early phase of HIV replication are quantified, and in step 2 the production of infectious virions in primary infected cells is determined. HIV infection was reduced in the presence of CPXV012 peptide, but not in the presence of DMSO vehicle control (Figure 2C). Both primary viral infection (step 1) and viral replication (step 2) were dose-dependently affected, with an inhibition of 82.7 ± 4.9% (step 1) and 62.2 ± 2.4% (step 2) at the highest peptide concentration (100 μg/mL).

For RVFV, an enveloped ssRNA virus of the Bunyavirus family, infection was monitored by viral eGFP expression 24 h post infection (Figure 2D). Infection was reduced by CPXV012 peptide in a concentration-dependent manner (65.5 ± 2.3% at 160 μg/mL).

In contrast, CPXV012 peptide showed no effect on infection by the measles virus, an enveloped ssRNA virus of the paramyxovirus family (Figure 3A). The formation of measles virus-induced syncytia was unchanged. Replication of measles virus and cell-to-cell spread was assessed by quantification of viral titers present in infected cultures using plaque assays. No difference was found between CPXV012 peptide-treated samples, the DMSO vehicle control, or the control peptide (data not shown). For VSV, an enveloped ssRNA virus of the rhabdovirus family, virus-encoded Renilla luciferase activity was measured 18 h after infection (Figure 3A). Luciferase activity was unchanged in the presence of CPXV012 peptide.

Next, the effect of CPXV012 on the infection of two non-enveloped viruses was tested. The infectivity of adenovirus, a dsDNA virus, was not altered in the presence of CPXV012 peptide, as measured by viral eGFP expression (Figure 3B). For the coxsackie B3 virus, a non-enveloped ssRNA virus, infection was monitored by the activity of the virus-encoded Renilla luciferase. No change in Renilla luciferase activity was observed in the presence of CPXV012 or control peptide (Figure 3B).

These results identify the CPXV012 peptide as an antiviral peptide acting on enveloped viruses.

3.2. CPXV012 May Bind to Viral Particles

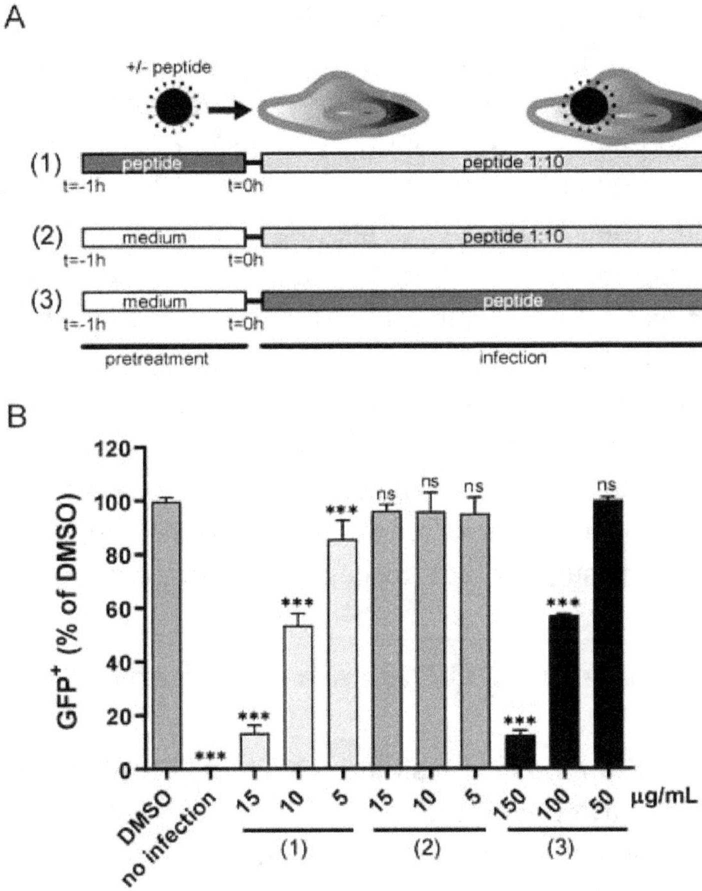

Figure 4. CPXV012 peptide inhibits infection by binding to viral particles. (A) Timeline of infection experiment as shown in (B). (1) MVA-eGFP was pretreated with 50, 100, or 150 µg/mL peptide or DMSO in culture medium at 37 °C for 1 h. Subsequently, the virus-peptide pre-incubation mixture was diluted ten times and used to infect MJS cells (corresponding to an MOI of 10), resulting in a final concentration of 5, 10, and 15 µg/mL peptide in the culture medium. (2–3) MVA-eGFP was incubated in culture medium only at 37 °C for 1 h. The virus mixture was added to MJS cells in the presence of 5, 10, or 15 µg/mL peptide (2) or 50, 100, or 150 µg/mL peptide (3) in the culture medium. (B) After 18 to 20 h of infection as shown in (A), the amount of infected eGFP-positive cells was quantified

using flow cytometry. Mean ± S.E.M. of three independent experiments is shown using different batches of peptide. Data were analyzed with one-way ANOVA followed by multiple comparisons Dunnett's test (the mean of each column was compared to that of the DMSO control). (* $p < 0.001$).**

To find out whether CPXV012 peptide mediates its inhibitory effect by interacting directly with viral particles, we performed a modified inhibition assay using MJS cells and MVA-eGFP. Instead of simultaneously adding CPXV012 peptide and virus to the cells, viral particles were pretreated with CPXV012 peptide for 1 h at 37 °C. After this preincubation, the virus-CPXV012 peptide mixture was diluted ten times and used to infect cells. Thus, the final peptide concentration during infection was 5 to 15 μg/mL (Figure 4A). Pretreating the virus with a high concentration of CPXV012 peptide before infection affected infectivity to a similar level as the presence of a high concentration of CPXV012 peptide during infection (Figure 4B). This effect was not due to the remaining CPXV012 peptide in the culture medium, as the diluted peptide alone was not sufficient to block infection when added directly to the cells. These data suggest that the peptide can directly act on viral particles.

3.3. The peptide variants CPXV012 have varying effects on virus infection

To determine the amino acid residues within the CPXV012 peptide crucial for virus inhibition, alanine substitution variants of the CPXV012 peptide were synthesized. These variants had small stretches of amino acid residues replaced by alanine residues (Figure 5A). To test the inhibitory capacity of these variants, MJS cells were infected with MVA-eGFP in the presence of 100 μg/mL of each of these peptides (Figure 5B). CPXV012 variants with the N-terminal five amino acid residues (Ala1) or the C-terminal four amino acid residues substituted by alanine (Ala7) inhibited MVA infection to similar levels as wild-type CPXV012 peptide. Substituting amino acid residues 6 to 20 for alanine (CPXV012-Ala2/Ala3/Ala4) significantly affected the inhibitory capacity of the peptides. Substituting amino acid residues 20 to 30 for alanine (Ala5 and Ala6) completely abolished the inhibitory effect of CPXV012 peptide.

Interestingly, the CPXV012 peptide variants with lower or no inhibitory capacity also had a lower net positive charge at pH 7.4 compared to the wild type peptide, due to the substitution of the charged amino acid residues lysine, arginine, or histidine, the latter being only partially charged. To test the role of these amino acid residues in virus inhibition, a peptide was synthesized with the charged amino acid residues replaced for alanine (Ala8). Indeed, this CPXV012 peptide variant lost its inhibitory capacity, thus confirming the role of basic amino acid residues in inhibiting virus infection.

The preferential inhibition of enveloped viruses suggests that CPXV012 peptide interacts with a common structure within these viral particles. A major constituent of these virions is the phospholipids forming the lipid bilayer of the viral envelope. To test the interaction between CPXV012 peptide and phospholipids, Langmuir monolayers were formed using the zwitterionic phospholipid phosphatidylcholine (PC) and the anionic phosphatidylserine (PS) on an aqueous buffer. Upon stabilization of the monolayer at an initial surface pressure of 25 mN/m, CPXV012 peptide was injected into the monolayer aqueous subphase and the surface pressure was measured for ~25 min (Figure 6A). A change in surface pressure is interpreted as the integration of CPXV012 peptide into the lipid monolayer. Although the surface pressure of monolayers formed by the zwitterionic PC changed rapidly upon the addition of CPXV012 peptide, the shift in surface pressure was much higher for monolayers formed by a 7:3 mixture of PC and the negatively charged PS (Figure 6B). These data suggest a preferred interaction between CPXV012 and the anionic PS. This is further supported by the circular dichroism (CD) experiments used to determine the secondary structure of CPXV012 peptide (Figure 6C). The CD spectrum of CPXV012 was measured in the presence of large unilamellar vesicles (LUVs), lipid vesicles consisting of PC only, or of PC and PS. In aqueous MES buffer or in the presence of PC, CD spectra of CPXV012 peptide were rather similar and becoming more negative at a shorter wavelength, suggesting the presence of a significant amount of random coil structure. However, upon addition of PS to PC lipids using a similar ratio as for the monolayer experiments, the CPXV012 peptide acquired a different CD spectrum with minima around 208 and 222 nm, typical for an alpha-helical structure. Thus, CPXV012 peptide adopts different secondary structures, depending on the presence of PS.

A

WT: QEGISRFKICPYHWYKQHMSLLFRRYYHKLDSII
Ala1: **AAAAA**RFKICPYHWYKQHMSLLFRRYYHKLDSII
Ala2: QEGIS**AAAAA**PYHWYKQHMSLLFRRYYHKLDSII
Ala3: QEGISRFKIC**AAAAA**KQHMSLLFRRYYHKLDSII
Ala4: QEGISRFKICPYHWY**AAAAA**LLFRRYYHKLDSII
Ala5: QEGISRFKICPYHWYKQHMS**AAAAA**YYHKLDSII
Ala6: QEGISRFKICPYHWYKQHMSLLFRR**AAAAA**DSII
Ala7: QEGISRFKICPYHWYKQHMSLLFRRYYHKL**AAAA**
Ala8: QEGIS**A**F**A**ICPY**A**WY**A**Q**A**MSLLF**AA**YY**AA**LDSII

B

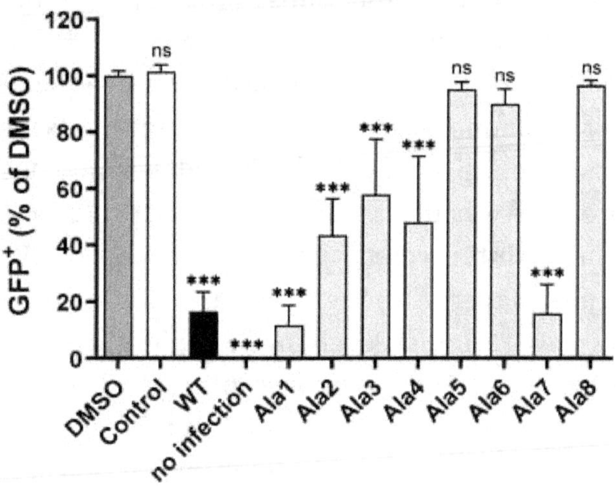

Figure 5. CPXV012 peptide variants differentially affect virus infection. (A) Amino acid sequence of CPXV012 peptide variants used in the experiment (B). (B) MJS cells were infected with MVA-eGFP at an MOI of 20 in the presence of 100 μg/mL of the peptide indicated or DMSO as vehicle control. After 20 h, cells were harvested, and the number of eGFP-positive cells was quantified using flow cytometry. Mean ± S.E.M. of three independent experiments is shown using different batches of peptide. Data were analyzed with one-way ANOVA followed by multiple comparisons Dunnett's test (the mean of each column was compared to that of the DMSO control). (* $p < 0.001$).**

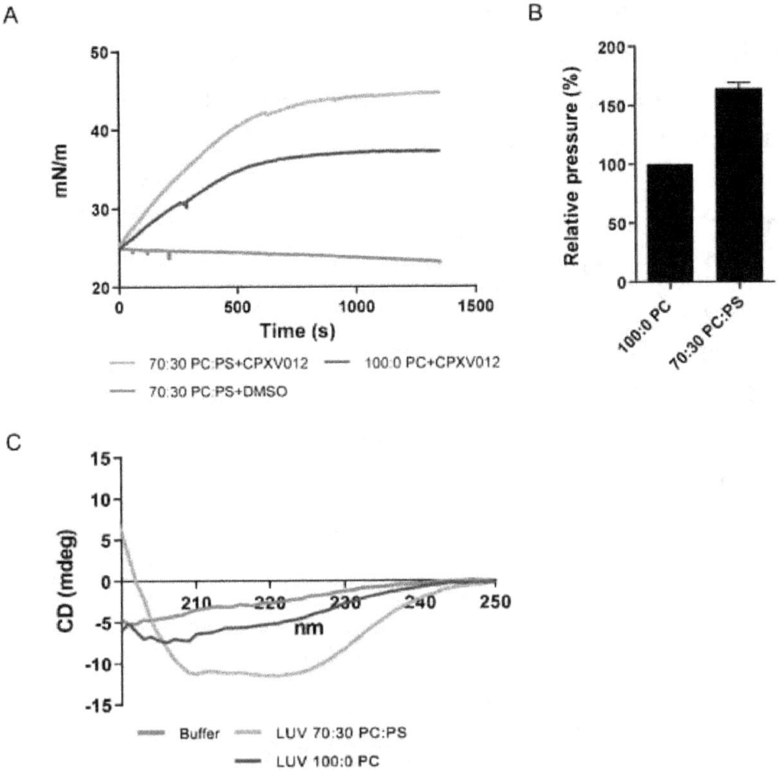

Figure 6. CPXV012 peptide interacts with PS. (A) CPXV012 peptide preferentially integrates into lipid monolayers composed of PC:PS (70:30). Langmuir monolayers of PC or PC:PS (70:30) with an initial surface pressure of 25 mN/m were formed over an aqueous subphase. Peptide was injected into the subphase at t = 0 s. As vehicle control, DMSO was injected in the subphase of PC:PS (70:30) monolayers. Results are representative of three independent experiments. (B) Quantification of CPXV012 peptide integration into lipid monolayers composed of PC:PS (70:30), as compared to monolayers composed of PC only (set at 100%). Mean ± S.E.M. of three independent experiments is shown. (C) CPXV012 peptide adopts different secondary structures depending on the presence of PS. CD spectrum of CPXV012 peptide was determined in the presence of aqueous buffer or LUVs composed of PC or PC:PS (70:30). Results are representative of three independent experiments.

3.4. The CPXV012 peptide exhibits an interaction with phospholipids that carry an electric charge

To gain further insight into the interaction between CPXV012 peptide and lipid membranes, the effect of the peptide on the integrity of lipid membrane vesicles was evaluated by membrane leakage assays using LUVs. These LUVs were composed of PC:PS and contained the self-quenching fluorophore calcein in their lumen. As soon as vesicle integrity is compromised, calcein leakage induces an increase in fluorescence, which can be measured in the supernatant using a fluorometer. As described previously, addition of human IAPP, but not murine IAPP, induced leakage of LUVs composed of PC and PS, whereas LUVs composed of PC alone were more resistant to leakage by human IAPP (Figure 7A,B) [31]. The addition of 25 μg/mL CPXV012 peptide disrupted the LUV membranes composed of both PC and PS, whereas membranes containing PC alone were unaffected. As described for human IAPP [32], membrane leakage by CPXV012 peptide is not instant but is preceded by a lag phase of several hours.

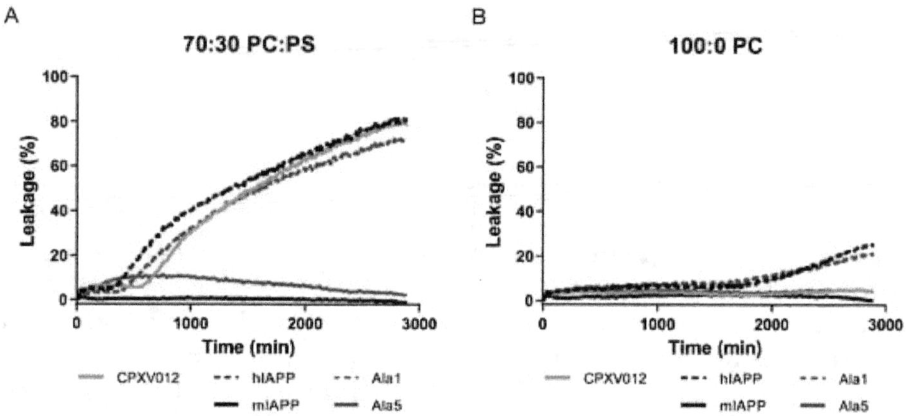

Figure 7. CPXV012 peptide disrupts membranes composed of PS. Calcein-containing PC:PS (7:3) LUVs (A) or PC LUVs (B) were incubated with human IAPP (hIAPP) as the positive control, murine IAPP (mIAPP) as the negative control, CPXV012 peptide, CPXV012 peptide variants Ala1 or Ala5 (see Figure 5A for amino acid sequences). Calcein leakage from the vesicle lumen was quantified using a microplate reader. Membrane leakage was compared to fully disintegrated detergent-treated LUVs (set at 100%). Results are representative of three independent experiments.

Two peptides with alanine substitutions, Ala1 and Ala5 (refer to Figure 5), were also subjected to testing. The CPXV012 peptide variation Ala1, which suppressed virus infection, caused membrane permeability of PC:PS LUVs at comparable levels and rates as the wildtype CPXV012 peptide. Unlike CPXV012, this variation was capable of causing a small amount of leakage in LUVs consisting solely of PC. The presence of Ala5 did not impact virus infection and did not damage the integrity of either of the LUV membranes (Figure 7B). Ultimately, the effectiveness of CPXV012 peptide in preventing virus infections is directly related to its potential to break lipid membranes that contain PS.

4. DISCUSSION

The CPXV012 peptide was discovered in this work as a potent antiviral drug with broad-spectrum activity, likely through its interaction with the anionic phospholipid PS. The antiviral efficacy of the CPXV012 peptide is quite low, however, we believe that this peptide has potential for biochemical enhancement. The CPXV012 peptide and PS presumably interact through opposing electrostatic forces, which are governed by the presence of cationic residues in the peptide and anionic head groups in PS. Furthermore, the presence of hydrophobic and aromatic amino acid residues, which constitute over 30% of the CPXV012 peptide, can potentially enhance the peptide's ability to enter the lipid portion of the membrane. Similar to what has been shown with other peptides, these interactions with the membrane can enhance the development of an alpha-helical secondary structure in peptides that are otherwise lacking a defined structure.

[33]. Due to the distribution of charged and hydrophobic residues over the length of the CPXV012 peptide, the alpha-helix can be expected to have somewhat of an amphipathic character. This may allow it to bind with an orientation that is roughly parallel to the membrane surface [34] or to assemble into oligomeric transmembrane pores [35]. These biophysical properties of CPXV012 peptide are shared with many antimicrobial host defense peptides [36], that form a crucial innate immune barrier by targeting the negatively charged surfaces of bacteria and certain fungi. Like CPXV012 peptide, these small peptides (up to 50 residues) have a net positive charge and high affinity for anionic phospholipids. The structural rearrangements occurring upon

membrane binding are crucial for the antimicrobial activity and can trigger pore formation, membrane depolarization, and disruption of the bacterial membrane [37,38].

Although less studied, antimicrobial peptides may also affect virus infection by different mechanisms. These mechanisms include interference with viral entry (binding to receptors, fusion), inhibition of viral replication, reduction/suppression of viral gene expression, immunomodulation, viral aggregation, and disruption of the integrity of the viral membrane (reviewed in [6,7]). While a list of peptides is reported to disrupt viral envelopes, their mode of action has not been formally tested. Still, a number of them were shown to directly affect viral membranes, although the specific interacting partners remain elusive. Examples include the human cathelicidin LL-37 and its murine equivalent mCRAMP, which disrupt the membranes of RSV, influenza A virus and VACV, and can prevent virus infection in vivo [39,40,41]. Although their effect on the infectivity of other viruses is not known, the structural similarities between these cationic peptides and CPXV012 peptide may suggest that these and other antimicrobial peptides possess a broader antiviral activity. Electron microscopy has also provided indications of envelope disruption with the cationic peptides Tachyplesin (VSV) [7], Temporin B (HSV-1) [42], and more recently Urumin (H1-bearing influenza A viruses) [43].

Membrane leakage induced by CPXV012 peptide was preceded by a lag phase, as was also observed for human IAPP. For human IAPP, this lag phase originates from the formation of transient oligomeric intermediates prior to the assembly of membrane-disruptive aggregates [32]. Likewise, disruption of membranes by CPXV012 peptides may be linked to the formation of (transient) oligomeric structures. The shorter lag time in the presence of PS is possibly due to more efficient binding of the peptide to the membrane: upon favorable electrostatic interactions, CPXV012 peptides achieve a higher local concentration and therefore the assembly process is supported. The lag phase of several hours appears to be inconsistent with the almost immediate effect observed in virus infection assays. There can be several reasons for this apparent discrepancy. First, the local peptide concentration at the membrane surface in the virus assays is likely to be much higher than in the leakage experiments, facilitating the oligomerization process. Second, antiviral activity may not require the formation of pores that are large enough to allow passage of calcein. Instead, binding to the viral membrane and/or insertion of hydrophobic amino acid side chains between the lipid acyl chains by itself already may be sufficient for an antiviral effect. Third, other factors may be involved in the virus infection assays, such as interactions with viral (glycol)proteins in addition to PS, that help accelerate peptide oligomerization on the viral surface.

Our data shows that the CPXV012 peptide effectively neutralizes the infectivity of viruses belonging to various families. Experiments conducted in laboratory conditions demonstrate that the CPXV012 peptide has a direct interaction with membranes. The CPXV012 peptide's disruptive activity on artificial membrane vesicles implies that it may have a similar effect on viral membranes. Due to the absence of a membrane repair mechanism, virus particles may be particularly destabilized and disrupted by this detrimental action. Significant membrane damage might hinder the fusion of the virus with the host cell membrane by affecting the flexibility and shapes of lipid membranes. Lipids, notably phosphatidylserine (PS), have a crucial function in causing these changes in membrane shape and the subsequent merging of membranes. [44,45]. Even in the absence of membrane damage, the interaction between CPXV012 peptide and PS in the viral envelope may interfere with virus infectivity by sterically blocking the interaction between viral PS and host cell PS receptors. A similar inhibitory mechanism was observed using the PS-specific antibody Bavituximab [3,46], or the PS-binding protein Annexin V [47,48]. CPXV012 peptide may also indirectly affect virus infection by binding to PS on the host cell membrane. Specific microdomains of the host cell plasma membrane are enriched for PS, including lipid rafts [49]. These sites are used by certain viruses as an entry point [50]. Although PS typically would reside in the inner leaflet of the plasma membrane, there are indications that some viruses induce redistribution of PS to the outer membrane to promote/

facilitate entry [50]. In addition, some viruses use lipid rafts as budding sites when exiting the host cell [51].

The effect of CPXV012 peptide on the viruses used in this study and potential inhibitory mechanism(s) are discussed in more detail.

4.1. Poxviruses

The CPXV012 peptide was able to inhibit infection with different orthopoxviruses (MVA, VACV, and CPXV) in several cell lines, using various detection methods. Notably, early gene expression, which starts immediately after virus entry, was highly reduced and indicates a block at the entry step. Viral particles of the poxviridae family adopt distinct structures and composition of the outer membrane, such as mature virions (MVs, also named IMVs, intracellular mature virus) [52]. MVs represent the majority of infectious progeny; they are very stable and largely responsible for viral spread. In viral preparations, MVs are the most abundant virion form. Importantly, membranes of MVs have been shown to contain large amounts of PS that is essential for entry [47,53]. We found that CPXV012 peptide binds to PS and thus likely mediates its inhibitory effect on poxvirus infection by interacting with PS in the viral membrane. This was confirmed by the observation that pretreating virions with CPXV012 peptide effectively blocked infection.

4.2. HSV-1

Inhibition of HSV-1 by cationic peptides has been previously reported, including the antimicrobial peptide LL-37 that presumably acts by targeting factors within the viral membrane [54,55,56]. It is believed that HSV-1 derives its final envelope from membranes of the late secretory pathway (e.g., trans-Golgi network) and/or endosomal pathway [57]. The inhibition of HSV-1 by CPXV012 peptide suggests that PS in the HSV-1 envelope is a potent target for antiviral therapy.

4.3. HBV

CPXV012 peptide affected HBV infection as measured by a decrease in viral DNA and viral antigen in the supernatant of infected cultures.

HBV particles are thought to acquire their envelope from multivesicular bodies [58], which are rich in PS [59]. The role of PS in HBV infection is underlined by studies showing that antibodies specific for PS block HBV infection [60]. Similar to poxviruses, PS may play a role in virus entry through apoptotic mimicry [61,62]. PS may also play a role in the proper folding of HBV envelope proteins [63].

4.4. HIV

The presence of PS in the membrane of HIV particles was shown to be an important cofactor for infection and blocking of PS led to a decrease in infection [64,65]. In addition, HIV infection was inhibited by the addition of lipid vesicles composed of PS that competed with the HIV particles for plasma membrane association, while lipid vesicles containing PC had no effect [64]. Therefore, the binding of CPXV012 peptide to PS in the viral envelope could explain the inhibitory effect observed in the EASY-HIT assay.

4.5. RVFV

The inhibition of RVFV infection by CPXV012 peptide suggests that PS is critical for RVFV infection. Although the lipid composition of RVFV is unknown, the viral envelope of the Uukuniemi virus, a related member of the bunyavirus family, is enriched for PS [66]. Like the Uukuniemi virus, RVFV obtains its membrane by budding into the Golgi membrane, suggesting that PS may also be enriched in these virus particles [67,68].

4.6. Viruses Not Affected by CPXV012 Peptide

No effects of CPXV012 peptide were observed on infection with adenovirus, coxsackie B3, measles virus, and VSV. Although the measles virus and VSV are enveloped viruses, the role of PS during infection is limited for these viruses. For the measles virus, different phospholipids were tested for their ability to reassemble with the viral envelope glycoproteins and to restore hemolytic activity. In contrast to phosphatidylethanolamine and PC, PS completely failed to restore hemagglutination activity [69]. For VSV, early studies considered PS expressed on the host cell important for cell entry [70]. In contrast, a more recent study found no correlation between PS surface levels and VSV binding. Moreover, masking of phosphatidylserine with annexin V during infection did not affect VSV binding to cells, or entry of virus particles pseudotyped with VSV-G [64,71]. Thus, the limited role of PS during entry of VSV and measles virus may explain the lack of inhibition by CPXV012 peptide.

Both adenovirus and Coxsackievirus B3 are non-enveloped viruses and the nucleocapsid does not contain any phospholipids (nor glycoproteins) CPXV012 peptide could bind to. CPXV012 peptide binding to the host cell membrane might not be sufficient for the steric impediment of viral ligand–cellular receptor interaction. Supporting this, the adenoviral elongated fiber protein responsible for binding to the host cell is flexible and large in size (9 to 30 nm), as is the coxsackie and adenovirus receptor (CAR) on the host cell [72].

CAR is also important for the infection of free CVB3 particles [73]. In addition, CVB3 may use a PS-dependent entry strategy for the bulk transmission of virus particles [74]. Infected cells release clusters of virus particles wrapped in PS-enriched membranes. These vesicles enhance subsequent virus entry in a process that highly depends on the presence of PS [74]. As CPXV012 peptide had no effect on CVB3 virus infection, PS may not play a significant role in our study. This discrepancy likely results from our virus production protocol that involves repeated freeze-thawing to liberate virus particles. Although necessary to obtain high virus titers, this method also releases CVB3 from PS-enriched vesicles [74].

Overall, the inhibition of virus infection by the CPXV012 peptide is closely associated with the reported interaction between the CPXV012 peptide and PS. The wide inhibitory range can thus be expanded to include other viruses that rely on PS for essential functions throughout their life cycle. The viruses mentioned are significant diseases with clinical and economic importance, including Ebola virus, Lassa virus, dengue virus, and poliovirus [61].

The distinct characteristics of viral and host membranes render PS vulnerable and susceptible to targeting by the CPXV012 peptide. It would be worthwhile to explore whether additional cationic peptides, such as specific antimicrobial peptides, can similarly selectively bind to PS located within the viral envelope, and potentially serve as broad-spectrum antiviral medicines. Examining the biophysical and physicochemical characteristics of the peptide can provide insights toward enhancing the limited effectiveness of CPXV012 and similar cationic peptides. Antiviral peptides with high efficiency are promising options for preventive treatment and antiviral therapy. They do not rely on active replication and specifically target viral components that are less prone to developing drug resistance [75].

The persistent risk of viral infections and the development of resistance to antiviral drugs necessitates a continuous pursuit of novel antiviral agents. Promising antiviral prospects are broadly-inhibiting medicines that target common components among several viruses. In this study, we have identified a peptide obtained from the protein CPXV012 of the cowpox virus that exhibits a wide range of antiviral activity. Our research revealed that the CPXV012 peptide inhibits the infection caused by various medically and economically significant enveloped viruses, such as poxviruses, herpes simplex virus-1, hepatitis B virus, HIV-1, and Rift Valley fever virus. Infections caused by non-enveloped viruses, such as Coxsackie B3 virus and adenovirus, remain unaffected. The findings further indicate that viral particles are deactivated by direct interactions with the CPXV012 peptide. Additionally, this cationic peptide may selectively attach to and break membranes containing the anionic phospholipid phosphatidylserine, which is a crucial element of various viral membranes. The collective findings strongly indicate that the CPXV012 peptide hinders virus infections through direct interactions with phosphatidylserine present in the viral envelope. These findings emphasize the potential of cationic peptides as effective inhibitors of a wide range of viruses.

REFERENCES

1. Marston, H.D.; Folkers, G.K.; Morens, D.M.; Fauci, A.S. Emerging viral diseases: Confronting threats with new technologies. Sci. Transl. Med. 2014, 6, 253ps10.
2. Zhu, J.D.; Meng, W.; Wang, X.J.; Wang, H.C. Broad-spectrum antiviral agents. Front. Microbiol. 2015, 6, 517.
3. Soares, M.M.; King, S.W.; Thorpe, P.E. Targeting inside-out phosphatidylserine as a therapeutic strategy for viral diseases. Nat. Med. 2008, 14, 1357–1362.
4. Vigant, F.; Lee, J.; Hollmann, A.; Tanner, L.B.; Akyol Ataman, Z.; Yun, T.; Shui, G.; Aguilar, H.C.; Zhang, D.; Meriwether, D.; et al. A mechanistic paradigm for broad-spectrum antivirals that target virus-cell fusion. PLoS Pathog. 2013, 9, e1003297.
5. Wolf, M.C.; Freiberg, A.N.; Zhang, T.; Akyol-Ataman, Z.; Grock, A.; Hong, P.W.; Li, J.; Watson, N.F.; Fang, A.Q.; Aguilar, H.C.; et al. A broad-spectrum antiviral targeting entry of enveloped viruses. Proc. Natl Acad. Sci. USA 2010, 107, 3157–3162.
6. Findlay, E.G.; Currie, S.M.; Davidson, D.J. Cationic host defence peptides: Potential as antiviral therapeutics. BioDrugs 2013, 27, 479–493.
7. Mulder, K.C.L.; Lima, L.A.; Miranda, V.J.; Dias, S.C.; Franco, O.L. Current scenario of peptide-based drugs: The key roles of cationic antitumor and antiviral peptides. Front. Microbiol. 2013, 4, 321.
8. Leventis, P.A.; Grinstein, S. The distribution and function of phosphatidylserine in cellular membranes. Annu. Rev. Biophys. 2010, 39, 407–427.
9. Moller-Tank, S.; Maury, W. Phosphatidylserine receptors: Enhancers of enveloped virus entry and infection. Virology 2014, 468–470, 565–580.
10. Andrews, N.W.; Almeida, P.E.; Corrotte, M. Damage control: Cellular mechanisms of plasma membrane repair. Trends Cell Biol. 2014, 24, 734–742.

11. Cooper, S.T.; McNeil, P.L. Membrane Repair: Mechanisms and Pathophysiology. Physiol. Rev. 2015, 95, 1205–1240.

12. Alzhanova, D.; Edwards, D.M.; Hammarlund, E.; Scholz, I.G.; Horst, D.; Wagner, M.J.; Upton, C.; Wiertz, E.J.; Slifka, M.K.; Früh, K. Cowpox Virus Inhibits the Transporter Associated with Antigen Processing to Evade T Cell Recognition. Cell Host Microbe 2009, 6, 433–445.

13. Byun, M.; Verweij, M.C.; Pickup, D.J.; Wiertz, E.J.H.J.; Hansen, T.H.; Yokoyama, W.M. Two Mechanistically Distinct Immune Evasion Proteins of Cowpox Virus Combine to Avoid Antiviral CD8 T Cells. Cell Host Microbe 2009, 6, 422–432.

14. Lin, J.; Eggensperger, S.; Hank, S.; Wycisk, A.I.; Wieneke, R.; Mayerhofer, P.U.; Tampé, R. A negative feedback modulator of antigen processing evolved from a frameshift in the cowpox virus genome. PLoS Pathog. 2014, 10, e1004554.

15. Luteijn, R.D.; Hoelen, H.; Kruse, E.; van Leeuwen, W.F.; Grootens, J.; Horst, D.; Koorengevel, M.; Drijfhout, J.W.; Kremmer, E.; Früh, K.; et al. Cowpox Virus Protein CPXV012 Eludes CTLs by Blocking ATP Binding to TAP. J. Immunol. 2014, 193, 1578–1589.

16. Praest, P.; Liaci, A.M.; Förster, F.; Wiertz, E.J.H.J. New insights into the structure of the MHC class I peptide-loading complex and mechanisms of TAP inhibition by viral immune evasion proteins. Mol. Immunol. 2019, 113, 103–114.

17. Kremb, S.; Helfer, M.; Heller, W.; Hoffmann, D.; Wolff, H.; Kleinschmidt, A.; Cepok, S.; Hemmer, B.; Durner, J.; Brack-Werner, R. EASY-HIT: HIV full-replication technology for broad discovery of multiple classes of HIV inhibitors. Antimicrob. Agents Chemother. 2010, 54, 5257–5268.

18. Gasteiger, G.; Kastenmuller, W.; Ljapoci, R.; Sutter, G.; Drexler, I. Cross-Priming of Cytotoxic T Cells Dictates Antigen Requisites for Modified Vaccinia Virus Ankara Vector Vaccines. J. Virol. 2007, 81, 11925–11936.

19. Staib, C.; Drexler, I.; Sutter, G. Construction and isolation of recombinant MVA. Methods Mol. Biol. 2004, 269, 77–100.
20. Lucifora, J.; Arzberger, S.; Durantel, D.; Belloni, L.; Strubin, M.; Levrero, M.; Zoulim, F.; Hantz, O.; Protzer, U. Hepatitis B virus X protein is essential to initiate and maintain virus replication after infection. J. Hepatol. 2011, 55, 996–1003.
21. Duprex, W.P.; McQuaid, S.; Roscic-Mrkic, B.; Cattaneo, R.; McCallister, C.; Rima, B.K. In vitro and in vivo infection of neural cells by a recombinant measles virus expressing enhanced green fluorescent protein. J. Virol. 2000, 74, 7972–7979.
22. Wohlleber, D.; Kashkar, H.; Gartner, K.; Frings, M.K.; Odenthal, M.; Hegenbarth, S.; Borner, C.; Arnold, B.; Hammerling, G.; Nieswandt, B.; et al. TNF-induced target cell killing by CTL activated through cross-presentation. Cell Rep. 2012, 2, 478–487.
23. Kortekaas, J.; Oreshkova, N.; Cobos-Jimenez, V.; Vloet, R.P.; Potgieter, C.A.; Moormann, R.J. Creation of a nonspreading Rift Valley fever virus. J. Virol. 2011, 85, 12622–12630.
24. Lanke, K.H.; van der Schaar, H.M.; Belov, G.A.; Feng, Q.; Duijsings, D.; Jackson, C.L.; Ehrenfeld, E.; van Kuppeveld, F.J. GBF1, a guanine nucleotide exchange factor for Arf, is crucial for coxsackievirus B3 RNA replication. J. Virol. 2009, 83, 11940–11949.
25. Hiemstra, H.S.; Duinkerken, G.; Benckhuijsen, W.E.; Amons, R.; de Vries, R.R.; Roep, B.O.; Drijfhout, J.W. The identification of CD4+ T cell epitopes with dedicated synthetic peptide libraries. Proc. Natl. Acad. Sci. USA 1997, 94, 10313–10318.
26. Repetto, G.; del Peso, A.; Zurita, J.L. Neutral red uptake assay for the estimation of cell viability/cytotoxicity. Nat. Protoc. 2008, 3, 1125–1131.
27. Livak, K.J.; Schmittgen, T.D. Analysis of relative gene expression data using real-time quantitative PCR and the 2(-Delta Delta C(T)) Method. Methods 2001, 25, 402–408.

28. Pfaffl, M.W. A new mathematical model for relative quantification in real-time RT-PCR. Nucleic Acids Res. 2001, 29, e45.

29. Rouser, G.; Kritchevsky, G.; Simon, G.; Nelson, G.J. Quantitative analysis of brain and spinach leaf lipids employing silicic acid column chromatography and acetone for elution of glycolipids. Lipids 1967, 2, 37–40.

30. Van de Weijer, M.L.; Luteijn, R.D.; Wiertz, E.J.H.J. Viral immune evasion: Lessons in MHC class I antigen presentation. Semin. Immunol. 2015, 27, 125–137.

31. Engel, M.F.; Khemtemourian, L.; Kleijer, C.C.; Meeldijk, H.J.; Jacobs, J.; Verkleij, A.J.; de Kruijff, B.; Killian, J.A.; Hoppener, J.W. Membrane damage by human islet amyloid polypeptide through fibril growth at the membrane. Proc. Natl. Acad. Sci. USA 2008, 105, 6033–6038.

32. Buchanan, L.E.; Dunkelberger, E.B.; Tran, H.Q.; Cheng, P.N.; Chiu, C.C.; Cao, P.; Raleigh, D.P.; de Pablo, J.J.; Nowick, J.S.; Zanni, M.T. Mechanism of IAPP amyloid fibril formation involves an intermediate with a transient beta-sheet. Proc. Natl. Acad. Sci. USA 2013, 110, 19285–19290.

33. Eiriksdottir, E.; Konate, K.; Langel, U.; Divita, G.; Deshayes, S. Secondary structure of cell-penetrating peptides controls membrane interaction and insertion. Biochim. Biophys. Acta 2010, 1798, 1119–1128.

34. Shepherd, C.M.; Vogel, H.J.; Tieleman, D.P. Interactions of the designed antimicrobial peptide MB21 and truncated dermaseptin S3 with lipid bilayers: Molecular-dynamics simulations. Biochem. J. 2003, 370, 233–243.

35. Hwang, P.M.; Vogel, H.J. Structure-function relationships of antimicrobial peptides. Biochem. Cell Biol. 1998, 76, 235–246.

36. Zasloff, M. Antimicrobial peptides of multicellular organisms. Nature 2002, 415, 389–395.

37. Baltzer, S.A.; Brown, M.H. Antimicrobial peptides: Promising alternatives to conventional antibiotics. J. Mol. Microbiol. Biotechnol. 2011, 20, 228–235.

38. Shai, Y. Mode of action of membrane active antimicrobial peptides. Pept. Sci. 2002, 66, 236–248.

39. Currie, S.M.; Gwyer Findlay, E.; McFarlane, A.J.; Fitch, P.M.; Bottcher, B.; Colegrave, N.; Paras, A.; Jozwik, A.; Chiu, C.; Schwarze, J.; et al. Cathelicidins Have Direct Antiviral Activity against Respiratory Syncytial Virus In Vitro and Protective Function In Vivo in Mice and Humans. J. Immunol. 2016, 196, 2699–2710.

40. Howell, M.D.; Jones, J.F.; Kisich, K.O.; Streib, J.E.; Gallo, R.L.; Leung, D.Y. Selective killing of vaccinia virus by LL-37: Implications for eczema vaccinatum. J. Immunol. 2004, 172, 1763–1767.

41. Tripathi, S.; Tecle, T.; Verma, A.; Crouch, E.; White, M.; Hartshorn, K.L. The human cathelicidin LL-37 inhibits influenza A viruses through a mechanism distinct from that of surfactant protein D or defensins. J. Gen. Virol. 2013, 94, 40–49.

42. Marcocci, M.E.; Amatore, D.; Villa, S.; Casciaro, B.; Aimola, P.; Franci, G.; Grieco, P.; Galdiero, M.; Palamara, A.T.; Mangoni, M.L.; et al. The Amphibian Antimicrobial Peptide Temporin B Inhibits In Vitro Herpes Simplex Virus 1 Infection. Antimicrob. Agents Chemother. 2018, 62, e02367-17.

43. Holthausen, D.J.; Lee, S.H.; Kumar, V.T.; Bouvier, N.M.; Krammer, F.; Ellebedy, A.H.; Wrammert, J.; Lowen, A.C.; George, S.; Pillai, M.R.; et al. An Amphibian Host Defense Peptide Is Virucidal for Human H1 Hemagglutinin-Bearing Influenza Viruses. Immunity 2017, 46, 587–595.

44. Sanchez-Migallon, M.P.; Aranda, F.J.; Gomez-Fernandez, J.C. Role of phosphatidylserine and diacylglycerol in the fusion of chromaffin granules with target membranes. Arch. Biochem. Biophys. 1994, 314, 205–216.

45. Teissier, E.; Pecheur, E.I. Lipids as modulators of membrane fusion mediated by viral fusion proteins. Eur. Biophys. J. 2007, 36, 887–899.

46. Dowall, S.D.; Graham, V.A.; Corbin-Lickfett, K.; Empig, C.; Schlunegger, K.; Bruce, C.B.; Easterbrook, L.; Hewson, R. Effective binding of a phosphatidylserine-targeting antibody to Ebola virus infected cells and purified virions. J. Immunol. Res. 2015, 2015, 347903.

47. Mercer, J.; Helenius, A. Vaccinia virus uses macropinocytosis and apoptotic mimicry to enter host cells. Science 2008, 320, 531–535.

48. Moller-Tank, S.; Kondratowicz, A.S.; Davey, R.A.; Rennert, P.D.; Maury, W. Role of the phosphatidylserine receptor TIM-1 in enveloped-virus entry. J. Virol. 2013, 87, 8327–8341.

49. Pike, L.J.; Han, X.; Chung, K.N.; Gross, R.W. Lipid rafts are enriched in arachidonic acid and plasmenylethanolamine and their composition is independent of caveolin-1 expression: A quantitative electrospray ionization/mass spectrometric analysis. Biochemistry 2002, 41, 2075–2088.

50. Briggs, J.A.; Wilk, T.; Fuller, S.D. Do lipid rafts mediate virus assembly and pseudotyping? J. Gen. Virol. 2003, 84, 757–768.

51. Lorizate, M.; Krausslich, H.G. Role of lipids in virus replication. Cold Spring Harb. Perspect. Biol. 2011, 3, a004820.

52. Roberts, K.L.; Smith, G.L. Vaccinia virus morphogenesis and dissemination. Trends Microbiol. 2008, 16, 472–479.

53. Ichihashi, Y.; Oie, M. The activation of vaccinia virus infectivity by the transfer of phosphatidylserine from the plasma membrane. Virology 1983, 130, 306–317.

54. Daher, K.A.; Selsted, M.E.; Lehrer, R.I. Direct inactivation of viruses by human granulocyte defensins. J. Virol. 1986, 60, 1068–1074.

55. Gordon, Y.J.; Huang, L.C.; Romanowski, E.G.; Yates, K.A.; Proske, R.J.; McDermott, A.M. Human cathelicidin (LL-37), a multifunctional peptide, is expressed by ocular surface epithelia and has potent antibacterial and antiviral activity. Curr. Eye Res. 2005, 30, 385–394.

56. Yasin, B.; Pang, M.; Turner, J.S.; Cho, Y.; Dinh, N.N.; Waring, A.J.; Lehrer, R.I.; Wagar, E.A. Evaluation of the inactivation of infectious Herpes simplex virus by host-defense peptides. Eur. J. Clin. Microbiol. Infect. Dis. 2000, 19, 187–194.

57. Crump, C. Virus assembly and egress of HSV. Adv. Exp. Med. Biol. 2018, 1045, 23–44.

58. Jiang, B.; Himmelsbach, K.; Ren, H.; Boller, K.; Hildt, E. Subviral Hepatitis B Virus Filaments, like Infectious Viral Particles, Are Released via Multivesicular Bodies. J. Virol. 2016, 90, 3330–3341.

59. Gyorgy, B.; Szabo, T.G.; Pasztoi, M.; Pal, Z.; Misjak, P.; Aradi, B.; Laszlo, V.; Pallinger, E.; Pap, E.; Kittel, A.; et al. Membrane vesicles, current state-of-the-art: Emerging role of extracellular vesicles. Cell Mol. Life Sci. 2011, 68, 2667–2688.

60. De Meyer, S.; Gong, Z.; Depla, E.; Maertens, G.; Yap, S.H. Involvement of phosphatidylserine and non-phospholipid components of the hepatitis B virus envelope in human Annexin V binding and in HBV infection in vitro. J. Hepatol. 1999, 31, 783–790.

61. Amara, A.; Mercer, J. Viral apoptotic mimicry. Nat. Rev. Microbiol. 2015, 13, 461–469.

62. Vanlandschoot, P.; Leroux-Roels, G. Viral apoptotic mimicry: An immune evasion strategy developed by the hepatitis B virus? Trends Immunol. 2003, 24, 144–147.

63. Gomez-Gutierrez, J.; Rodriguez-Crespo, I.; Peterson, D.L.; Gavilanes, F. Reconstitution of hepatitis B surface antigen proteins into phospholipid vesicles. Biochim. Biophys. Acta 1994, 1192, 45–52.

64. Callahan, M.K.; Popernack, P.M.; Tsutsui, S.; Truong, L.; Schlegel, R.A.; Henderson, A.J. Phosphatidylserine on HIV envelope is a cofactor for infection of monocytic cells. J. Immunol. 2003, 170, 4840–4845.

65. Lorizate, M.; Sachsenheimer, T.; Glass, B.; Habermann, A.; Gerl, M.J.; Krausslich, H.G.; Brugger, B. Comparative lipidomics analysis of HIV-1 particles and their producer cell membrane in different cell lines. Cell Microbiol. 2013, 15, 292–304.

66. Renkonen, O.; Kaariainen, L.; Pettersson, R.; Oker-Blom, N. The phospholipid composition of Uukuniemi virus, a non-cubical tick-borne arbovirus. Virology 1972, 50, 899–901.

67. Ellis, D.S.; Shirodaria, P.V.; Fleming, E.; Simpson, D.I. Morphology and development of Rift Valley fever virus in Vero cell cultures. J. Med. Virol. 1988, 24, 161–174.

68. Kuismanen, E.; Hedman, K.; Saraste, J.; Pettersson, R.F. Uukuniemi virus maturation: Accumulation of virus particles and viral antigens in the Golgi complex. Mol. Cell Biol. 1982, 2, 1444–1458.

69. Hall, W.W.; Martin, S.J. Structure and function relationships of the envelope of measles virus. Med. Microbiol. Immunol. 1974, 160, 143–154.

70. Schlegel, R.; Tralka, T.S.; Willingham, M.C.; Pastan, I. Inhibition of VSV binding and infectivity by phosphatidylserine: Is phosphatidylserine a VSV-binding site? Cell 1983, 32, 639–646.

71. Coil, D.A.; Miller, A.D. Phosphatidylserine is not the cell surface receptor for vesicular stomatitis virus. J. Virol. 2004, 78, 10920–10926.

72. Bergelson, J.M.; Coyne, C.B. Picornavirus entry. Adv. Exp. Med. Biol. 2013, 790, 24–41.

73. Marjomaki, V.; Turkki, P.; Huttunen, M. Infectious Entry Pathway of Enterovirus B Species. Viruses 2015, 7, 6387–6399.

74. Chen, Y.H.; Du, W.; Hagemeijer, M.C.; Takvorian, P.M.; Pau, C.; Cali, A.; Brantner, C.A.; Stempinski, E.S.; Connelly, P.S.; Ma, H.C.; et al. Phosphatidylserine vesicles enable efficient en bloc transmission of enteroviruses. Cell 2015, 160, 619–630.

75. Ahmed, A.; Siman-Tov, G.; Hall, G.; Bhalla, N.; Narayanan, A. Human Antimicrobial Peptides as Therapeutics for Viral Infections. Viruses 2019, 11, 704.

04 "REPAIR ME IF YOU CAN": MEMBRANE DAMAGE, RESPONSE, AND CONTROL FROM THE VIRAL PERSPECTIVE

1. INTRODUCTION: AN OVERVIEW

Cellular membranes are selective, permeable barriers, consisting of a phospholipid bilayer. They operate by physically separating two compartments within the cell or separate the contents of the cell from the external environment, and they regulate the exchange of chemicals, ions, and biomolecules across them. Cellular membranes thus define organelles and delineate vesicles in intracellular cargo transport systems, such as the endo-lysosomal or the exocytosis pathway. They are both fluid and dynamic, as well as solid and impermeable if necessary [1]. Therefore, membrane integrity is an essential part of cellular homeostasis, which is constantly challenged by pathogens (bacteria, virus, fungi), protein aggregates, or chemicals. Cells have evolutionary conserved surveillance systems to detect and respond to membrane damage and secure cell survival. As outlined in this review, a hallmark of membrane damage is the transient exposure of complex glycosylated proteins in the cytosolic compartment, which are detected by galectins (Gal). Galectins are small cytosolic carbohydrate binding proteins, which cluster at the site of membrane damage. Galectins also coordinate differential responses such as autophagy, to remove membrane remnants [2,3], and the activation of the endosomal sorting complex required for transport (ESCRT) machinery, favoring membrane repair [4,5]. There is increasing evidence that pathogens, including viruses, take control of the cell response to membrane damage and the associated machinery for their own benefit. Therefore, provoking membrane damage upon infection may not be just an accidental by-product of the pathogen entry process. Instead, eliciting membrane damage could be purposeful and part of a sophisticated entry strategy. This review will discuss this possibility and provide an overview of current knowledge on the relationship between membrane damage, cellular response, and pathogen entry. While briefly summarizing observations made for bacterial pathogens and membrane damage in general, our focus is on the entry of the membrane lytic adenovirus (Ad), and other non-enveloped viruses. We discuss the limited existing knowledge about how viruses manipulate and exploit the cellular membrane damage response. We apologize in advance to the many authors who contribute to the emerging field of membrane damage and whose work we cannot cite due to limited space.

1.1. Virus Inflicted Membrane Damage

Most viruses are taken up by receptor-mediated endocytosis and enter cells through the endo-lysosomal compartment. From here, they have to penetrate into the cytosol to reach their site of replication. For most RNA viruses, replication takes place in the cytosol, while most DNA viruses replicate in the nucleus. The endosome is a dynamic entry compartment, and while it provides some protection from cytosolic innate immune sensors, viruses need to escape before they are either sorted to the lysosome or recycled back to the cell surface [6].

Most enveloped viruses fuse their lipid shell (viral envelope) with the endosomal membrane and release their capsid to the cytosol. This strategy is efficient and topologically simple: it does, a priori, not require any kind of membrane rupture [7]. In contrast, non-enveloped viruses, with their hydrophilic capsids, have to cross cell membranes by inflicting membrane damage. This approach risks exposure of the luminal content of the endosome to the cytosol, which can trigger inflammation. To penetrate the cellular membrane, most non-enveloped viruses undergo conformational changes and/or proteolytic processing that allow them to use membrane lytic/modulating factors. Strategies include lipid modification, pore insertion, or large-scale membrane disruption (Figure 1). Still, not much is known about the structural and mechanistic details of the process how non-enveloped viruses inflict membrane damage to cross membranes and even less is known how they engage in the cellular response [8,9].

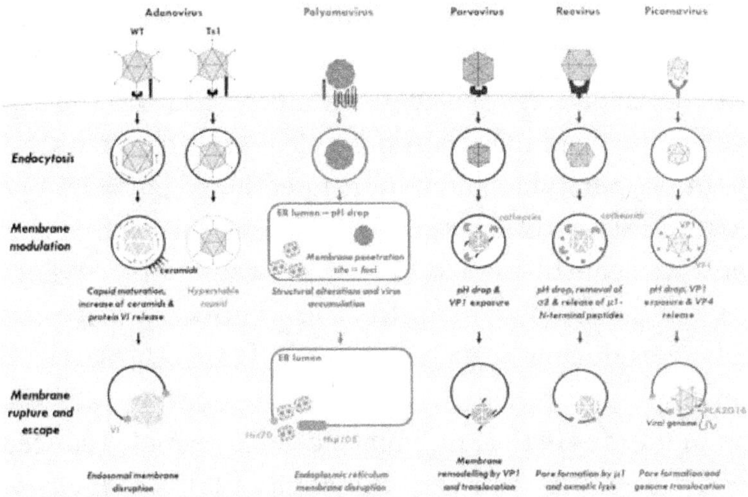

Figure 1. Virus-inflicted membrane damage. After binding to cell-surface receptors, viruses are internalized through endocytosis. Once in the endosome, adenovirus capsid undergoes partial disassembly and releases protein VI. The increase of ceramide concentration enhances the binding of protein VI to the endosomal membrane and its subsequent rupture. Polyomavirus-containing endosomes are targeted to the endoplasmic reticulum (ER) where the virus undergoes conformational changes to penetrate the ER-membrane and escape to the cytosol. Parvovirus and reovirus require a pH drop and the action of endosomal cathepsins to induce conformational rearrangements, disrupt the endosome, and reach the cytosol. After endocytosis and conformational changes, picornaviruses rely on a cellular lipid-modifying enzyme (PLA2G16) to facilitate the translocation of its genome via selective pores across the endosomal membrane. See Section 1 for further details. Abbreviations: ER, endoplasmic reticulum; Hsc70, Heat shock cognate 71 kDa protein; Hsp105, Heat shock protein 105 kDa; PLA2G16, phospholipase A2 group XVI.

1.1.1. Adenovirus

Adenoviruses (Ad) are among the best-studied non-enveloped viruses for membrane penetration. Their ~90 nm capsids enter target cells following receptor binding through clathrin-mediated endocytosis. Most Ads rapidly escape the endosomal compartment to avoid lysosomal sorting and use cytosolic motors to reach the nucleus for replication. Adenovirus endosome penetration was shown through co-uptake of non-membrane permeable substrates [10,11]. Observations made with the thermosensitive mutant virus Ad2/5 ts1 showed that membrane penetration is an essential step in the infection process [12,13,14]. This ts1 mutant fails to package the adenoviral protease into the capsid. This failure prevents capsid maturation by cleavage of precursor proteins, resulting in hyperstable and non-infectious particles [15,16,17]. Adenovirus capsid maturation is required for the release of the internal membrane lytic capsid protein VI upon virus entry. What triggers protein VI release from the capsid is not known, but it may involve disassembly cues during initial cell binding [14,18,19,20,21]. Protein VI encodes an amphipathic helix with membrane binding and lytic activity [13,22] and mutations in the amphipathic helix strongly reduced both its membrane lytic activity and viral infectivity [23]. Adenovirus inflicted membrane damage creates openings large enough for cytosolic delivery of 70 kDa dextrans or 25 nm parvovirus particles [10,11]. Electron microscopy images of Ad particles in partially disrupted endosomes show large physical openings in the endosome plugged by the virus [3]. High-resolution fluorescence microscopy images suggest localized protein VI release from the capsid at the membrane damage site [24]. Ceramides have been proposed to increase protein VI membrane affinity, showing the importance of the local lipid composition at the membrane penetration site [19]. Protein VI release also plays a role in virus escape from endosomes by counteracting cellular autophagy, which is normally mounted in response to membrane damage as detailed in Section 2 [3].

1.1.2. Polyomavirus

Polyomaviruses (e.g., BK virus and simian virus 40) are small non-enveloped DNA viruses with a diameter of ~45 nm. Particles enter by endocytosis and are transported to the endoplasmic reticulum (ER), from where they reach the cytosol. To penetrate the membrane, particles undergo conformational changes triggered by pH drop and recruited cellular chaperones [25]. Exposure of the N-terminus of the major capsid protein VP1 uncovers a hidden myristylated domain in the internal capsid protein VP2. This structural alteration makes the virus particle significantly more hydrophobic and primes the capsid for membrane binding and particle translocation [26]. Penetration sites are located at the ER membrane and form distinct, virus-induced foci. Several cellular proteins, including chaperones Heat shock cognate 71 kDa protein (Hsc70) and human heat shock protein 105 kDa (Hsp105), accumulate at these foci likely assisting viral particles to escape into the cytosol [27,28]. The exact nature of the penetration foci and the mechanism of translocation are largely unknown. There are no reports on the size of the resulting membrane damage (if any), but the translocated capsid appears to remain intact, therefore requiring the formation of large openings [29]. However, a more recent report suggests that capsid translocation could be coupled to capsid disassembly for genome release, raising the possibility that the inflicted membrane damage is smaller than initially thought [30].

1.1.3. Parvovirus

With a ~25 nm diameter, parvoviruses (e.g., adeno-associated virus (AAV) and canine parvovirus (CPV)) are rather small, non-enveloped DNA viruses that enter cell by endocytosis. Acidification in the endosome is a crucial step to induce conformational changes required for endosome penetration. The drop in pH allows the deployment of the N-terminus of capsid protein VP1, which has a phospholipase type 2 (PLA2) activity [11,31]. This PLA2 activity is essential for endosome penetration, most likely by transient and localized lipid modification [11,32,33]. For example, CPV entry allows the release of 3 kDa dextrans, but not 10 kDa dextrans from endosomes. The absence of co-release of larger molecules during parvovirus endosomal escape suggests that the resulting membrane damage is limited and does not involve complete endosome lysis [31]. However, the exact mechanism of parvovirus membrane translocation is not known.

1.1.4. Reovirus

Reoviruses (e.g., rotavirus) have non-enveloped capsids with a diameter of ~75 nm and contain RNA genomes. The exact entry of reoviruses is unclear, but it is likely to occur through endocytosis. After uptake, endosomal cathepsins are activated by a drop in pH. The outer capsid is then proteolytically processed and capsid protein σ3 is removed [34]. This uncovers the membrane lytic capsid protein µ1 and autoproteolytic processing renders it membrane-lytic [35,36]. The fully processed µ1 N-terminal peptides are then myristylated, released from the capsid and insert into the endosomal membrane, where they form size selective pores [37,38]. While these pores are too small (estimated to 4–9 nm) to permit reovirus translocation, it was suggested that endosome lysis could involve osmotic lysis [39,40].

1.1.5. Picornavirus

Picornaviruses (PV, e.g., polioviruses or rhinoviruses) enter their host cells by endocytosis. Unlike other non-enveloped viruses, PV do not translocate their capsid to the cytosol. Instead, they perforate the endosome membrane by creating a pore to translocate their RNA genome to initiate cytosolic replication [41]. Thus, contrary to viruses like Ad, PV inflicted membrane damage has a small diameter [42]. Receptor binding and a drop in pH provide cues of conformational changes in the virus particle. As a result the amphipathic helix of the capsid protein, VP1 is externalized and VP4 is released to allow membrane binding and pore formation [43,44]. The enzymatic activity of the cellular protein phospholipase A2 group XVI (PLA2G16) then facilitates genome translocation into the cytoplasm, but the exact mechanism of pore formation remains unknown [45]. Some PV encode the 2A protein, which are homologues of the cellular PLA2G16, suggesting an essential and conserved role in membrane penetration [46].

1.2. Bacteria Induced Membrane Damage

Pathogen-induced membrane damage has been much more extensively studied in bacterial infections than for viruses. Many intracellular bacteria enter their host cell by phagocytosis, ending up in vacuoles connected with the endo-lysosomal system. To avoid degradation and to proliferate, a number of bacteria escape from their vacuole to the cytosol. Others, such as Salmonella typhimurium, first proliferate inside vacuoles and may occasionally escape at a later time point to reach the cytoplasm for hyperproliferation [47]. In all cases, they must destabilize and break the membrane of their vacuole. Numerous studies have shown that this involves bacterial effectors (or their toxins) cooperating with cellular factors. Shigella flexneri, for example, is a gram-negative bacteria known to infect epithelial cells. It uses a type 3 secretion system (T3SS) to inject effectors into the cell, resulting in phagocytic uptake. After internalization, S. flexneri must escape the vacuole for cytoplasmic proliferation. Initially believed to be mediated via the bacterial T3SS secretion system [48], a subsequent small interfering RNA (siRNA) screening identified the cellular small GTPase Rab11 to be a crucial factor required for vacuole breakdown through the formation of macropinosomes at the invasion site [49,50].

S. typhimurium is another bacterial example for membrane damage and probably the best documented. S. typhimurium enters its target cells by endocytosis and establishes itself in a replication vacuole (SCV). A small but significant fraction of the bacteria escapes to reach the cytosol, through their T3SS [51]. TANK binding kinase 1 (TBK1) was suspected to control the integrity of the SCV membrane because in the absence of TBK1, S. typhimurium replication was more efficient [52]. More recently, it was shown that TBK1 instead coordinates an autophagic response against bacteria after cytosolic exposure (see Section 2) [53,54]. Some evidence suggests that the accumulation of COPII complexes on the SCV membrane destabilizes the membrane through a mechanism that remains to be deciphered [55].

Listeria monocytogenes encodes Listeriolysin O (LLO), a toxin it uses for vacuole lysis. Listeriolysin O inserts into the vacuole membrane by binding to cholesterol and forms membrane- disrupting pores. The LLO effect is potentiated by cellular factors including gamma-interferon-inducible lysosomal thiol reductase (GILT) and cystic fibrosis transmembrane conductance regulator (CFTR) [56,57]. Once in the cytoplasm, L. monocytogenes replicates and controls the polymerization of actin to be able to propagate from cell-to-cell in secondary vesicles. Unlike the first endosomal escape, escaping from secondary vesicles needs destabilization of two membranes. L monocytogenes accomplishes this task using LLO as well as the phospholipases PIcA and PIcB [58,59]. The use of a bacterial toxin for membrane rupture is not unique to L. monocytogenes and other bacterial toxins, such as perfringolysin (Clostridium difficile), pneumolysin (Streptococcus pneumonia), and VacA (Helicobacter pylori) may be deployed from bacteria to cause membrane penetration for the same purpose [60,61,62]. Escape from vacuoles has been demonstrated for several other bacteria including Mycobacterium tuberculosis, Rickettsia prowazekii, Burkholderia pseudomallei, and Francisella tularensis [63,64,65,66,67,68,69,70,71,72,73]. The underlying mechanisms are less clear, but generally involve the use of endogenous bacterial genes in conjunction with cellular factors, such as phospholipases, to breach vacuolar membranes. Interestingly for all described examples, these bacterial genes are often associated with pathogenicity linking membrane damage with disease [68].

1.3. Membrane Damage Induced by Other Pathogens

Membrane damage is not restricted to bacteria and viruses. There is increasing evidence for a role of membrane damage during fungal and parasite infection. Candida albicans is a polymorphic fungus that can change from a yeast form to a hyphal form [74]. After phagocytosis in the form of yeast, conversion into the hyphal form inside the phagosome involves a change in morphology through stretching, which ultimately leads to the rupture of the phagosomal membrane [75]. This is likely to occur through mechanical pressure (filamentation) rather via a biochemical process [76]. As part of a complex life cycle, the protozoa Trypanosoma cruzi invades red blood cells and escapes its entry vacuole to reach the cytoplasm [77]. This vacuolar escape depends in part on the parasite PFP TcTOX protein, which seems to have a similar role as the LLO bacterial toxin of L. monocytogenes. Additional factors may be involved in breaching the membrane and the precise mechanism needs further clarification.

1.4. Non-Pathogenic Membrane Damage

Many neurodegenerative diseases and cancer cells are associated with the induction of sterile membrane damage [78]. In neurodegenerative diseases, membrane damage is often a consequence of the formation of large amyloid aggregates, propagating from cell-to-cell. Studies on α-synuclein showed the ability of these amyloid aggregates to induce endosome rupture after endocytosis [79,80]. This capacity is not restricted to α-synuclein, but also extends to Tau and Huntingtin assemblies [80]. The exact reason why cell invasion by amyloid assemblies and their ability to break the membrane of the endosome causes disease is not fully understood, but it is likely to involve dysregulation of autophagy associated with irreparable damage [80]. Lysosomotropic compounds have emerged to artificially trigger lysosome damage and induce cell death, e.g., as therapeutic concept in cancer [81]. Conceptually, the acidic pH inside makes lysosomes susceptible to the accumulation of weak bases that can penetrate them. Once inside, they get protonated and stay trapped within the lysosome where they can induce specific membrane damage [81]. One compound, the L-leucyl-L-leucine methyl ester (LLOMe), is often used as a positive control in endolysosomal rupture experiments in combination with detection of membrane damage markers [82,83]. Endosomal membrane damage can also be caused via a mechanical/physical trigger, especially in the context of intracellular drug delivery, where it is important to achieve efficient endosomal escape. Some lipid formulations used in transfection of siRNAs were shown to be particularly efficient in causing endosomal membrane damage [84,85]. Gold nanoparticles and nanodiamonds were also employed to physically disrupt the endosomal membrane [86,87].

The non-exhaustive summary of possible pathogenic and non-pathogenic membrane insults presented in this section provides insight into the diversity of threats to cellular membrane integrity. This listing also highlights the importance for cellular surveillance of membrane integrity and the ability to respond to membrane damage to secure cell survival, which is discussed in the next section.

2. MEMBRANE DAMAGE RECOGNITION AND CELL RESPONSE

Extensive damage in endolysosomal vesicles can leak their content and flood the cell with reactive oxygen species (ROS), acid hydrolases (cathepsins), and/or calcium ions provoking injuries to several organelles and ultimately resulting in cell death through necrosis or pyroptosis. A key question, when trying to understand how viruses and other pathogens inflict membrane damage, is how cells sense their membranes have been actively penetrated and how they mount a response. A critical role in the detection of membrane damage is done by cytosolic lectins belonging to the family of Gal. Galectins recognize complex glycans present on lipids or glycosylated proteins. These glycosylated substrates are mainly localized at the cell surface or protruding from the intra-luminal membrane leaflet, including the interior of endosomes and lysosomes. In contrast, they are virtually absent from the cytosol. Hence, membrane damage exposes these glycans to the cytosol and they become easy targets for Gal sensing.

2.1. Galectins and Their Role in Membrane Damage Recognition

Galectins are characterized by the presence of one or two carbohydrate recognition domains (CRD) [88]. They can be monovalent (with a single CRD: Gal-1, 2, 5, 7, 10, 11, 13, 14, 15), bivalent (with two CRDs, Gal -4, 6, 8, 9 and 12) or even chimeric (Gal3) [89,90]. The diversity of Gals allows them to have different affinities for a plethora of substrates, making them ideal pathway mediators. Their cellular functions include regulating cell adhesion, organizing membrane domains, signaling and trafficking, apoptosis and cell cycle regulation. Here, we focus on their role as danger sensors since they quickly accumulate on galactosides exposed to the cytosol, which are recognized as danger signals at sites of membrane breach [91,92]. Although mammals have 15 different forms of Gal, to date only Gal1, 3, 8 and 9 have been described to sense damaged vacuoles [2,92]. They accumulate around membranes ruptured by both pathogen and non-pathogen stimuli, such as protein aggregates [80], liposome transfection [85] or lysosomotropic agents [82]. Upon membrane rupture, glycans and Gals form clusters (puncta) that can be easily detected by immunofluorescence using Gal-specific antibodies or cells expressing Gals tagged with a fluorescent protein [82,92,93]. Most Gals (e.g., Gal1/3) are widely expressed, making them useful targets to assess membrane damage [82,93].

Most observations and functional studies using Gals were made using invasive bacteria models. Galectin3 recognizes the vacuoles from which S. flexneri escaped and was the first example of a Gal recruited to pathogen-induced membrane damage [92]. A systematic screen including all Gals identified found Gal3, 8, and 9 as binding to S. typhimurium ruptured vacuoles [2]. Galectin 8 was also the first Gal shown to restrict bacterial proliferation and thus to have a direct antimicrobial effect [2]. To date, very few studies have addressed if virus membrane penetration also triggers Gal recruitment. Initial studies with Ad used Gal3 as a marker to show the subcellular localization and the timing of Ad endosome penetration [93,94]. In a subsequent study, it was shown that Ad also recruit Gal8 to the endosome penetration site and that this is essential for the cellular autophagy response [3]. Picornavirus pore formation in the endosomal membrane also recruits Gal8-mediating cellular autophagy and remains so far the only other example showing the recruitment of Gals to virus inflicted membrane damage [45]. Whether cells employ Gals in response to entry of other (non-enveloped) viruses remains to be shown. In any case and as discussed below, Ad and PV found ways to counteract Gal8 mediated antiviral autophagy to efficiently propagate.

2.2. Galectins and Ubiquitin Coordinate Autophagy for Membrane Damage Removal

The easiest way for a cell to deal with damage is to remove it via degradation pathways. An important part in membrane damage removal has been attributed to autophagy [78]. Autophagy is a conserved cytosolic degradation process induced under stress conditions, such as starvation, hypoxia or during membrane damage and pathogen infection [95], which allows the removal and degradation of damaged organelles or proteins from the cytosol into double membrane vesicles called autophagosomes.

These autophagosomes subsequently fuse with lysosomes for cargo degradation and recycling [96]. Autophagy is driven by proteins of the "ATG family" (for autophagy-related), a core set of proteins that coordinates formation, elongation and maturation of autophagosomes [96,97,98]. It is now clearly established that autophagic degradation is highly selective. In mammals, this selectivity is provided by a family of proteins called "autophagic receptors" (including, e.g., the nuclear dot protein 52 kD (NDP52) and p62, reviewed in [99]), which allow targeting of distinct substrates for degradation. Each autophagic degradation process is specified by a "cargo-defined" name such as mitophagy defining the selective degradation of mitochondria, xenophagy for specific degradation of intracellular pathogens or lysophagy naming the degradation of lysosomal membranes upon endolysosomal damage [99].

Membrane damage removal by autophagy can be separated into two steps (Figure 2, lysophagy). In a first step, the cell has to sense and discriminate intact vs. damaged vacuoles. The second step is to assemble and recruit the autophagic machinery to degrade membrane remnants. Key elements of the first step are Gals and ubiquitin. Both are so called "eat-me signals", a name playing with their function to mark membrane remnants for autophagic removal. Certain Gals recognize membrane damage by virtue of complex glycan exposure e.g., during pathogen infection [2,92,93]. To enlist autophagy, Gals have to bind downstream effectors. Thurston et al. reported that Gal8 recruited after SCV rupture forms a complex with the autophagy receptor NDP52 [2]. NDP52 binds FAK family kinase-interacting protein of 200 kDa (FIP200), which serves as seed assembly platform for autophagy initiation [100]. FIP200 assembles a primary kinase complex, the Unc-51 like autophagy activating kinase (ULK1/2) complex, ATG13, and ATG101. Once assembled, this complex activates a downstream Class III PI3-kinase complex with VPS34 and Beclin-1 [100,101]. The Beclin-1 complex modifies membranes and creates a phospholipid patch (also termed phagophore) to support binding of WD-repeat protein interacting with phosphoinositides (WIPI) proteins, creating a membrane anchored landing platform. WIPI proteins recruit the ATG5-ATG12/ATG16L E3 conjugation complex to the phagophore

[102]. The ATG5-complex conjugates soluble ATG8 (better known as LC3B) with phosphatidylethanolamine (PE) to be incorporated into the expanding autophagosomal membranes [103]. Additional interactions, such as NDP52-LC3 and FIP200-ATG16L stabilize the developing cargo-autophagosome complex to promote membrane elongation. Gal8 and Gal3 present at the site of membrane damage can also interact with other effectors such as the tripartite motif protein 16 (TRIM16), as shown for ruptured lysosomes using LLOMe or bacteria [104,105]. TRIM16 serves as another assembly hub to initiate autophagy by recruiting the core autophagic machinery (ATG16L1, ULK1, and Beclin-1) directly on damaged membranes [104]. Interestingly, the recruitment of Gal3 to the membrane damage side can also be pro-bacterial by impeding on the autophagic response [106]. Silencing of Gal3 during L. monocytogenes infection in murine macrophages increased LC3 recruitment to the vacuoles and reduced bacterial replication. Moreover, treatment of cells with sialidase (which removes sialic acid from glycans) increased Gal3 and decreased Gal8 on the damaged phagosomes. This change in Gal preference directly influenced the autophagic response [106], showing that cytosolic Gals can discriminate the origin of the damaged membrane and orchestrate an adapted cellular response. Deciphering the glycosylation pattern driving Gal specificity could thus be a valuable point for therapeutic intervention. Furthermore, Gal3 also directs other antibacterial functions including the recruitment of interferon (IFN)-inducible guanylate binding proteins (GBPs) to pathogen-containing vacuoles [107].

Next to Gals, ubiquitin is the second most important early danger signal of many membrane damage sites following bacterial [2,108,109,110], viral [3,111], and apathogenic insults [112]. Ubiquitin appears shortly after Gals, and like them, is involved in the recruitment of a variety of autophagic receptors (also known as sequestosome like receptors or SLRs), which contain ubiquitin binding motifs. All SLRs also contain LIR-domains (LC3-interacting region) and bridge the ubiquitylated cargo and LC3 to link the cargo into the growing autophagosomal membrane [99]. Moreover, ubiquitin can directly bind the autophagic machinery through ATG16L1, which may enhance its anchoring to the surface of ubiquitylated endosomes [113]. Two types of ubiquitin chains, K48- and K63 linked, are predominantly found on damaged lysosomes [113].

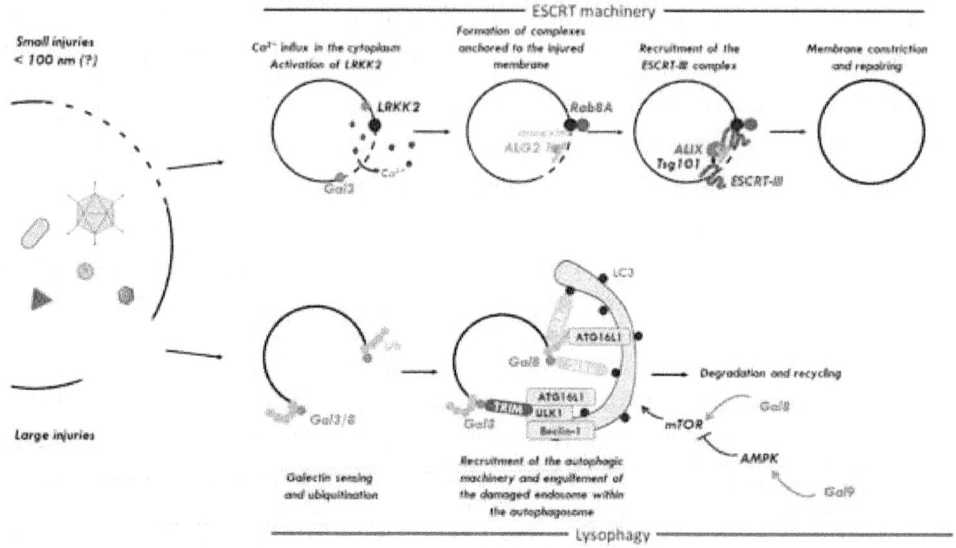

Figure 2. Membrane damage repair and removal. The endosome is constantly challenged by pathogens (bacteria, virus, and fungi), protein aggregates, or chemicals that can disrupt its membrane and provoke injuries of different sizes. Small disruptions (<100 nm) trigger a leakage of Ca2+ into the cytoplasm and activate LRKK2. LRKK2 and Ca2+ effectors mediate the recruitment of the ESCRT machinery to promote repair of the injured organelle. Galectin-3 (Gal3) is recruited to damage sites and may promote ESCRT assembly. If the injury is too large, the cell will trigger a process of degradation called lysophagy. During lysophagy, damaged vacuoles are sensed and tagged by galectins (Gal) and ubiquitin (Ub). Both signals mediate the recruitment of the autophagic machinery (either directly via autophagic receptors, such as the sequestosome like receptors (SLRs) or indirectly via TRIMs). The membrane remnant is engulfed in a double-membrane vesicle called autophagosome, which fuses with lysosomes for content degradation and recycling. Autophagy is also controlled by metabolic kinase mechanistic target of rapamycin (mTOR) through Gal8. Moreover, Gal9 can also control autophagy induction by directly activating AMP-activated protein kinase (AMPK) in response to endosomal damage to inhibit mTOR. See text for details. Abbreviations: ESCRT, endosomal sorting complexes required for transport; Gal8, galectin8; LRRK2, Leucine-rich repeat kinase 2; ROS, reactive oxygen species; Ub, ubiquitin.

K63 chains are targeted by SLRs and recruit the autophagy machinery [113,114]. The enzymes driving ubiquitylation of membrane damage, and how the chain type influences the cell response, are not entirely clear, and may involve more than one player. Some of them have been characterized in the context of bacterial infection, where they allow the ubiquitylation of both the bacteria and the membrane remnants to turn them into signaling platforms. The first E3 ubiquitin ligase described to target bacterial membrane damage was the LRR-containing RING E3 ligase (LRSAM) and is required for autophagic clearance of S. typhimurium [115]. Another E3 ligase is linear ubiquitin chain assembly complex (LUBAC), which mediates the formation of M1-linked ubiquitin chains and is also involved in autophagic degradation of S. typhimurium [116]. The aforementioned TRIM16, recruited via Gal3/8 to endomembrane damage, is also an E3 ubiquitin ligase promoting K63-linkage to ATG proteins and to ubiquitylate lysosome substrates e.g., after LLOMe treatment [104]. FBXO27 (F-Box Protein 27, part of the SCF ubiquitin ligase SKP1/CUL1/F-box protein complex) ubiquitylates several glycoproteins present on the surface of damaged lysosomes (e.g., lysosomal-associated membrane protein LAMP2), which in turn promotes the recruitment of the autophagic machinery [112]. A recent siRNA screen identified a crucial role for Ubiquitin Conjugating Enzyme E2 Q Family Like 1 (UBE2QL1), an E2 ubiquitin conjugating enzyme, in directing lysophagy [117]. Depletion of UBE2QL1 increased steady-state lysosomal membrane damage and prevented efficient ubiquitylation and SLR recruitment, essentially curtailing the autophagy response. It also prevented recruitment of the AAA-ATPase valosin-containing protein, VCP/p97 complex. The VCP/p97 complex is likely to extract membrane proteins to facilitate downstream lysophagy and cooperates with YOD1/OTU1, a K48-specific deubiquitylating enzyme. This observation suggests that selective membrane deubiquitylation, possibly of K48 labeled VCP/p97 substrates, maybe necessary before lysophagy proceeds [78,118]. A similar mechanism appears active in neurodegenerative diseases, where protein aggregates rupture endo-lysosomes and Gal3/8 recruits NDP52 and p62 suggesting that the response might be cell intrinsic [80,119].

Some receptors including NDP52 and p62 are phosphorylated by TBK1, to enhance their ability to bind ubiquitin following recruitment to the membrane damage site [120,121]. TBK1 also phosphorylates WIPI proteins and enhances autophagosome formation in response to bacterial membrane penetration [54,122]. Thus, TBK1 plays an important role in promoting and fine-tuning the antimicrobial autophagy response at the site of membrane damage.

On a larger scale, autophagy is under the control of two different kinases, the metabolic Ser/Thr kinases mechanistic target of rapamycin (mTOR) and AMP-activated protein kinase (AMPK). Both kinases are major regulators of autophagy. They act in an antagonistic way in that mTOR suppresses autophagy by phosphorylating inhibitory sites on autophagy regulators, such as ULK1/2, while phosphorylation by AMPK activates autophagy regulators [123]. Active mTOR can be found on the outer lysosomal membrane and upon lysosome damage translocates to the cytosol and becomes inactivated. Keeping mTOR inactive involves inhibitory phosphorylation by AMPK, connecting both kinases. It was recently shown that Gal8/9 regulate mTOR and AMPK on damaged lysosomes identifying new functions beyond glycan recognition [4]. The study shows that interaction of Gal8 with the lysosome membrane integral neutral amino acid transporter Solute Carrier Family 38 Member 9 (SLC38A9) upon membrane damage reorganizes a signaling complex (the Ragulator-Rag complex) on the lysosomal membrane. The Ragulator-Rag complex normally retains mTOR in an active state bound to the lysosome surface due to nutrient sensing. The reorganization through Gal8 (e.g., following membrane damage) promotes mTOR dissociation from lysosomes. Inactivation of mTOR is then reinforced by the action of another Gal, Gal9, which activates AMPK [4]. Taken together, this work offers the interesting perspective that membrane damage can be controlled by Gals at two different levels. Locally, at the physical site of membrane damage through recruitment of SLRs and assembly of the autophagic machinery. In addition, Gals also act at the cellular level by coordinating the activities of mTOR and AMPK. This hierarchical organization could be used to amplify and disseminate the local response, e.g., dependent on the extent of the damage by modifying major upstream autophagy regulators. Widening the response via mTOR inactivation has several effects on the cell, including the nuclear translocation of transcription factor EB (TFEB), a mTOR target, and master regulator normally sequestered in the cytosol, which drives gene expression programs for neosynthesis of genes involved in the biogenesis of lysosomes [104], in autophagy [124], and in inflammation [125].

2.3. Viral Control of the Autophagy Response to Membrane Damage Removal

Most of the investigations concerning the cellular response to membrane damage use invasive bacteria or models of induced lysosome damage. Although details are still emerging, there seems to be consistency since cells respond in all cases with autophagy directed against the membrane damage site. Viral membrane damage is no exception, although only two viral systems have been investigated. Adenovirus and PV both cause different types of membrane damage upon entry (Figure 3). Depending on the cell system used, Ads penetrate the endosome within 15–20 min and trigger fast selective autophagy in the infected cell. Adenovirus damaged endosomal membranes are detected by Gal3 [93,94,111] and Gal8 [3]. Using co-detection of viral particles and/or the exposed membrane lytic protein VI revealed that the viral association with ruptured membrane is transient, and that the Gal3/8-positive membrane associated with NDP52, p62, ubiquitin and LC3 [3,111]. Gal3 positive membrane remnants were cleared from the cell within 3 h of entry presumably via autophagy [111]. Interestingly wild type Ad was not cleared by autophagy and suppressing autophagy with genetic or pharmacological tools did not affect viral infectivity [3]. Using the mutant Ad-M1, the authors showed that being refractive to autophagic clearance is not simply due to a rapid escape from the endosome, but also involves active control of autophagosome maturation. The mutant Ad-M1 has strongly reduced infectivity compared to wild type Ad. It lacks a PPxY motif encoded in the membrane lytic protein VI, which is required to recruit the ubiquitin ligase neural precursor cell expressed developmentally down-regulated 4-like (NEDD4.2) upon membrane penetration [18]. Unlike wild type Ad, Ad-M1 is unable to escape from endosomes and becomes susceptible to autophagic degradation [3]. Comparing both viruses revealed that wild type viruses do not prevent autophagy initiation. Instead, they use the PPxY motif to prevent autophagosomes from maturing and fusing to lysosomes, a process that was also observed upon NEDD4.2 depletion from cells using viral and non-viral autophagy stimuli [3]. The exact role for NEDD4.2 is not yet clear and may involve regulating autophagy effectors such as ULK1 or protein VI itself [18,126]. However, this example shows how Ad has evolved a short peptide motif that recruits an essential cellular factor to stall the cell response to membrane damage, thereby buying enough time to reach the safety of the cytosol. The PPxY motif is strategically located in the membrane lytic factor making sure it is exposed at the right time and place. Consequently, eliminating membrane damage recognition by depleting either Gal8 or autophagy (e.g., in ATG5 KO cells), fully restores viral infectivity of the PPxY-deficient Ad-M1. In contrast, this does not accelerate viral escape from the endosome, showing that escape mechanism and autophagy suppression are separate processes, and that one favors the other [3].

Unlike Ad, which generates large openings in the endosomal membrane, PV creates pores of limited size within endosomes to translocate their RNA genome across the membrane. A haploid cell-based genetic screen identified the small phospholipase PLA2G16 (Group 16 phospholipase A2) as an essential factor required for genome translocation [45]. Using a counter screen in PLA2G16 deficient cells, the authors found that depletion of several autophagy genes (e.g., ATG5/7/12) and Gal8 restored viral infectivity. Using fluorescent viruses and alternative membrane rupture assays, they could show that Gal8 and PLA2G16 relocated after membrane damage independently of each other. This suggests that PLA2G16, like Gal8, senses virus-induced membrane rupture, but is recruited in a virus- and Gal8-independent manner. PLA2G16 has no effect on the autophagic flux itself, but facilitates the translocation of the genome into the cytoplasm and prevents its clearance by autophagy. When PLA2G16 is silenced, genomes are degraded via autophagic clearance together with the virus, a phenotype that can be reversed if Gal8 or ATG7 is also silenced [45].

This work thus identified PLA2G16 as cellular factor exploited by a virus to control antiviral autophagy directed against PV-induced membrane damage, a strategy analogous to the PPxY motif of Ad. Unlike for Ads, PV protection from autophagic degradation through accelerated cytosolic translocation concerns only the viral genome.

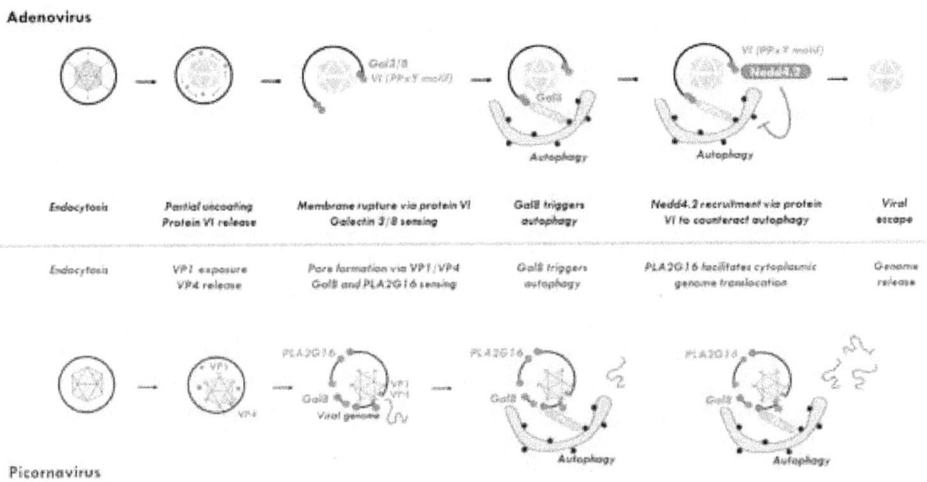

Figure 3. Viral control of the membrane damage response. Top: endocytosed adenoviruses partially uncoat and release the membrane lytic capsid protein VI for endosomal membrane lysis. Membrane damage is sensed by galectins 3 and 8. Galectin 8 recruits autophagic receptors and triggers autophagy. Adenoviruses stall autophagy through a short PPxY peptide motif in protein VI that recruits the ubiquitin ligase Nedd4.2. As a consequence, they avoid degradation and escape into the cytoplasm. Bottom: After endocytosis and acidification of the endosome, picornaviruses undergo conformational changes to expose capsid protein VP1 and release of VP4. Both proteins attach to the endosomal membrane creating membrane-penetrating pores. Membrane damage is then independently sensed by galectin 8 activating autophagy and PLA2G16. PLA2G16 facilitates genome translocation into the cytoplasm preventing autophagic clearance. See Section 2 and Section 3 for further details. Abbreviations: Gal, galectin; NDP52, Nuclear dot protein 52; Nedd4.2, neural precursor cell expressed, developmentally down-regulated 4.2; PLA2G16, phospholipase A2 group XVI.

It is conceivable that the enzymatic activity of PLA2G16 favors PV genome translocation over autophagic clearance of the membrane damage by local lipid modification, although the exact mode of action is currently unclear. Interesting in both cases, neither Ad nor PV have evolved mechanisms to suppress autophagy initiation suggesting that there is additional benefit in triggering autophagy, possibly linked to counteracting inflammatory signaling (Section 4).

2.4. ESCRT Machinery for Membrane Repair

Cells use autophagy to remove and degrade damaged membranes such as lysosomes (lysophagy) for cell survival [2,104,112,113]. However, the cell response can be more restrained and small membrane injuries were shown to trigger a membrane repair mechanism via the ESCRT system rather than removal [127,128,129] (Figure 2, ESCRT machinery). The ESCRT complex is divided into five sub-complexes (ESCRT-0, I, II, III, and disassembly proteins) and it was first discovered for its role in regulating endosomal trafficking, but is now known to be involved in numerous other cellular processes such as vesicle formation, vesicle budding and cytokinesis as detailed elsewhere [5,130]. Small membrane lesions (<100 nm), especially at the plasma membrane, trigger calcium ion influx and lead to a rapid recruitment of cytosolic annexin7 and the calcium sensor ALG2 at the membrane damage site. This complex recruits subsequently the ALG2 interactors ALIX and Tsg101, both ESCRT-I proteins that orchestrate the recruitment of the ESCRT-III complex to repair the membrane. ESCRT-III has the ability to form filaments that constrict the membrane and shed the damaged membrane [5,127,128]. A recent report showed that the ESCRT machinery was also quickly recruited to damaged endo-lysosomes to allow their repair following treatment with either LLOMe or silica crystals [131]. Even if the initial recruitment of ALG2 (apoptosis-linked gene-2) is still debated [129,131], endosome repair requires recruitment of the ESCRT-I components ALIX (ALG-2-interacting protein X) and Tsg101 (tumor-susceptibility gene 101), and the subsequent action of the ESCRT-III complex. Repair also involves transient inactivation of lysosomal hydrolases followed by re-acidification [132]. Even with minor damage, Gal3 is still recruited to damaged endo-lysosomes, but follows a slower kinetic than ESCRT recruitment. The ESCRT-III response (monitored by recruitment of CHMP4B/ESCRT-III) occurs on the timescale of minutes preceding lysophagy that mounts within the hour [129,131]. Depletion of either ALIX or Tsg101 or both does not affect Gal3 recruitment, but delays membrane repair in favor of removal/lysophagy [129,131]. A partial explanation for this observation might be that Gal3 initially promotes interactions between ALIX and the downstream ESCRT-III effector CHMP4. At later times, however, Gal3 controls the autophagic response via TRIM16 supported by Gal8/9 that regulate mTOR/AMPK [133]. A very recent study showed that macrophages challenged with either invasive bacteria or LLOMe activates the Parkinson's disease related kinase leucine-rich repeat kinase 2 (LRKK2), which in turn recruits the Rab GTPase Rab8A. Both coordinate the activity of the ESCRT-III complex for membrane repair. In contrast, depletion of LRKK2 and Rab8A change the damage response phenotype from membrane repair to lysophagy [134]. The interplay of Gal3/8/9 with the ESCRT machinery, autophagy and metabolic signaling is also observed during membrane damage upon M. tuberculosis and Coxiella burnetii infection [129,133]. Interestingly, mycobacterial effectors EsxG/TB9.8 and EsxH/TB10.4 secreted by the ESX-3 T7SS secretion system antagonize the ESCRT response with a kinetic that matches the speed with which the cell responds, showing that bacteria have also developed efficient countermeasures against membrane repair mechanisms [135]. It is still possible that ESCRT and autophagy have overlapping roles in endosome repair because ATG5 was shown to be involved in repairing the endosomal membrane damaged by the type-1 secretion system T1SS after infection with S. typhimurium [136]. ESCRT and autophagy also cooperate in the maintenance of the bacterial vacuole in Dictyostelium discoideum infected with Mycobacterium marinum [137]. Taken together the cell reaction to membrane damage appears conserved, fast, and flexible. This implies an active and hierarchical organization, where the ESCRT-mediated membrane repair is transient and precedes lysophagy [129,133].

2.5. Viral Control of the ESCRT-Response to Membrane Repair

If there is an ESCRT-response against membrane penetration by non-enveloped viruses upon entry has not been investigated. The ESCRT machinery plays a major role in the release process of most enveloped viruses and mediates their non-lytic release from cells (including some non-enveloped viruses). Many viruses use a conserved PPxY peptide motif or the related P(S/T)AP motif encoded in their capsid proteins to recruit ESCRT-I proteins or NEDD4 ubiquitin ligases (for review see [138,139,140]). These domains are still present in entering virus particles and could easily target the ESCRT machinery during entry. As discussed above, such a conserved PPxY domain able to recruit NEDD4-family ubiquitin ligases is encoded in the Ad membrane lytic protein VI [18], which suppress Gal8-mediated antiviral autophagy upon endosome penetration likely through a NEDD4.2 mediated process [3]. It is not clear if this is functionally related to the ESCRT machinery. A different study showed that upstream of endosomal escape, the binding of Ad to the cell surface results in small (transient) lesions at the plasma membrane in a protein VI-dependent manner that provokes extracellular calcium ion influx and dye penetration into the cell [128]. This local increase in calcium ions triggers exocytosis and local fusion of secretory lysosomes as part of a repair mechanism [128]. Upon membrane fusion, these lysosomes are suggested to release lipid converting acidic sphingomyelinase (ASM) catalyzing sphingomyelin thereby generating high local concentrations of ceramides. Ceramides, when present in the endosomal membrane, increase the affinity of protein VI, helping Ad to penetrate the endosome [19]. Such a mechanism suggests that Ad membrane penetration is a two-step process involving small and large sized membrane damage [19]. As shown recently, sphingomyelins play an important role in initiating the membrane collapse in bacterial vacuoles preceding Gal8 recruitment to exposed glycans [141]. It would be interesting to know if the small lesions observed for Ad in addition to the exocytosis of lysosomes have a similar function and also trigger an ESCRT-response similar to the one described above [127,128]. In this case, capsid-encoded late domains, such as the PPxY motif in protein

VI or perhaps a second motif present in the viral penton protein [142], may exert additional control functions upstream of viral restriction of Gal8-mediated antiviral autophagy in response to endosome damage [3]. More generally, if the ESCRT-response directed against small lesions in endosomes or lysosomes is a cell intrinsic response, several other viruses would have to face and counteract such a response.

2.6. Inflammation to Signal Membrane Damage

Membrane damage, especially in the endo-lysosomal compartment, can be repaired or removed. However, it also enables leakage of soluble contents from the endocytic compartments into the cytosol including protons, ROS, calcium ions, and soluble acid hydrolases such as Cathepsin B. This leakage eventually damages mitochondria further increasing ROS production, which can trigger inflammasome activation [143,144] (Figure 4). Inflammasomes are important cytosolic signaling platform, mediating caspase-1 activation and the secretion of pro-inflammatory cytokines including IL-1β and IL-18 [145]. The best-studied inflammasome is the Nod-like receptor family, pyrin domain-containing 3 (NLRP3)-inflammasome, which can be activated by multiple stimuli, including membrane damage (Figure 4). Its activation can result in cell death by pyroptosis unless cell survival pathways, such as autophagy are activated [146,147]. Not surprisingly, pro-inflammatory signaling and autophagy are an important part of the cellular stress response and are partially overlapping (summarized in [148]).

Recognizing invading pathogens involves pathogen-associated molecular patterns (PAMPs) intrinsic to the invading microbes and damage-associated molecular patterns (DAMPs). DAMPs are best described as out-of-place detection of cell components due to pathogens or stress. The understanding of pro-inflammatory signaling responding to pathogen-induced membrane damage is limited, partly because it is difficult to discriminate membrane damage signaling from microbial PAMP-activated signals. The latter trigger pathogen recognition receptors including lumenal Toll-like receptors (TLRs), cytosolic nucleic acid sensing receptors such as Cyclic GMP-AMP synthase/Stimulator of interferon genes protein (cGAS/STING), retinoic acid-inducible gene I (RIG-I) like receptors (RLRs) and Nod-like receptors (NLRs).

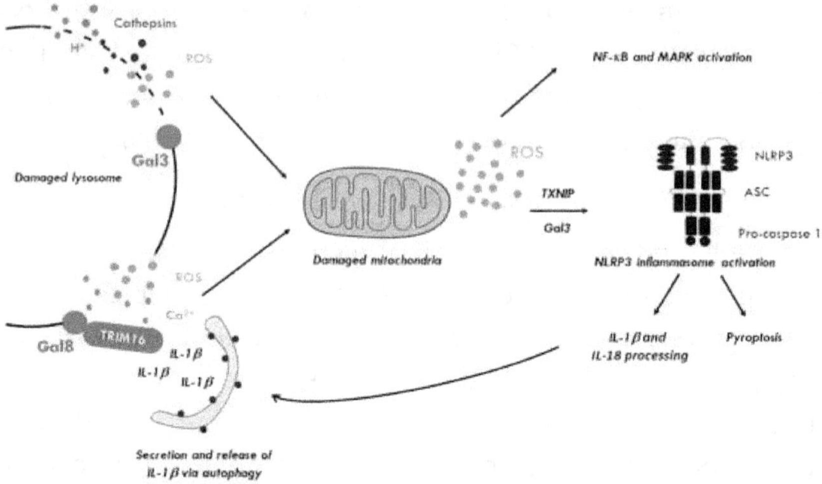

Figure 4. Membrane damage signaling. Endolysosomal rupture triggers the leakage of harmful components (e.g., ROS, H+, cathepsins or Ca2+), resulting in mitochondrial damage and increased ROS production. Increasing levels of ROS activate nuclear factor-kappa B (NF-κB) and mitogen-activated protein kinase (MAPK) signaling and triggers the release of the thioredoxin-interacting protein (TXNIP) from thioredoxin. Soluble TXNIP mediates Nod-like receptor family, pyrin domain-containing 3 (NLRP3)-inflammasomes activation and association with their adapters (ASC), thereby triggering pro-caspase conversion [149,150]. Activated inflammasomes process pro-inflammatory cytokines (IL-1β and IL-18) and may trigger cell death through pyroptosis. Processed IL-1β in turn can be recruited by Gal8 and TRIM16, that coordinate the autophagic machinery to secrete IL-1β via an unconventional secretory pathway [105,151]. Abbreviations: Gal3, galectin3; Gal8, galectin8; ROS, reactive oxygen species.

The different receptors exert their pathogen sensing role in association with the endocytic entry compartment [152]. To complicate matters even further, different sensing pathways can converge in overlapping responses such as IFN or pro-inflammatory cytokine expression [153]. As discussed above, cytosolic glycan exposure is considered an important and unique DAMP for membrane damage. Thus, inflammatory responses upon non-microbial membrane damage provides some insight into cause and consequences and can be studied e.g., in the context of neuro-degenerative diseases and neuroinflammation [154]. Some, neurodegenerative diseases are caused by the formation of pathogenic protein aggregates. These include amyloid-β or tau in Alzheimer's disease, α-synuclein in Parkinson disease or aberrant polyglutamine stretches in mutated huntingtin in Huntington's disease (reviewed in [155]). During cell-to-cell transmission, aggregates reach the cytosol by damaging the endocytic compartment, which results in Gal3 accumulation on endosomes but mostly lysosomes [80]. Furthermore, α-synuclein mediated lysosome rupture produces ROS e.g., due to mitochondrial damage upon cathepsin release [79,149]. Unbalanced ROS accumulating in the cytosol can trigger the inflammasome, but also other inflammatory pathways including the nuclear factor-kappa B (NF-κB) pathway ([156,157], reviewed in [158]). Thus, it is important to understand which pathway is activated upon membrane breach. A recent study using several Huntington's disease models showed a role for Gal3 in the regulation of the inflammatory response towards membrane damage caused by the huntingtin mutant. Pharmacological inhibition or depletion of Gal3 in respective microglia cells or a mouse model attenuated pro-inflammatory cytokines including interleukins (IL-1β, IL-6) and TNFα. This reduction correlated with reduced NLRP3 inflammasome levels suggesting a direct regulation via Gal3 [159]. Mechanistically, the authors discuss a role for Gal3 in inhibiting the clearing of damaged lysosomes and a possible, Gal3-mediated crosstalk with the NF-κB pathway [160]. Based on these results, it is possible that Gal3 plays part in maintaining the inflammatory signaling upon membrane damage by keeping the inflammasome activated until overruled by autophagy (see above). Other Gals, such as Gal8/9, also contribute to inflammation upon membrane damage. A recent study showed that Gal8 recruited to damaged lysosomes binds to TRIM16, which sequesters processed IL-1β (Figure 4). TRIM16 coordinates autophagosome assembly and transfers IL-1β to secretory autophagic vesicles to enhance IL-1β secretion [105,145]. If Gal3/Trim16 has a similar role is not known. These examples indicate that Gals may have a dual role by regulating inflammatory responses on top off membrane damage control.

2.7. Inflammation upon Virus Membrane Penetration

Detecting viral nucleic acids such as double stranded RNA or cytosolic DNA is crucial for cells to sense invading viruses. Toll-like receptors are the first line of sensors (10 in humans) and can detect different danger signals including nucleic acids (DNA and RNA) [161]. They localize either to the plasma membrane (TLR1/2/4/5/6/10) or inside the endosomal compartment (TLR3/7/8/9) [162]. Pathogen-associated molecular pattern detection induces a change in TLR conformation, allowing interaction with adapter molecules, such as Myeloid differentiation primary response protein (Myd88) or TIR Domain-Containing Adapter-Inducing Beta Interferon (TRIF), which activate in turn IRF transcription factors and induce IFN-I production. The cytosolic presence of short double-stranded RNA with a 5′ triphosphate end activates RIG-I [163,164], while long double-stranded RNA activates the Melanoma differentiation-associated protein 5 (MDA5) [165]. These two RLRs recruit MAVS adapter molecules located on the mitochondrial surface were they activate signaling kinases inhibitor of nuclear factor kappa-B kinase subunit alpha (IKKα) and TBK1 [166,167]. Both kinases phosphorylate transcription factors IRF3/7 and NF-κB, which drive expression of inflammatory genes and interferons [167,168]. Reovirus and PV, with their RNA genomes, were shown to activate the RIG-I/MDA5 pathway although probably not during entry [169,170,171,172,173]. In addition, the encephalomyocarditis virus viroporin 2B was shown to activate the NLRP3 inflammasome and to trigger IL-1β release [174,175]. Other PV like human rhinovirus (HRV) or expression of the HRV 2B protein also activated the NLRP3 inflammasome. However, because UV-inactivation of PV strongly decreased inflammasome activation it is unlikely that the PV membrane penetration step plays a dominant role [176]. Moreover, cGAS is a ubiquitous sensor of cytosolic DNA, which recognizes nucleic acids e.g., from entering DNA viruses. In the presence of DNA, it catalyzes the local formation of cGAMP, which binds to the ER associated adapter protein STING. This induces a conformational change in STING, which will then translocate to the Golgi where it serves as phosphorylation platform by recruiting TBK1 and its IRF3 substrate, again resulting in expression of pro-inflammatory cytokines and IFN. The Ad ts1 mutant was instrumental in identifying inflammatory signaling linked to Ad membrane penetration. As discussed above, Ad ts1 is endocytosed, but does not rupture the endosome. Wild type virus (but not ts1) induces a pro-inflammatory response upon cell entry, characterized by increasing the activity of p38 and Extracellular signal-regulated kinase (ERK) [177]. Likewise, rupture of the endosomal membrane with high doses of wild type Ad in murine macrophages induced the expression of IFN via activation of TBK1/IRF3 in a STING dependent manner [178,179]. Immune-complexed Ad in monocyte-derived DC (MoDC) was shown to release protein VI and to accumulate Gal3 indicative of membrane rupture and to activate the absent in melanoma 2 (AIM2) inflammasome [180]. Using a macrophage model, it was shown that high doses of Ad leads to the rapid production of ROS, release of cathepsins into the cytosol and mitochondrial damage. This resulted in secretion of IL-1β following activation of the inflammasome NLRP3. In their studies neither the Ad-ts1 mutant nor reovirus was capable of inducing this kind of response [181,182,183]. The data show that inflammasome activation by Ads requires membrane damage and cathepsin release. However, endosome rupture and cytosolic accumulation also exposes the viral DNA to cGAS/STING and RIG-I detection triggering additional immune responses [184,185,186]. The immune activation mechanisms induced by Ad membrane rupture are therefore not easy to dissect and it remains to be established if there is signaling hierarchy or cooperativity. Further evidence of membrane modulation in the direct activation of the antiviral response have also been found in enveloped viruses discussed elsewhere [187,188].

3. INTERPLAY AMONG MEMBRANE INJURY, AUTOPHAGY, AND INFLAMMATORY REACTION

As previously mentioned, membrane breakdown triggers both autophagy and inflammation. It is reasonable to infer that autophagy and inflammation have a reciprocal regulatory effect on each other. Notably, a number of the mechanisms that cause inflammation when the cell membrane is damaged or when a virus enters also lead to an overall increase in autophagy. Unsurprisingly, autophagy has been proposed to play a role in regulating and constraining both innate and adaptive immunity. (reviewed in [189]). An early indication of the direct link between autophagy and inflammation was shown by studying Crohn's disease in ATG16L1 deficient mice [190]. These mice had high levels of IL-1β and IL-18, which are indicative of elevated inflammasome activation. Similarly, IL-1β production was increased in macrophages treated with the autophagy inhibitor 3-MA or macrophages deficient for several ATGs [190,191]. The absence of ATG7 [192] or ATG5 [193,194] also caused an increase in IL-1β production in murine (alveolar) macrophages and induced pyroptosis showing how important autophagy is in limiting inflammation at least in immune cells. Mechanistically autophagy either removes inflammasome activators, inflammasome substrates or removes the activated inflammasome itself. For example, autophagy selectively degrades damaged lysosomes (lysophagy) to remove a source of cytosolic cathepsins, which can cause depolarization of mitochondria (Figure 4). Damaged mitochondria produce large amounts of ROS or oxidized mitochondrial DNA, both strong inflammasome activators. The selective removal of damaged mitochondria (mitophagy), thus, further restricts inflammasome activity [143,194]. Like lysophagy, mitophagy is a process of selective autophagy and uses selective autophagy receptors such as p62 [195] or Fanconi anemia complementation group C (FANCC) [196]. These receptors are recruited to damaged mitochondria by virtue of ubiquitin tags placed by the ubiquitin ligase Parkin [197] described in detail elsewhere [198]. Autophagy also sequesters and removes the immature form of pro-inflammatory IL-1β, an inflammasome substrate [199]. Autophagy also directly degrades activated inflammasomes. For example, the AIM2 inflammasome is ubiquitylated via the TRIM11 E3 ligase upon activation. The SLR, p62 recognizes the ubiquitin signal on AIM2 and mediates selective autophagy degradation [200,201]. A different TRIM, TRIM20, was shown to ubiquitylate pro-caspase 1 as well as the NLRP1/3 receptor [202], suggesting a more universal degradation mechanism of activated immune regulators by TRIMs (reviewed in [203]). In contrast, several inflammasome receptors interact with Beclin-1 and block the induction of autophagy [204]. This observation suggests that NLR receptors may have reciprocal regulatory functions through direct interactions with ATGs controlling autophagy and inflammasome activation [204]. Autophagic processes are not systematically anti-inflammatory. In macrophages and probably other immune cells, Gal8/TRIM16 sequestration of mature IL-1β to autophagic membranes regulates unconventional secretion of IL-1β [205,206]. Consequently, inhibiting autophagy decreased inflammation in response to activators of AIM2 and NLRP3 inflammasomes in some studies [205,207].

Inflammasomes are not the only signaling platforms degraded by autophagy. Virus sensing in the endosome via TLR, cytosolic RIG-I/MDA5 sensing of viral RNAs or cGAS/STING sensing of viral DNAs all assemble specific signaling platforms that activate interferon and/or proinflammatory cytokine expression. Whether some (or all) of these pathogen sensing and signaling platforms are linked to the surveillance of membrane integrity is not clear. It would make sense for cells to aim to understand what penetrates the membrane and establish some mechanistic link between membrane damage detection, sensing, signaling, and removal by integrating autophagic and immunity signaling. This idea is supported by several observations. Autophagy can be induced following the activation of certain TLRs [208,209]. For example, TLR7 and 3 (involved in innate responses to viruses) induce autophagy in vitro after their activation by single-stranded and double-stranded RNA present in viral genomes [210,211]. The precise mechanism is not known, but involves recruitment of Beclin-1 [211]. Autophagy in turn limits TLR signaling by degrading several mediators such as IKK/TBK1 and IRF3/7 (see below, [203]). ATGs (and autophagy) are also involved in preventing unprovoked RIG-I pathway activation by blocking the interaction between RIG-I and mitochondrial antiviral-signaling protein (MAVS). For example, ATG12-ATG5 inhibits RIG-I and MAVS by binding to their caspase activation and recruitment (CARD) domain [212]. A similar function was shown for Beclin-1 blocking RIG-I/MAVS also by binding the CARD domain of MAVS [213]. In contrast, once activated, RIG-I/MAVS trigger the production of IFN and the expression of interferon inducible genes (ISGs). Two ISGs, tetherin and ISG-15, provide a negative feedback loop targeting respectively MAVS and RIG-I for selective autophagy [213,214,215]. Autophagic restriction of cGAS/STING works in a similar fashion. ATG9 prevents activation of TBK1 by controlling STING [216], while Beclin-1 binds cGAS to prevent the production of cGAMP and thereby IFN production [217]. In addition, at steady state, cGAS is rapidly turned over via K48 ubiquitylation, recognized by p62 and targeted to selective autophagy [218]. This constant turnover is counteracted upon cGAS stimulation by the ISG TRIM14 [218]. STING can also be ubiquitylated presumably by TRIM56 [219] resulting, like for cGAS, in degradation via p62 and selective autophagy [220]. Both pathways, RIG-I/MDA5 and cGAS/STING recruit kinases TBK1 and IKK respectively to phosphorylate transcription factor IRF3/7. Like the different signaling platforms (including the inflammasome), the kinases and transcription factors substrates are also subject to autophagic control. TRIM21 is capable of recruiting the autophagic machinery and initiating de novo autophagy to target IKKβ [221], while TBK1 is degraded via a TRIM27 mediated process [222]. Furthermore, TRIM21 regulates the process of autophagy degradation of active IRF3 dimers [202]. This list provides a limited number of examples that demonstrate the complex and close relationship between innate immunity and the autophagic degradation mechanism, which both contribute to both intensifying and limiting the inflammatory response.

4. CONCISE SUMMARY

Our comprehension of the cellular membrane damage response has significantly expanded over the years. The majority of our knowledge originates from membrane injury produced by invasive pathogens or from mechanical disruption due to protein aggregation or lysosomal destruction. The research of virus-induced membrane damage has led to further advancements in the field, revealing that the cell response is conserved and connected to pro-inflammatory signaling.

Regardless of the stimulus, cells possess the ability to ascertain the magnitude and extent of membrane injury. The initial cellular response to membrane damage, known as the "repair me if you can" response, involves the ESCRT machinery. However, in cases of more extensive and prolonged damage, autophagy is activated, along with potential pro-inflammatory signals. According to this review, Gals, SLRs, ATGs, and TRIMs are four protein families that are closely related and play a crucial role in the cellular response to membrane damage. They are responsible for regulating autophagy and modulating the resulting inflammatory response [148,203]. Galectins identify the presence of membrane damage and direct the ATGs through SLRs to initiate the autophagic process. Currently, only one protein called TRIM16 has been shown to directly contribute to the cellular response to membrane damage caused by Gals and autophagy. However, over 60 TRIMs that were screened using two separate methods, more than half of them were identified as positive regulators of autophagy ([223,224]) or virophagy (such as TRIM21/23/41) ([223]). Furthermore, numerous papers demonstrate that multiple TRIMs have a role in innate immunity, either by controlling the activity of TLRs (such as TRIM8/30a/31/32/38/56) and/or the nucleic acid sensors RIG-1/MDA5 and cGAS/STING (such as TRIM13/14/25/29/31/32/38/56), or by directly attacking viruses (TRIM 5α/21/22). The TRIM targets are consistently exposed to autophagic degradation, as discussed in [203] and its associated references. Curiously, a range of TRIMs that are involved in the regulation of the immune system or autophagy were also found to bind Gal3/8 in a pull down experiment. These TRIMs include TRIM5α, TRIM6, TRIM17, TRIM20, TRIM22, and TRIM23 [104]. Currently, it remains uncertain whether Gal binding and autophagy, or the immune regulation mediated by these TRIMs, effectively connects the identification of membrane damage with immune regulation and/or pathogen sensing. Currently, there is a scarcity of information regarding the damage caused to viral membranes, particularly by non-enveloped viruses, during the process of cell entry. A distinctive characteristic of virus entry is that, unlike bacteria, invading viruses are incapable of expressing and utilizing effector molecules to counteract the cell's defenses until their genomes are delivered and expressed. Therefore, viral capsid components are the sole available method to counteract the cellular reaction. Both Ad and PV, the two instances analyzed in this paper, share the characteristic of impeding the membrane damage response by postponing autophagy until they have transported either the capsid (Ad) or their genome (PV) to the cytosol. Furthermore, both viruses elicit a cellular membrane damage response, as demonstrated through the use of a mutant virus or depletion studies. This implies that there may be a beneficial connection between the cell response and virus penetration. This observation is significant because any harm to the membrane or the process of antiviral autophagy can be detrimental to the virus. After being taken up by human macrophages, Ad particles that were coated with antibodies were shown to rupture lysosomes and activate inflammasomes. Meanwhile, cytosolic TRIM21 would identify any virus that was coated with antibodies and had entered the cytosol, and then transfer it to cGAS/STING sensing [225]. As previously mentioned, autophagy downregulates the activity of inflammasome and cGAS/STING. Similarly, Ad-M1 mutant viruses are incapable of halting the membrane damage response upon entry due to their absence of a crucial PPxY peptide sequence. Ultimately, Ad-M1 undergoes degradation by autophagy and is subsequently presented with more efficiency to MHC class II compared to wild type Ad in a murine model [3]. In addition, TBK1 phosphorylates SLRs during the cell response to bacteria, regardless of its role in inducing IFN through IRF3 phosphorylation [122]. PV genomes are destroyed along with their capsids without any delay in autophagy [45].

On the other hand, it is widely recognized that PV utilizes autophagy to establish a membrane compartment for replicating the genome in the cytosol (as discussed in [226]). Delayed activation of autophagy upon genome translocation may promote the initiation of PV replication. In a broader context, it seems plausible that non-enveloped viruses deliberately exploit the cellular membrane damage response after entry in order to restrict the cell's inherent immune response (Figure 5). Certain viruses have perhaps developed the ability to evade the endosome by utilizing small openings that can be healed, causing only limited and temporary harm without triggering a substantial immune response.

Activating autophagy through prolonged membrane damage and regulating it, such as with Ad and PV, may present a potential trade-off in the opposite scenario. On one hand, it enables the release from endosomes and the utilization of autophagic membranes for replication, whereas autophagy may be utilized to expedite the degradation of molecules implicated in immunological detection and pro-inflammatory signaling. Attaining an optimal equilibrium by carefully choosing, delaying, and coordinating the cell's response could serve as a potent alternative for the inability to convey effector molecules. Although this hypothesis is intriguing and deserves additional exploration, there is currently a dearth of experimental data supporting a direct connection between virus-induced membrane damage response and the management of the inflammatory response by autophagy.

The potential for several TRIMs to bind Gals suggests that a greater number of TRIM substrates could be enlisted at regions of membrane degradation. The objective of inducing membrane damage in invading viruses is to eliminate a significant number of cell effector molecules and dampen the cell's response to viral entrance.

Taken together, there remains much to be discovered concerning our understanding how viruses penetrate cell membranes and how cells respond to this insult. Using the few existing examples presented in this review about the cell response to virus inflicted membrane damage, we invite the reader to reflect on the wider role of membrane damage for the virus.

Cells face ongoing challenges from pathogens (such as bacteria, viruses, and fungi), as well as protein aggregates or chemicals. These challenges can cause damage to the cell membrane either at the plasma membrane or within the endo-lysosomal compartments. Glycan exposure to the cytoplasm due to membrane disruption is detected by a group of lectins called galectins, which are sugar-binding proteins. Galectins, along with other factors, coordinate distinct responses to membrane damage, including as the recruitment of the endosomal sorting complex needed for transport (ESCRT) machinery.

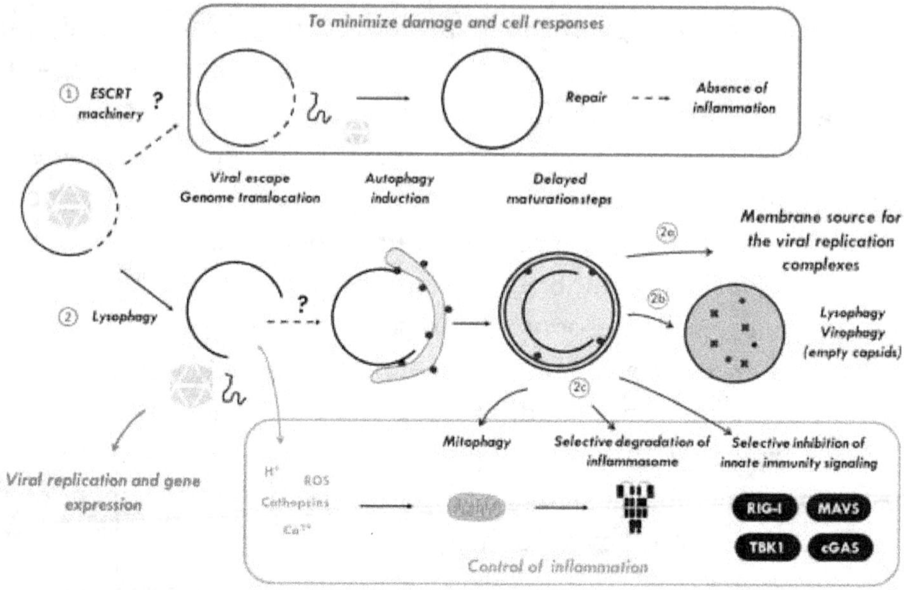

Figure 5. Model for non-enveloped virus membrane penetration. Non-enveloped viruses breach endo-lysosomal membranes for particle escape into the cytosol or cytosolic genome translocation. (1) If the membrane damage is small enough, Ca2+-dependent recruitment of the ESCRT machinery will repair the endosomes limiting time and extend of the damage. This strategy could provide enough time for the virus/genome to reach the cytosol without activating the inflammatory pathways. (2) If this repair fails or the inflicted damage is too large, "eat-me signals" (galectins and ubiquitin) will accumulate at the damage site and initiate autophagy. Viruses use viral factors (e.g., Ad) or cellular factors (e.g., PV) to delay the autophagy response until capsids/viral genomes reach the safety of the cytosol. The ensuing autophagy response than may continue and work in favor of the virus. Activating autophagy via membrane damage could (2a) provide a membranes source for viral replication, and (2b) remove and recycle membrane remnants, damaged lysosomes, and empty viral capsids to avoid excessive cells responses. Additionally, autophagy could (2c) limit excessive inflammation through autophagic degradation of damaged mitochondria and by removing activated inflammasomes and effectors of the innate immunity signaling pathways (see text for details).

This can either facilitate the repair of damaged membranes or trigger the activation of autophagy to eliminate residual membrane fragments.

Unregulated membrane damage leads to the release of detrimental substances such as protons, reactive oxygen species, or cathepsins, which trigger inflammation. This article presents a comprehensive summary of the existing understanding of membrane injury and the subsequent cellular responses. Specifically, our focus lies on the harm caused to the endo-lysosomal system by non-enveloped viruses, such adenovirus. We also examine the methods employed by these viruses to manipulate the cellular response to membrane damage. Ultimately, we examine the connection between autophagy and inflammation in this particular situation and explore the potential that autophagy triggered by a virus upon entry restricts inflammation.

REFERENCES

1. Lombard, J. Once upon a time the cell membranes: 175 years of cell boundary research. Biol. Direct 2014, 9, 1–35.
2. Thurston, T.L.M.M.; Wandel, M.P.; Von Muhlinen, N.; Foeglein, Á.; Randow, F.; Foeglein, A.; Randow, F.; Foeglein, Á.; Randow, F. Galectin 8 targets damaged vesicles for autophagy to defend cells against bacterial invasion. Nature 2012, 482, 414–418.
3. Montespan, C.; Marvin, S.A.; Austin, S.; Burrage, A.M.; Roger, B.; Rayne, F.; Faure, M.; Campell, E.M.; Schneider, C.; Reimer, R.; et al. Multi-layered control of Galectin-8 mediated autophagy during adenovirus cell entry through a conserved PPxY motif in the viral capsid. PLoS Pathog. 2017, 13.
4. Jia, J.; Abudu, Y.P.; Claude-Taupin, A.; Gu, Y.; Kumar, S.; Choi, S.W.; Peters, R.; Mudd, M.H.; Allers, L.; Salemi, M.; et al. Galectins Control mTOR in Response to Endomembrane Damage. Mol. Cell 2018, 70, 120–135.e8.
5. Vietri, M.; Radulovic, M.; Stenmark, H. The many functions of ESCRTs. Nat. Rev. Mol. Cell Biol. 2020, 21, 25–42.
6. Lozach, P.Y.; Huotari, J.; Helenius, A. Late-penetrating viruses. Curr. Opin. Virol. 2011, 1, 35–43.

7. Ketter, E.; Randall, G. Virus Impact on Lipids and Membranes. Annu. Rev. Virol. 2019, 6, 319–340.

8. Moyer, C.L.; Nemerow, G.R. Viral weapons of membrane destruction: Variable modes of membrane penetration by non-enveloped viruses. Curr. Opin. Virol. 2011, 1, 44–49.

9. Stewart, P.L.; Dermody, T.S.; Nemerow, G.R. Structural basis of nonenveloped virus cell entry. Adv. Protein Chem. 2003, 64, 455–491.

10. Prchla, E.; Plank, C.; Wagner, E.; Blaas, D.; Fuchs, R. Virus-mediated release of endosomal content in vitro: Different behavior of adenovirus and rhinovirus serotype 2. J. Cell Biol. 1995, 131, 111–123.

11. Farr, G.A.; Zhang, L.G.; Tattersall, P. Parvoviral virions deploy a capsid-tethered lipolytic enzyme to breach the endosomal membrane during cell entry. Proc. Natl. Acad. Sci. USA 2005, 102, 17148–17153.

12. Greber, U.F.; Webster, P.; Weber, J.; Helenius, A. The role of the adenovirus protease in virus entry into cells. EMBO J. 1996, 15, 1766–1777.

13. Wiethoff, C.M.; Wodrich, H.; Gerace, L.; Nemerow, G.R. Adenovirus Protein VI Mediates Membrane Disruption following Capsid Disassembly. J. Virol. 2005, 79, 1992–2000.

14. Martinez, R.; Schellenberger, P.; Vasishtan, D.; Aknin, C.; Austin, S.; Dacheux, D.; Rayne, F.; Siebert, A.; Ruzsics, Z.; Gruenewald, K.; et al. The amphipathic helix of adenovirus capsid protein VI contributes to penton release and postentry sorting. J. Virol. 2015, 89, 2121–2135.

15. Weber, J. Genetic analysis of adenovirus type 2 III. Temperature sensitivity of processing viral proteins. J. Virol. 1976, 17, 462–471.

16. Cotten, M.; Weber, J.M. The adenovirus protease is required for virus entry into host cells. Virology 1995, 213, 494–502.

17. Martín, C.S. Latest insights on adenovirus structure and assembly. Viruses 2012, 4, 847–877.

18. Wodrich, H.; Henaff, D.; Jammart, B.; Segura-Morales, C.; Seelmeir, S.; Coux, O.; Ruzsics, Z.; Wiethoff, C.M.; Kremer, E.J. A capsid-encoded PPxY-motif facilitates adenovirus entry. PLoS Pathog. 2010, 6.

19. Luisoni, S.; Suomalainen, M.; Boucke, K.; Tanner, L.B.; Wenk, M.R.; Guan, X.L.; Grzybek, M.; Coskun, U.; Greber, U.F. Co-option of membrane wounding enables virus penetration into cells. Cell Host Microbe 2015, 18, 75–85.

20. Wiethoff, C.M.; Nemerow, G.R. Adenovirus membrane penetration: Tickling the tail of a sleeping dragon. Virology 2015, 479–480, 591–599.

21. Hernando-Pérez, M.; Martín-González, N.; Pérez-Illana, M.; Suomalainen, M.; Condezo, G.N.; Ostapchuk, P.; Gallardo, J.; Menéndez, M.; Greber, U.F.; Hearing, P.; et al. Dynamic competition for hexon binding between core protein VII and lytic protein VI promotes adenovirus maturation and entry. Proc. Natl. Acad. Sci. USA 2020, 117, 13699–13707.

22. Maier, O.; Galan, D.L.; Wodrich, H.; Wiethoff, C.M. An N-terminal domain of adenovirus protein VI fragments membranes by inducing positive membrane curvature. Virology 2010, 402, 11–19.

23. Moyer, C.L.; Wiethoff, C.M.; Maier, O.; Smith, J.G.; Nemerow, G.R. Functional Genetic and Biophysical Analyses of Membrane Disruption by Human Adenovirus. J. Virol. 2011, 85, 2631–2641.

24. Pied, N.; Wodrich, H. Imaging the adenovirus infection cycle. FEBS Lett. 2019, 593, 3419–3448.

25. Dupzyk, A.; Tsai, B. How polyomaviruses exploit the ERAD machinery to cause infection. Viruses 2016, 8, 242.

26. Magnuson, B.; Rainey, E.K.; Benjamin, T.; Baryshev, M.; Mkrtchian, S.; Tsai, B. ERp29 triggers a conformational change in polyomavirus to stimulate membrane binding. Mol. Cell 2005, 20, 289–300.

27. Guerrero, C.A.; Bouyssounade, D.; Zárate, S.; Iša, P.; López, T.; Espinosa, R.; Romero, P.; Méndez, E.; López, S.; Arias, C.F. Heat Shock Cognate Protein 70 Is Involved in Rotavirus Cell Entry. J. Virol. 2002, 76, 4096–4102.

28. Walczak, C.P.; Ravindran, M.S.; Inoue, T.; Tsai, B. A Cytosolic Chaperone Complexes with Dynamic Membrane J-Proteins and Mobilizes a Nonenveloped Virus out of the Endoplasmic Reticulum. PLoS Pathog. 2014, 10, e1004007.

29. Inoue, T.; Tsai, B. A large and intact viral particle penetrates the endoplasmic reticulum membrane to reach the cytosol. PLoS Pathog. 2011, 7, 1002037.

30. Spriggs, C.C.; Badieyan, S.; Verhey, K.J.; Cianfrocco, M.A.; Tsai, B. Golgi-associated BICD adaptors couple ER membrane penetration and disassembly of a viral cargo. J. Cell Biol. 2020, 219.

31. Suikkanen, S.; Antila, M.; Jaatinen, A.; Vihinen-Ranta, M.; Vuento, M. Release of canine parvovirus from endocytic vesicles. Virology 2003, 316, 267–280.

32. Canaan, S.; Zádori, Z.; Ghomashchi, F.; Bollinger, J.; Sadilek, M.; Moreau, M.E.; Tijssen, P.; Gelb, M.H. Interfacial Enzymology of Parvovirus Phospholipases A2. J. Biol. Chem. 2004, 279, 14502–14508.

33. Dorsch, S.; Liebisch, G.; Kaufmann, B.; von Landenberg, P.; Hoffmann, J.H.; Drobnik, W.; Modrow, S. The VP1 Unique Region of Parvovirus B19 and Its Constituent Phospholipase A2-Like Activity. J. Virol. 2002, 76, 2014–2018.

34. Ebert, D.H.; Deussing, J.; Peters, C.; Dermody, T.S. Cathepsin L and cathepsin B mediate reovirus disassembly in murine fibroblast cells. J. Biol. Chem. 2002, 277, 24609–24617.

35. Chandran, K.; Farsetta, D.L.; Nibert, M.L. Strategy for Nonenveloped Virus Entry: A Hydrophobic Conformer of the Reovirus Membrane Penetration Protein μ1 Mediates Membrane Disruption. J. Virol. 2002, 76, 9920–9933.

36. Odegard, A.L.; Chandran, K.; Zhang, X.; Parker, J.S.L.; Baker, T.S.; Nibert, M.L. Putative Autocleavage of Outer Capsid Protein μ1, Allowing Release of Myristoylated Peptide μ1N during Particle Uncoating, Is Critical for Cell Entry by Reovirus. J. Virol. 2004, 78, 8732–8745.

37. Liemann, S.; Chandran, K.; Baker, T.S.; Nibert, M.L.; Harrison, S.C. Structure of the reovirus membrane-penetration protein, μ1, in a complex with its protector protein, σ3. Cell 2002, 108, 283–295.

38. Zhang, L.; Agosto, M.A.; Ivanovic, T.; King, D.S.; Nibert, M.L.; Harrison, S.C. Requirements for the Formation of Membrane Pores by the Reovirus Myristoylated μ1N Peptide. J. Virol. 2009, 83, 7004–7014.

39. Agosto, M.A.; Ivanovic, T.; Nibert, M.L. Mammalian reovirus, a nonfusogenic nonenveloped virus, forms size-selective pores in a model membrane. Proc. Natl. Acad. Sci. USA 2006, 103, 16496–16501.

40. Ivanovic, T.; Agosto, M.A.; Zhang, L.; Chandran, K.; Harrison, S.C.; Nibert, M.L. Peptides released from reovirus outer capsid form membrane pores that recruit virus particles. EMBO J. 2008, 27, 1289–1298.

41. Tuthill, T.J.; Groppelli, E.; Hogle, J.M.; Rowlands, D.J. Picornaviruses. In Current Topics in Microbiology and Immunology; Springer: Berlin/Heidelberg, Germany, 2010; Volume 343, pp. 43–89.

42. Schober, D.; Kronenberger, P.; Prchla, E.; Blaas, D.; Fuchs, R. Major and Minor Receptor Group Human Rhinoviruses Penetrate from Endosomes by Different Mechanisms. J. Virol. 1998, 72, 1354–1364.

43. Panjwani, A.; Strauss, M.; Gold, S.; Wenham, H.; Jackson, T.; Chou, J.J.; Rowlands, D.J.; Stonehouse, N.J.; Hogle, J.M.; Tuthill, T.J. Capsid Protein VP4 of Human Rhinovirus Induces Membrane Permeability by the Formation of a Size-Selective Multimeric Pore. PLoS Pathog. 2014, 10, e1004294.

44. Staring, J.; Raaben, M.; Brummelkamp, T.R. Viral escape from endosomes and host detection at a glance. J. Cell Sci. 2018, 131.

45. Staring, J.; Von Castelmur, E.; Blomen, V.A.; Van Den Hengel, L.G.; Brockmann, M.; Baggen, J.; Thibaut, H.J.; Nieuwenhuis, J.; Janssen, H.; Van Kuppeveld, F.J.M.M.; et al. PLA2G16 represents a switch between entry and clearance of Picornaviridae. Nature 2017, 541, 412–416.

46. Hughes, P.J.; Stanway, G. The 2A proteins of three diverse picornaviruses are related to each other and to the H-rev107 family of proteins involved in the control of cell proliferation. J. Gen. Virol. 2000, 81, 201–207.

47. Knodler, L.A.; Vallance, B.A.; Celli, J.; Winfree, S.; Hansen, B.; Montero, M.; Steele-Mortimer, O. Dissemination of invasive Salmonella via bacterial-induced extrusion of mucosal epithelia. Proc. Natl. Acad. Sci. USA 2010, 107, 17733–17738.

48. Mellouk, N.; Enninga, J. Cytosolic access of intracellular bacterial pathogens: The Shigella paradigm. Front. Cell. Infect. Microbiol. 2016, 6, 1–8.

49. Mellouk, N.; Weiner, A.; Aulner, N.; Schmitt, C.; Elbaum, M.; Shorte, S.L.; Danckaert, A.; Enninga, J. Shigella subverts the host recycling compartment to rupture its vacuole. Cell Host Microbe 2014, 16, 517–530.

50. Weiner, A.; Mellouk, N.; Lopez-Montero, N.; Chang, Y.-Y.Y.; Souque, C.; Schmitt, C.; Enninga, J. Macropinosomes are Key Players in Early Shigella Invasion and Vacuolar Escape in Epithelial Cells. PLoS Pathog. 2016, 12, e1005602.

51. Birmingham, C.L.; Smith, A.C.; Bakowski, M.A.; Yoshimori, T.; Brumell, J.H. Autophagy controls Salmonella infection in response to damage to the Salmonella-containing vacuole. J. Biol. Chem. 2006, 281, 11374–11383.

52. Radtke, A.L.; Delbridge, L.M.; Balachandran, S.; Barber, G.N.; O'Riordan, M.X.D. TBK1 protects vacuolar integrity during intracellular bacterial infection. PLoS Pathog. 2007, 3.

53. Boyle, K.B.; Thurston, T.L.M.; Randow, F. TBK1 directs WIPI2 against Salmonella. Autophagy 2016, 12, 2508–2509.

54. Thurston, T.L.; Boyle, K.B.; Allen, M.; Ravenhill, B.J.; Karpiyevich, M.; Bloor, S.; Kaul, A.; Noad, J.; Foeglein, A.; Matthews, S.A.; et al. Recruitment of TBK 1 to cytosol-invading Salmonella induces WIPI 2-dependent antibacterial autophagy. EMBO J. 2016, 35, 1779–1792.

55. Santos, J.C.; Duchateau, M.; Fredlund, J.; Weiner, A.; Mallet, A.; Schmitt, C.; Matondo, M.; Hourdel, V.; Chamot-Rooke, J.; Enninga, J. The COPII complex and lysosomal VAMP7 determine intracellular Salmonella localization and growth. Cell. Microbiol. 2015, 17, 1699–1720.

56. Singh, R.; Jamieson, A.; Cresswell, P. GILT is a critical host factor for Listeria monocytogenes infection. Nature 2008, 455, 1244–1247.

57. Radtke, A.L.; Anderson, K.L.; Davis, M.J.; DiMagno, M.J.; Swanson, J.A.; O'Riordan, M.X. Listeria monocytogenes exploits cystic fibrosis transmembrane conductance regulator (CFTR) to escape the phagosome. Proc. Natl. Acad. Sci. USA 2011, 108, 1633–1638.

58. Gedde, M.M.; Higgins, D.E.; Tilney, L.G.; Portnoy, D.A. Role of listeriolysin O in cell-to-cell spread of Listeria monocytogenes. Infect. Immun. 2000, 68, 999–1003.

59. Alberti-Segui, C.; Goeden, K.R.; Higgins, D.E. Differential function of Listeria monocytogenes listeriolysin O and phospholipases C in vacuolar dissolution following cell-to-cell spread. Cell. Microbiol. 2007, 9, 179–195.

60. Malet, J.K.; Cossart, P.; Ribet, D. Alteration of epithelial cell lysosomal integrity induced by bacterial cholesterol-dependent cytolysins. Cell. Microbiol. 2017, 19, 1–11.

61. Palframan, S.L.; Kwok, T.; Gabriel, K. Vacuolating cytotoxin A (VacA), a key toxin for Helicobacter pylori pathogenesis. Front. Cell. Infect. Microbiol. 2012, 2, 92.

62. Li, F.Y.; Weng, I.C.; Lin, C.H.; Kao, M.C.; Wu, M.S.; Chen, H.Y.; Liu, F.T. Helicobacter pylori induces intracellular galectin-8 aggregation around damaged lysosomes within gastric epithelial cells in a host O-glycan-dependent manner. Glycobiology 2018, 29, 151–162.

63. van der Wel, N.; Hava, D.; Houben, D.; Fluitsma, D.; Van Zon, M.; Pierson, J.; Brenner, M.; Peters, P.J.M. tuberculosis and M. leprae Translocate from the Phagolysosome to the Cytosol in Myeloid Cells. Cell 2007, 129, 1287–1298.

64. De Jonge, M.I.; Pehau-Arnaudet, G.; Fretz, M.M.; Romain, F.; Bottai, D.; Brodin, P.; Honoré, N.; Marchal, G.; Jiskoot, W.; England, P.; et al. ESAT-6 from Mycobacterium tuberculosis dissociates from its putative chaperone CFP-10 under acidic conditions and exhibits membrane-lysing activity. J. Bacteriol. 2007, 189, 6028–6034.

65. Akimana, C.; Al-Khodor, S.; Kwaik, Y.A. Host factors required for modulation of phagosome biogenesis and proliferation of francisella tularensis within the cytosol. PLoS ONE 2010, 5.

66. De Leon, J.; Jiang, G.; Ma, Y.; Rubin, E.; Fortune, S.; Sun, J. Mycobacterium tuberculosis ESAT-6 exhibits a unique membrane-interacting activity that is not found in its ortholog from non-pathogenic Mycobacterium smegmatis. J. Biol. Chem. 2012, 287, 44184–44191.

67. Peng, X.; Sun, J. Mechanism of ESAT-6 membrane interaction and its roles in pathogenesis of Mycobacterium tuberculosis. Toxicon 2016, 116, 29–34.

68. Pilatz, S.; Breitbach, K.; Hein, N.; Fehlhaber, B.; Schulze, J.; Brenneke, B.; Eberl, L.; Steinmetz, I. Identification of Burkholderia pseudomallei genes required for the intracellular life cycle and in vivo virulence. Infect. Immun. 2006, 74, 3576–3586.

69. Whitworth, T.; Popov, V.L.; Yu, X.J.; Walker, D.H.; Bouyer, D.H. Expression of the Rickettsia prowazekii pld or tlyC gene in Salmonella enterica serovar typhimurium mediates phagosomal escape. Infect. Immun. 2005, 73, 6668–6673.

70. Renesto, P.; Dehoux, P.; Gouin, E.; Touqui, L.; Cossart, P.; Raoult, D. Identification and Characterization of a Phospholipase D-Superfamily Gene in Rickettsiae. J. Infect. Dis. 2003, 188, 1276–1283.

71. Clemens, D.L.; Lee, B.Y.; Horwitz, M.A. Virulent and avirulent strains of Francisella tularensis prevent acidification and maturation of their phagosomes and escape into the cytoplasm in human macrophages. Infect. Immun. 2004, 72, 3204–3217.

72. Clemens, D.L.; Ge, P.; Lee, B.Y.; Horwitz, M.A.; Zhou, Z.H. Atomic structure of T6SS reveals interlaced array essential to function. Cell 2015, 160, 940–951.

73. Chong, A.; Wehrly, T.D.; Nair, V.; Fischer, E.R.; Barker, J.R.; Klose, K.E.; Celli, J. The early phagosomal stage of Francisella tularensis determines optimal phagosomal escape and Francisella pathogenicity island protein expression. Infect. Immun. 2008, 76, 5488–5499.

74. Berman, J. Morphogenesis and cell cycle progression in Candida albicans. Curr. Opin. Microbiol. 2006, 9, 595–601.

75. Westman, J.; Moran, G.; Mogavero, S.; Hube, B.; Grinsteina, S. crossm Candida albicans Hyphal Expansion Causes Phagosomal. MBio 2018, 9, 1–14.

76. Vylkova, S.; Carman, A.J.; Danhof, H.A.; Collette, J.R.; Zhou, H.; Lorenz, M.C. The fungal pathogen candida albicans autoinduces hyphal morphogenesis by raising extracellular pH. MBio 2011, 2.

77. De Souza, W.; De Carvalho, T.M.U.; Barrias, E.S. Review on Trypanosoma cruzi: Host cell interaction. Int. J. Cell Biol. 2010, 2010.

78. Papadopoulos, C.; Kirchner, P.; Bug, M.; Grum, D.; Koerver, L.; Schulze, N.; Poehler, R.; Dressler, A.; Fengler, S.; Arhzaouy, K.; et al. VCP/p97 cooperates with YOD 1, UBXD 1 and PLAA to drive clearance of ruptured lysosomes by autophagy. EMBO J. 2017, 36, 135–150.

79. Freeman, D.; Cedillos, R.; Choyke, S.; Lukic, Z.; McGuire, K.; Marvin, S.; Burrage, A.M.; Sudholt, S.; Rana, A.; O'Connor, C.; et al. Alpha-Synuclein Induces Lysosomal Rupture and Cathepsin Dependent Reactive Oxygen Species Following Endocytosis. PLoS ONE 2013, 8.

80. Flavin, W.P.; Bousset, L.; Green, Z.C.; Chu, Y.; Skarpathiotis, S.; Chaney, M.J.; Kordower, J.H.; Melki, R.; Campbell, E.M. Endocytic vesicle rupture is a conserved mechanism of cellular invasion by amyloid proteins. Acta Neuropathol. 2017, 134, 629–653.

81. Villamil Giraldo, A.M.; Appelqvist, H.; Ederth, T.; Öllinger, K. Lysosomotropic agents: Impact on lysosomal membrane permeabilization and cell death. Biochem. Soc. Trans. 2014, 42, 1460–1464.

82. Aits, S.; Kricker, J.; Liu, B.; Ellegaard, A.-M.M.; Hämälistö, S.; Tvingsholm, S.; Corcelle-Termeau, E.; Høgh, S.; Farkas, T.; Jonassen, A.H.; et al. Sensitive detection of lysosomal membrane permeabilization by lysosomal galectin puncta assay. Autophagy 2015, 11, 1408–1424.

83. Uchimoto, T.; Nohara, H.; Kamehara, R.; Iwamura, M.; Watanabe, N.; Kobayashi, Y. Mechanism of apoptosis induced by a lysosomotropic agent, L-leucyl-L- leucine methyl ester. Apoptosis 1999, 4, 357–362.

84. Du Rietz, H.; Hedlund, H.; Wilhelmson, S.; Nordenfelt, P.; Wittrup, A. Imaging small molecule-induced endosomal escape of siRNA. Nat. Commun. 2020, 11, 1809.

85. Wittrup, A.; Ai, A.; Liu, X.; Hamar, P.; Trifonova, R.; Charisse, K.; Manoharan, M.; Kirchhausen, T.; Lieberman, J. Visualizing lipid-formulated siRNA release from endosomes and target gene knockdown. Nat. Biotechnol. 2015, 33, 870–876.

86. Fraire, J.C.; Houthaeve, G.; Liu, J.; Raes, L.; Vermeulen, L.; Stremersch, S.; Brans, T.; García-Díaz Barriga, G.; De Keulenaer, S.; Van Nieuwerburgh, F.; et al. Vapor nanobubble is the more reliable photothermal mechanism for inducing endosomal escape of siRNA without disturbing cell homeostasis. J. Control. Release 2020, 319, 262–275.

87. Chu, Z.; Miu, K.; Lung, P.; Zhang, S.; Zhao, S.; Chang, H.C.; Lin, G.; Li, Q. Rapid endosomal escape of prickly nanodiamonds: Implications for gene delivery. Sci. Rep. 2015, 5, 11661.

88. Barondes, S.H.; Castronovo, V.; Cooper, D.N.W.; Cummings, R.D.; Drickamer, K.; Felzi, T.; Gitt, M.A.; Hirabayashi, J.; Hughes, C.; Kasai, K.I.; et al. Galectins: A family of animal β-galactoside-binding lectins. Cell 1994, 76, 597–598.

89. Hirabayashi, J.; Kasai, K.I. The family of metazoan metal-independent β-galactoside-binding lectins: Structure, function and molecular evolution. Glycobiology 1993, 3, 297–304.

90. Wang, W.H.; Lin, C.Y.; Chang, M.R.; Urbina, A.N.; Assavalapsakul, W.; Thitithanyanont, A.; Chen, Y.H.; Liu, F.T.; Wang, S.F. The role of galectins in virus infection—A systemic literature review. J. Microbiol. Immunol. Infect. 2019.

91. Baum, L.G.; Garner, O.B.; Schaefer, K.; Lee, B. Microbe-host interactions are positively and negatively regulated by galectin-glycan interactions. Front. Immunol. 2014, 5, 284.

92. Paz, I.; Sachse, M.; Dupont, N.; Mounier, J.; Cederfur, C.; Enninga, J.; Leffler, H.; Poirier, F.; Prevost, M.C.; Lafont, F.; et al. Galectin-3, a marker for vacuole lysis by invasive pathogens. Cell. Microbiol. 2010, 12, 530–544.

93. Maier, O.; Marvin, S.A.; Wodrich, H.; Campbell, E.M.; Wiethoff, C.M. Spatiotemporal Dynamics of Adenovirus Membrane Rupture and Endosomal Escape. J. Virol. 2012, 86, 10821–10828.

94. Martinez, R.; Burrage, A.M.; Wiethoff, C.M.; Wodrich, H. High temporal resolution imaging reveals endosomal membrane penetration and escape of adenoviruses in real time. Methods Mol. Biol. 2013, 1064, 211–226.

95. Klionsky, D.J.; Codogno, P. The mechanism and physiological function of macroautophagy. J. Innate Immun. 2013, 5, 427–433.

96. Bento, C.F.; Renna, M.; Ghislat, G.; Puri, C.; Ashkenazi, A.; Vicinanza, M.; Menzies, F.M.; Rubinsztein, D.C. Mammalian Autophagy: How Does It Work? Annu. Rev. Biochem. 2016, 85, 685–713.

97. Suzuki, K.; Noda, T.; Ohsumi, Y. Interrelationships among Atg proteins during autophagy inSaccharomyces cerevisiae. Yeast 2004, 21, 1057–1065.

98. Klionsky, D.J.; Abdelmohsen, K.; Abe, A.; Abedin, M.J.; Abeliovich, H.; Acevedo Arozena, A.; Adachi, H.; Adams, C.M.; Adams, P.D.; Adeli, K.; et al. Guidelines for the use and interpretation of assays for monitoring autophagy (3rd edition). Autophagy 2016, 12, 1–222.

99. Johansen, T.; Lamark, T. Selective autophagy mediated by autophagic adapter proteins. Autophagy 2011, 7, 279–296.

100. Ravenhill, B.J.; Boyle, K.B.; von Muhlinen, N.; Ellison, C.J.; Masson, G.R.; Otten, E.G.; Foeglein, A.; Williams, R.; Randow, F. The Cargo Receptor NDP52 Initiates Selective Autophagy by Recruiting the ULK Complex to Cytosol-Invading Bacteria. Mol. Cell 2019, 74, 320–329.e6.

101. Vargas, J.N.S.; Wang, C.; Bunker, E.; Hao, L.; Maric, D.; Schiavo, G.; Randow, F.; Youle, R.J. Spatiotemporal Control of ULK1 Activation by NDP52 and TBK1 during Selective Autophagy. Mol. Cell 2019, 74, 347–362.e6.

102. Roberts, R.; Ktistakis, N.T. Omegasomes: PI3P platforms that manufacture autophagosomes. Essays Biochem. 2013, 55, 17–27.

103. Tanida, I.; Ueno, T.; Kominami, E. LC3 conjugation system in mammalian autophagy. Int. J. Biochem. Cell Biol. 2004, 36, 2503–2518.

104. Chauhan, S.; Kumar, S.; Jain, A.; Ponpuak, M.; Mudd, M.H.; Kimura, T.; Choi, S.W.; Peters, R.; Mandell, M.; Bruun, J.A.; et al. TRIMs and Galectins Globally Cooperate and TRIM16 and Galectin-3 Co-direct Autophagy in Endomembrane Damage Homeostasis. Dev. Cell 2016, 39, 13–27.

105. Kimura, T.; Jia, J.; Kumar, S.; Choi, S.W.; Gu, Y.; Mudd, M.; Dupont, N.; Jiang, S.; Peters, R.; Farzam, F.; et al. Dedicated SNARE s and specialized TRIM cargo receptors mediate secretory autophagy. EMBO J. 2017, 36, 42–60.

106. Weng, I.C.; Chen, H.L.; Lo, T.H.; Lin, W.H.; Chen, H.Y.; Hsu, D.K.; Liu, F.T. Cytosolic galectin-3 and -8 regulate antibacterial autophagy through differential recognition of host glycans on damaged phagosomes. Glycobiology 2018, 28, 392–405.

107. Feeley, E.M.; Pilla-Moffett, D.M.; Zwack, E.E.; Piro, A.S.; Finethy, R.; Kolb, J.P.; Martinez, J.; Brodsky, I.E.; Coers, J. Galectin-3 directs antimicrobial guanylate binding proteins to vacuoles furnished with bacterial secretion systems. Proc. Natl. Acad. Sci. USA 2017, 114, E1698–E1706.

108. Herhaus, L.; Dikic, I. Regulation of Salmonella-host cell interactions via the ubiquitin system. Int. J. Med. Microbiol. 2018, 308, 176–184.

109. Dupont, N.; Lacas-Gervais, S.; Bertout, J.; Paz, I.; Freche, B.; Van Nhieu, G.T.; van der Goot, F.G.; Sansonetti, P.J.; Lafont, F. Shigella Phagocytic Vacuolar Membrane Remnants Participate in the Cellular Response to Pathogen Invasion and Are Regulated by Autophagy. Cell Host Microbe 2009, 6, 137–149.

110. Mostowy, S.; Sancho-Shimizu, V.; Hamon, M.A.; Simeone, R.; Brosch, R.; Johansen, T.; Cossart, P. p62 and NDP52 proteins target intracytosolic Shigella and Listeria to different autophagy pathways. J. Biol. Chem. 2011, 286, 26987–26995.

111. Luisoni, S.; Bauer, M.; Prasad, V.; Boucke, K.; Papadopoulos, C.; Meyer, H.; Hemmi, S.; Suomalainen, M.; Greber, U. Endosomophagy clears disrupted early endosomes but not virus particles during virus entry into cells. Matters 2016, 1–9.

112. Yoshida, Y.; Yasuda, S.; Fujita, T.; Hamasaki, M.; Murakami, A.; Kawawaki, J.; Iwai, K.; Saeki, Y.; Yoshimori, T.; Matsuda, N.; et al. Ubiquitination of exposed glycoproteins by SCFFBXO27 directs damaged lysosomes for autophagy. Proc. Natl. Acad. Sci. USA 2017, 114, 8574–8579.

113. Fujita, N.; Morita, E.; Itoh, T.; Tanaka, A.; Nakaoka, M.; Osada, Y.; Umemoto, T.; Saitoh, T.; Nakatogawa, H.; Kobayashi, S.; et al. Recruitment of the autophagic machinery to endosomes during infection is mediated by ubiquitin. J. Cell Biol. 2013, 203, 115–128.

114. Tan, J.M.M.; Wong, E.S.P.; Kirkpatrick, D.S.; Pletnikova, O.; Ko, H.S.; Tay, S.P.; Ho, M.W.L.; Troncoso, J.; Gygi, S.P.; Lee, M.K.; et al. Lysine 63-linked ubiquitination promotes the formation and autophagic clearance of protein inclusions associated with neurodegenerative diseases. Hum. Mol. Genet. 2008, 17, 431–439.

115. Huett, A.; Heath, R.J.; Begun, J.; Sassi, S.O.; Baxt, L.A.; Vyas, J.M.; Goldberg, M.B.; Xavier, R.J. The LRR and RING domain protein LRSAM1 is an E3 ligase crucial for ubiquitin-dependent autophagy of intracellular salmonella typhimurium. Cell Host Microbe 2012, 12, 778–790.

116. Noad, J.; Von Der Malsburg, A.; Pathe, C.; Michel, M.A.; Komander, D.; Randow, F. LUBAC-synthesized linear ubiquitin chains restrict cytosol-invading bacteria by activating autophagy and NF-κB. Nat. Microbiol. 2017, 2, 1–10.

117. Koerver, L.; Papadopoulos, C.; Liu, B.; Kravic, B.; Rota, G.; Brecht, L.; Veenendaal, T.; Polajnar, M.; Bluemke, A.; Ehrmann, M.; et al. The ubiquitin-conjugating enzyme UBE 2 QL 1 coordinates lysophagy in response to endolysosomal damage. EMBO Rep. 2019, 20.

118. Seczynska, M.; Dikic, I. Removing the waste bags: How p97 drives autophagy of lysosomes. EMBO J. 2017, 36, 129–131.

119. Falcon, B.; Noad, J.; McMahon, H.; Randow, F.; Goedert, M. Galectin-8-mediated selective autophagy protects against seeded tau aggregation. J. Biol. Chem. 2018, 293, 2438–2451.

120. Richter, B.; Sliter, D.A.; Herhaus, L.; Stolz, A.; Wang, C.; Beli, P.; Zaffagnini, G.; Wild, P.; Martens, S.; Wagner, S.A.; et al. Phosphorylation of OPTN by TBK1 enhances its binding to Ub chains and promotes selective autophagy of damaged mitochondria. Proc. Natl. Acad. Sci. USA 2016, 113, 4039–4044.

121. Thurston, T.L.M.; Ryzhakov, G.; Bloor, S.; von Muhlinen, N.; Randow, F. The TBK1 adaptor and autophagy receptor NDP52 restricts the proliferation of ubiquitin-coated bacteria. Nat. Immunol. 2009, 10, 1215–1221.

122. Pilli, M.; Arko-Mensah, J.; Ponpuak, M.; Roberts, E.; Master, S.; Mandell, M.A.; Dupont, N.; Ornatowski, W.; Jiang, S.; Bradfute, S.B.; et al. TBK-1 Promotes Autophagy-Mediated Antimicrobial Defense by Controlling Autophagosome Maturation. Immunity 2012, 37, 223–234.

123. Kim, J.; Kundu, M.; Viollet, B.; Guan, K.L. AMPK and mTOR regulate autophagy through direct phosphorylation of Ulk1. Nat. Cell Biol. 2011, 13, 132–141.

124. Settembre, C.; Fraldi, A.; Medina, D.L.; Ballabio, A.; Children, T. Signals from the lysosome. Nat. Rev. Mol. Cell Biol. 2013, 14, 283–296.

125. Brady, O.A.; Martina, J.A.; Puertollano, R. Emerging roles for TFEB in the immune response and inflammation. Autophagy 2018, 14, 181–189.

126. Nazio, F.; Carinci, M.; Valacca, C.; Bielli, P.; Strappazzon, F.; Antonioli, M.; Ciccosanti, F.; Rodolfo, C.; Campello, S.; Fimia, G.M.; et al. Fine-tuning of ULK1 mRNA and protein levels is required for autophagy oscillation. J. Cell Biol. 2016, 215, 841–856.

127. Jimenez, A.J.; Maiuri, P.; Lafaurie-Janvore, J.; Divoux, S.; Piel, M.; Perez, F. ESCRT machinery is required for plasma membrane repair. Science 2014, 343.

128. Scheffer, L.L.; Sreetama, S.C.; Sharma, N.; Medikayala, S.; Brown, K.J.; Defour, A.; Jaiswal, J.K. Mechanism of $Ca2^{+}$-triggered ESCRT assembly and regulation of cell membrane repair. Nat. Commun. 2014, 5, 5646.

129. Radulovic, M.; Schink, K.O.; Wenzel, E.M.; Nähse, V.; Bongiovanni, A.; Lafont, F.; Stenmark, H. ESCRT-mediated lysosome repair precedes lysophagy and promotes cell survival. EMBO J. 2018, 37, 1–15.

130. Hurley, J.H. ESCRTs are everywhere. Embo J. 2015, 34, 2398–2407.

131. Skowyra, M.L.; Schlesinger, P.H.; Naismith, T.V.; Hanson, P.I. Triggered recruitment of ESCRT machinery promotes endolysosomal repair. Science 2018, 360.

132. Repnik, U.; Distefano, M.B.; Speth, M.T.; Wui Ng, M.Y.; Progida, C.; Hoflack, B.; Gruenberg, J.; Griffiths, G. L-leucyl-L-leucine methyl ester does not release cysteine cathepsins to the cytosol but inactivates them in transiently permeabilized lysosomes. J. Cell Sci. 2017, 130, 3124–3140.

133. Jia, J.; Abudu, Y.P.; Claude-Taupin, A.; Gu, Y.; Kumar, S.; Choi, S.W.; Peters, R.; Mudd, M.H.; Allers, L.; Salemi, M.; et al. Galectins Control MTOR and AMPK in Response to Lysosomal Damage to Induce Autophagy; Taylor and Francis Inc.: Abingdon, UK, 2019; Volume 15, pp. 169–171.

134. Herbst, S.; Campbell, P.; Harvey, J.; Bernard, E.M.; Papayannopoulos, V.; Wood, N.W.; Morris, H.R.; Gutierrez, M.G. LRRK2 activation controls the repair of damaged endomembranes in macrophages. EMBO J. 2020, 44, 1–14.

135. Mittal, E.; Skowyra, M.L.; Uwase, G.; Tinaztepe, E.; Mehra, A.; Köster, S.; Hanson, P.I.; Philips, J.A. Mycobacterium tuberculosis type VII secretion system effectors differentially impact the ESCRT endomembrane damage response. MBio 2018, 9.

136. Kreibich, S.; Emmenlauer, M.; Fredlund, J.; Dehio, C. Autophagy Proteins Promote Repair of Endosomal Membranes Damaged by the Salmonella Type Three Secretion System 1 Graphical Abstract Highlights d RNAi screen implicates autophagy in Salmonella-containing vacuole (SCV) maintenance d Autophagy promotes repair of TTSS-1-damaged SCV membranes and TTSS-2 expression d In parallel, autophagy eliminates cytosolic Salmonella d Autophagy could play a general role in maintaining endosomal membrane integrity. Cell Host Microbe 2015, 18, 527–537.

137. López-Jiménez, A.T.; Cardenal-Muñoz, E.; Leuba, F.; Gerstenmaier, L.; Barisch, C.; Hagedorn, M.; King, J.S.; Soldati, T. The ESCRT and autophagy machineries cooperate to repair ESX-1-dependent

damage at the Mycobacterium-containing vacuole but have opposite impact on containing the infection. PLOS Pathog. 2018, 14, e1007501.

138. Hurley, J.H.; Emr, S.D. The ESCRT complexes: Structure and mechanism of a membrane-trafficking network. Annu. Rev. Biophys. Biomol. Struct. 2006, 35, 277–298.

139. Robinson, M.; Schor, S.; Barouch-Bentov, R.; Einav, S. Viral journeys on the intracellular highways. Cell. Mol. Life Sci. 2018, 75, 3693–3714.

140. Votteler, J.; Sundquist, W.I. Virus budding and the ESCRT pathway. Cell Host Microbe 2013, 14, 232–241.

141. Ellison, C.J.; Kukulski, W.; Boyle, K.B.; Munro, S.; Randow, F. Transbilayer Movement of Sphingomyelin Precedes Catastrophic Breakage of Enterobacteria-Containing Vacuoles. Curr. Biol. 2020, 30, 2974–2983.e6.

142. Gout, E.; Gutkowska, M.; Takayama, S.; Reed, J.C.; Chroboczek, J. Co-chaperone BAG3 and adenovirus penton base protein partnership. J. Cell. Biochem. 2010, 111, 699–708.

143. Zhao, M.; Antunes, F.; Eaton, J.W.; Brunk, U.T. Lysosomal enzymes promote mitochondrial oxidant production, cytochrome c release and apoptosis. Eur. J. Biochem. 2003, 270, 3778–3786.

144. Deus, C.M.; Yambire, K.F.; Oliveira, P.J.; Raimundo, N. Mitochondria–Lysosome Crosstalk: From Physiology to Neurodegeneration. Trends Mol. Med. 2020, 26, 71–88.

145. Christgen, S.; Place, D.E.; Kanneganti, T.D. Toward targeting inflammasomes: Insights into their regulation and activation. Cell Res. 2020, 30, 315–327.

146. Boya, P.; Kroemer, G. Lysosomal membrane permeabilization in cell death. Oncogene 2008, 27, 6434–6451.

147. Aits, S.; Jäättelä, M. Lysosomal cell death at a glance. J. Cell Sci. 2013, 126, 1905–1912.

148. Deretic, V.; Saitoh, T.; Akira, S. Autophagy in infection, inflammation and immunity. Nat Rev Immunol 2013, 13, 722–737.

149. Zhou, R.; Tardivel, A.; Thorens, B.; Choi, I.; Tschopp, J. Thioredoxin-interacting protein links oxidative stress to inflammasome activation. Nat. Immunol. 2010, 11, 136–140.

150. Zhou, R.; Yazdi, A.S.; Menu, P.; Tschopp, J. A role for mitochondria in NLRP3 inflammasome activation. Nature 2011, 469, 221–226.

151. Kimura, T.; Jia, J.; Claude-Taupin, A.; Kumar, S.; Won Choi, S.; Gu, Y.; Mudd, M.; Dupont, N.; Jiang, S.; Peters, R.; et al. Cellular and molecular mechanism for secretory autophagy. Autophagy 2017, 13, 1084–1085.

152. Brubaker, S.W.; Bonham, K.S.; Zanoni, I.; Kagan, J.C. Innate Immune Pattern Recognition: A Cell Biological Perspective. Annu. Rev. Immunol. 2015, 33, 257–290.

153. Kawai, T.; Akira, S. Toll-like Receptors and Their Crosstalk with Other Innate Receptors in Infection and Immunity. Immunity 2011, 34, 637–650.

154. Gabandé-Rodríguez, E.; Pérez-Cañamás, A.; Soto-Huelin, B.; Mitroi, D.N.; Sánchez-Redondo, S.; Martínez-Sáez, E.; Venero, C.; Peinado, H.; Ledesma, M.D. Lipid-induced lysosomal damage after demyelination corrupts microglia protective function in lysosomal storage disorders. EMBO J. 2019, 38, 1–22.

155. Davis, A.A.; Leyns, C.E.G.; Holtzman, D.M. Intercellular Spread of Protein Aggregates in Neurodegenerative Disease. Annu. Rev. Cell Dev. Biol. 2018, 34, 545–568.

156. Schoonbroodt, S.; Ferreira, V.; Best-Belpomme, M.; Boelaert, J.R.; Legrand-Poels, S.; Korner, M.; Piette, J. Crucial Role of the Amino-Terminal Tyrosine Residue 42 and the Carboxyl-Terminal PEST Domain of IκBα in NF-κB Activation by an Oxidative Stress. J. Immunol. 2000, 164, 4292–4300.

157. Takada, Y.; Mukhopadhyay, A.; Kundu, G.C.; Mahabeleshwar, G.H.; Singh, S.; Aggarwal, B.B. Hydrogen peroxide activates NF-κB through

tyrosine phosphorylation of IκBα and serine phosphorylation of p65. Evidence for the involvement of IκBα kinase and Syk protein-tyrosine kinase. J. Biol. Chem. 2003, 278, 24233–24241.

158. Anderson, F.L.; Coffey, M.M.; Berwin, B.L.; Havrda, M.C. Inflammasomes: An Emerging Mechanism Translating Environmental Toxicant Exposure Into Neuroinflammation in Parkinson's Disease. Toxicol. Sci. 2018, 166, 3–15.

159. Siew, J.J.; Chen, H.M.; Chen, H.Y.; Chen, H.L.; Chen, C.M.; Soong, B.W.; Wu, Y.R.; Chang, C.P.; Chan, Y.C.; Lin, C.H.; et al. Galectin-3 is required for the microglia-mediated brain inflammation in a model of Huntington's disease. Nat. Commun. 2019, 10, 1–18.

160. Streetly, M.J.; Maharaj, L.; Joel, S.; Schey, S.A.; Gribben, J.G.; Cotter, F.E. GCS-100, a novel galectin-3 antagonist, modulates MCL-1, NOXA, and cell cycle to induce myeloma cell death. Blood 2010, 115, 3939–3948.

161. Kawasaki, T.; Kawai, T. Toll-like receptor signaling pathways. Front. Immunol. 2014, 5, 1–8.

162. O'Neill, L.A.J.; Golenbock, D.; Bowie, A.G. The history of Toll-like receptors-redefining innate immunity. Nat. Rev. Immunol. 2013, 13, 453–460.

163. Pichlmair, A.; Schulz, O.; Tan, C.P.; Näslund, T.I.; Liljeström, P.; Weber, F.; Reis E Sousa, C. RIG-I-mediated antiviral responses to single-stranded RNA bearing 5′-phosphates. Science 2006, 314, 997–1001.

164. Takahasi, K.; Yoneyama, M.; Nishihori, T.; Hirai, R.; Kumeta, H.; Narita, R.; Gale, M.; Inagaki, F.; Fujita, T. Nonself RNA-Sensing Mechanism of RIG-I Helicase and Activation of Antiviral Immune Responses. Mol. Cell 2008, 29, 428–440.

165. Kato, H.; Takeuchi, O.; Sato, S.; Yoneyama, M.; Yamamoto, M.; Matsui, K.; Uematsu, S.; Jung, A.; Kawai, T.; Ishii, K.J.; et al. Differential roles of MDA5 and RIG-I helicases in the recognition of RNA viruses. Nature 2006, 441, 101–105.

166. Zeng, W.; Sun, L.; Jiang, X.; Chen, X.; Hou, F.; Adhikari, A.; Xu, M.; Chen, Z.J. Reconstitution of the RIG-I pathway reveals a signaling role of unanchored polyubiquitin chains in innate immunity. Cell 2010, 141, 315–330.

167. Loo, Y.M.; Gale, M. Immune Signaling by RIG-I-like Receptors. Immunity 2011, 34, 680–692.

168. Pichlmair, A.; Reis e Sousa, C. Innate Recognition of Viruses. Immunity 2007, 27, 370–383.

169. Goubau, D.; Schlee, M.; Deddouche, S.; Pruijssers, A.J.; Zillinger, T.; Goldeck, M.; Schuberth, C.; Van Der Veen, A.G.; Fujimura, T.; Rehwinkel, J.; et al. Antiviral immunity via RIG-I-mediated recognition of RNA bearing 59-diphosphates. Nature 2014, 514, 372–375.

170. Gitlin, L.; Barchet, W.; Gilfillan, S.; Cella, M.; Beutler, B.; Flavell, R.A.; Diamond, M.S.; Colonna, M. Essential role of mda-5 in type I IFN responses to polyriboinosinic: Polyribocytidylic acid and encephalomyocarditis picornavirus. Proc. Natl. Acad. Sci. USA 2006, 103, 8459–8464.

171. Feng, Q.; Hato, S.V.; Langereis, M.A.; Zoll, J.; Virgen-Slane, R.; Peisley, A.; Hur, S.; Semler, B.L.; van Rij, R.P.; van Kuppeveld, F.J.M. MDA5 Detects the Double-Stranded RNA Replicative Form in Picornavirus-Infected Cells. Cell Rep. 2012, 2, 1187–1196.

172. Wang, J.P.; Cerny, A.; Asher, D.R.; Kurt-Jones, E.A.; Bronson, R.T.; Finberg, R.W. MDA5 and MAVS Mediate Type I Interferon Responses to Coxsackie B Virus. J. Virol. 2010, 84, 254–260.

173. Slater, L.; Bartlett, N.W.; Haas, J.J.; Zhu, J.; Message, S.D.; Walton, R.P.; Sykes, A.; Dahdaleh, S.; Clarke, D.L.; Belvisi, M.G.; et al. Co-ordinated role of TLR3, RIG-I and MDA5 in the innate response to rhinovirus in bronchial epithelium. PLoS Pathog. 2010, 6.

174. Ito, M.; Yanagi, Y.; Ichinohe, T. Encephalomyocarditis Virus Viroporin 2B Activates NLRP3 Inflammasome. PLoS Pathog. 2012, 8.

175. Rajan, J.V.; Rodriguez, D.; Miao, E.A.; Aderem, A. The NLRP3 Inflammasome Detects Encephalomyocarditis Virus and Vesicular Stomatitis Virus Infection. J. Virol. 2011, 85, 4167–4172.

176. da Costa, L.S.; Outlioua, A.; Anginot, A.; Akarid, K.; Arnoult, D. RNA viruses promote activation of the NLRP3 inflammasome through cytopathogenic effect-induced potassium efflux. Cell Death Dis. 2019, 10.

177. Smith, J.S.; Xu, Z.; Tian, J.; Palmer, D.J.; Ng, P.; Byrnes, A.P. The role of endosomal escape and mitogen-activated protein kinases in adenoviral activation of the innate immune response. PLoS ONE 2011, 6, e26755.

178. Nociari, M.; Ocheretina, O.; Murphy, M.; Falck-Pedersen, E. Adenovirus Induction of IRF3 Occurs through a Binary Trigger Targeting Jun N-Terminal Kinase and TBK1 Kinase Cascades and Type I Interferon Autocrine Signaling. J. Virol. 2009, 83, 4081–4091.

179. Stein, S.C.; Falck-Pedersen, E. Sensing Adenovirus Infection: Activation of Interferon Regulatory Factor 3 in RAW 264.7 Cells. J. Virol. 2012, 86, 4527–4537.

180. Eichholz, K.; Bru, T.; Tran, T.T.P.; Fernandes, P.; Welles, H.; Mennechet, F.J.D.; Manel, N.; Alves, P.; Perreau, M.; Kremer, E.J. Immune-Complexed Adenovirus Induce AIM2-Mediated Pyroptosis in Human Dendritic Cells. PLoS Pathog. 2016, 12, e1005871.

181. Barlan, A.U.; Griffin, T.M.; Mcguire, K.A.; Wiethoff, C.M. Adenovirus Membrane Penetration Activates the NLRP3 Inflammasome. J. Virol. 2011, 85, 146–155.

182. McGuire, K.A.; Barlan, A.U.; Griffin, T.M.; Wiethoff, C.M. Adenovirus Type 5 Rupture of Lysosomes Leads to Cathepsin B-Dependent Mitochondrial Stress and Production of Reactive Oxygen Species. J. Virol. 2011, 85, 10806–10813.

183. Tibbles, L.A.; Spurrell, J.C.L.; Bowen, G.P.; Liu, Q.; Lam, M.; Zaiss, A.K.; Robbins, S.M.; Hollenberg, M.D.; Wickham, T.J.; Muruve, D.A. Activation of p38 and ERK Signaling during Adenovirus Vector Cell

Entry Lead to Expression of the C-X-C Chemokine IP-10. J. Virol. 2002, 76, 1559–1568.

184. Anghelina, D.; Lam, E.; Falck-Pedersen, E. Diminished Innate Antiviral Response to Adenovirus Vectors in cGAS/STING-Deficient Mice Minimally Impacts Adaptive Immunity. J. Virol. 2016, 90, 5915–5927.

185. Maler, M.D.; Nielsen, P.J.; Stichling, N.; Cohen, I.; Ruzsics, Z.; Wood, C.; Engelhard, P.; Suomalainen, M.; Gyory, I.; Huber, M.; et al. Key role of the scavenger receptor MARCO in mediating adenovirus infection and subsequent innate responses of macrophages. MBio 2017, 8.

186. Watkinson, R.E.; McEwan, W.A.; Tam, J.C.H.; Vaysburd, M.; James, L.C. TRIM21 Promotes cGAS and RIG-I Sensing of Viral Genomes during Infection by Antibody-Opsonized Virus. PLoS Pathog. 2015, 11, e1005253.

187. Hare, D.; Mossman, K.L. Novel paradigms of innate immune sensing of viral infections. Cytokine 2013, 63, 219–224.

188. Collins, S.E.; Mossman, K.L. Danger, diversity and priming in innate antiviral immunity. Cytokine Growth Factor Rev. 2014, 25, 525–531.

189. Deretic, V. Autophagy as an innate immunity paradigm: Expanding the scope and repertoire of pattern recognition receptors. Curr. Opin. Immunol. 2012, 24, 21–31.

190. Saitoh, T.; Fujita, N.; Jang, M.H.; Uematsu, S.; Yang, B.G.; Satoh, T.; Omori, H.; Noda, T.; Yamamoto, N.; Komatsu, M.; et al. Loss of the autophagy protein Atg16L1 enhances endotoxin-induced IL-1β production. Nature 2008, 456, 264–268.

191. Santeford, A.; Wiley, L.A.; Park, S.; Bamba, S.; Nakamura, R.; Gdoura, A.; Ferguson, T.A.; Rao, P.K.; Guan, J.L.; Saitoh, T.; et al. Impaired autophagy in macrophages promotes inflammatory eye disease. Autophagy 2016, 12, 1876–1885.

192. Pu, Q.; Gan, C.; Li, R.; Li, Y.; Tan, S.; Li, X.; Wei, Y.; Lan, L.; Deng, X.; Liang, H.; et al. Atg7 Deficiency Intensifies Inflammasome

Activation and Pyroptosis in Pseudomonas Sepsis. J. Immunol. 2017, 198, 3205–3213.

193. Bechelli, J.; Vergara, L.; Smalley, C.; Buzhdygan, T.P.; Bender, S.; Zhang, W.; Liu, Y.; Popov, V.L.; Wang, J.; Garg, N.; et al. Atg5 supports Rickettsia australis infection in macrophages in vitro and in vivo. Infect. Immun. 2019, 87.

194. Nakahira, K.; Haspel, J.A.; Rathinam, V.A.K.; Lee, S.J.; Dolinay, T.; Lam, H.C.; Englert, J.A.; Rabinovitch, M.; Cernadas, M.; Kim, H.P.; et al. Autophagy proteins regulate innate immune responses by inhibiting the release of mitochondrial DNA mediated by the NALP3 inflammasome. Nat. Immunol. 2011, 12, 222–230.

195. Zhong, Z.; Umemura, A.; Sanchez-Lopez, E.; Liang, S.; Shalapour, S.; Wong, J.; He, F.; Boassa, D.; Perkins, G.; Ali, S.R.; et al. NF-κB Restricts Inflammasome Activation via Elimination of Damaged Mitochondria. Cell 2016, 164, 896–910.

196. Sumpter, R.; Sirasanagandla, S.; Fernández, Á.F.; Wei, Y.; Dong, X.; Franco, L.; Zou, Z.; Marchal, C.; Lee, M.Y.; Clapp, D.W.; et al. Fanconi Anemia Proteins Function in Mitophagy and Immunity. Cell 2016, 165, 867–881.

197. Nguyen, T.N.; Padman, B.S.; Lazarou, M. Deciphering the Molecular Signals of PINK1/Parkin Mitophagy. Trends Cell Biol. 2016, 26, 733–744.

198. Chen, G.; Kroemer, G.; Kepp, O. Mitophagy: An Emerging Role in Aging and Age-Associated Diseases. Front. Cell Dev. Biol. 2020, 8, 1–15.

199. Harris, J.; Hartman, M.; Roche, C.; Zeng, S.G.; O'Shea, A.; Sharp, F.A.; Lambe, E.M.; Creagh, E.M.; Golenbock, D.T.; Tschopp, J.; et al. Autophagy controls IL-1β secretion by targeting Pro-IL-1β for degradation. J. Biol. Chem. 2011, 286, 9587–9597.

200. Shi, C.-S.S.; Shenderov, K.; Huang, N.-N.N.; Kabat, J.; Abu-Asab, M.; Fitzgerald, K.A.; Sher, A.; Kehrl, J.H. Activation of autophagy

by inflammatory signals limits IL-1β production by targeting ubiquitinated inflammasomes for destruction. Nat. Immunol. 2012, 13, 255–263.

201. Liu, T.; Tang, Q.; Liu, K.; Xie, W.; Liu, X.; Wang, H.; Wang, R.F.; Cui, J. TRIM11 Suppresses AIM2 Inflammasome by Degrading AIM2 via p62-Dependent Selective Autophagy. Cell Rep. 2016, 16, 1988–2002.
202. Kimura, T.; Jain, A.; Choi, S.W.; Mandell, M.A.; Schroder, K.; Johansen, T.; Deretic, V. TRIM-mediated precision autophagy targets cytoplasmic regulators of innate immunity. J. Cell Biol. 2015, 210, 973–989.
203. Van Gent, M.; Sparrer, K.M.J.; Gack, M.U. TRIM proteins and their roles in antiviral host defenses. Annu. Rev. Virol. 2018, 5, 385–405.
204. Jounai, N.; Kobiyama, K.; Shiina, M.; Ogata, K.; Ishii, K.J.; Takeshita, F. NLRP4 Negatively Regulates Autophagic Processes through an Association with Beclin1. J. Immunol. 2011, 186, 1646–1655.
205. Dupont, N.; Jiang, S.; Pilli, M.; Ornatowski, W.; Bhattacharya, D.; Deretic, V. Autophagy-based unconventional secretory pathway for extracellular delivery of IL-1β. EMBO J. 2011, 30, 4701–4711.
206. Zhang, M.; Kenny, S.J.; Ge, L.; Xu, K.; Schekman, R. Translocation of interleukin-1β into a vesicle intermediate in autophagy-mediated secretion. Elife 2015, 4.
207. Wang, L.J.; Huang, H.Y.; Huang, M.P.; Liou, W.; Chang, Y.T.; Wu, C.C.; Ojcius, D.M.; Chang, Y.S. The microtubule-associated protein EB1 links AIM2 inflammasomes with autophagy-dependent secretion. J. Biol. Chem. 2014, 289, 29322–29333.
208. Deretic, V. Multiple regulatory and effector roles of autophagy in immunity. Curr. Opin. Immunol. 2009, 21, 53–62.
209. Xu, Y.; Liu, X.-D.; Gong, X.; Eissa, N.T. Signaling pathway of autophagy associated with innate immunity. Autophagy 2008, 4, 110–112.

210. Delgado, M.A.; Elmaoued, R.A.; Davis, A.S.; Kyei, G.; Deretic, V. Toll-like receptors control autophagy. EMBO J. 2008, 27, 1110–1121.

211. Shi, C.-S.; Kehrl, J.H. MyD88 and Trif target Beclin 1 to trigger autophagy in macrophages. J. Biol. Chem. 2008, 283, 33175–33182.

212. Jounai, N.; Takeshita, F.; Kobiyama, K.; Sawano, A.; Miyawaki, A.; Xin, K.Q.; Ishii, K.J.; Kawai, T.; Akira, S.; Suzuki, K.; et al. The Atg5-Atg12 conjugate associates with innate antiviral immune responses. Proc. Natl. Acad. Sci. USA 2007, 104, 14050–14055.

213. Jin, S.; Tian, S.; Chen, Y.; Zhang, C.; Xie, W.; Xia, X.; Cui, J.; Wang, R. USP 19 modulates autophagy and antiviral immune responses by deubiquitinating Beclin-1. EMBO J. 2016, 35, 866–880.

214. Jin, S.; Tian, S.; Luo, M.; Xie, W.; Liu, T.; Duan, T.; Wu, Y.; Cui, J. Tetherin Suppresses Type I Interferon Signaling by Targeting MAVS for NDP52-Mediated Selective Autophagic Degradation in Human Cells. Mol. Cell 2017, 68, 308–322.e4.

215. Du, Y.; Duan, T.; Feng, Y.; Liu, Q.; Lin, M.; Cui, J.; Wang, R. LRRC25 inhibits type I IFN signaling by targeting ISG15-associated RIG-I for autophagic degradation. EMBO J. 2018, 37, 351–366.

216. Saitoh, T.; Fujita, N.; Hayashi, T.; Takahara, K.; Satoh, T.; Lee, H.; Matsunaga, K.; Kageyama, S.; Omori, H.; Noda, T.; et al. Atg9a controls dsDNA-driven dynamic translocation of STING and the innate immune response. Proc. Natl. Acad. Sci. USA 2009, 106, 20842–20846.

217. Liang, Q.; Seo, G.J.; Choi, Y.J.; Kwak, M.J.; Ge, J.; Rodgers, M.A.; Shi, M.; Leslie, B.J.; Hopfner, K.P.; Ha, T.; et al. Crosstalk between the cGAS DNA sensor and beclin-1 autophagy protein shapes innate antimicrobial immune responses. Cell Host Microbe 2014, 15, 228–238.

218. Chen, M.; Meng, Q.; Qin, Y.; Liang, P.; Tan, P.; He, L.; Zhou, Y.; Chen, Y.; Huang, J.; Wang, R.F.; et al. TRIM14 Inhibits cGAS Degradation Mediated by Selective Autophagy Receptor p62 to Promote Innate Immune Responses. Mol. Cell 2016, 64, 105–119.

219. Tsuchida, T.; Zou, J.; Saitoh, T.; Kumar, H.; Abe, T.; Matsuura, Y.; Kawai, T.; Akira, S. The ubiquitin ligase TRIM56 regulates innate immune responses to intracellular double-stranded DNA. Immunity 2010, 33, 765–776.

220. Prabakaran, T.; Bodda, C.; Krapp, C.; Zhang, B.; Christensen, M.H.; Sun, C.; Reinert, L.; Cai, Y.; Jensen, S.B.; Skouboe, M.K.; et al. Attenuation of c GAS-STING signaling is mediated by a p62/ SQSTM 1-dependent autophagy pathway activated by TBK1. EMBO J. 2018, 37.

221. Niida, M.; Tanaka, M.; Kamitani, T. Downregulation of active IKKβ by Ro52-mediated autophagy. Mol. Immunol. 2010, 47, 2378–2387.

222. Zheng, Q.; Hou, J.; Zhou, Y.; Yang, Y.; Xie, B.; Cao, X. Siglec1 suppresses antiviral innate immune response by inducing TBK1 degradation via the ubiquitin ligase TRIM27. Cell Res. 2015, 25, 1121–1136.

223. Sparrer, K.M.J.; Gableske, S.; Zurenski, M.A.; Parker, Z.M.; Full, F.; Baumgart, G.J.; Kato, J.; Pacheco-Rodriguez, G.; Liang, C.; Pornillos, O.; et al. TRIM23 mediates virus-induced autophagy via activation of TBK1. Nat. Microbiol. 2017, 2, 1543–1557.

224. Mandell, M.A.; Jain, A.; Arko-Mensah, J.; Chauhan, S.; Kimura, T.; Dinkins, C.; Silvestri, G.; Münch, J.; Kirchhoff, F.; Simonsen, A.; et al. TRIM proteins regulate autophagy and can target autophagic substrates by direct recognition. Dev. Cell 2014, 30, 394–409.

225. Labzin, L.I.; Bottermann, M.; Rodriguez-Silvestre, P.; Foss, S.; Andersen, J.T.; Vaysburd, M.; Clift, D.; James, L.C. Antibody and DNA sensing pathways converge to activate the inflammasome during primary human macrophage infection. EMBO J. 2019, 38.

226. Klein, K.A.; Jackson, W.T. Picornavirus subversion of the autophagy pathway. Viruses 2011, 3, 1549–1561.

05 THE TRANSPORT OF EBOLA VIRUS NUCLEOCAPSID-LIKE STRUCTURES INTO CELLS RELIES ON THE ARP2/3 SIGNALING PATHWAY FOR LONG-DISTANCE MOVEMENT

1. INTRODUCTION: AN OVERVIEW

The dynamic actin cytoskeleton is commonly utilized by pathogens for entry, exit and their intracellular assembly [1]. While for some intracellular bacteria and DNA viruses, the mechanism of hijacking the actin cytoskeleton is described in detail, the remodeling of actin through highly pathogenic RNA viruses such as filoviruses remains poorly understood [1,2]. Recently, it has been demonstrated that the intracytoplasmic transport of the highly pathogenic Marburg virus (MARV) and the Ebola virus (EBOV) depends on actin polymerization, but the detailed mechanisms and cellular interaction partners remain largely unknown [3,4,5].

MARV and EBOV belong to the family of Filoviridae, which are filamentous, enveloped viruses containing a single strand, negative-sense RNA genome, and which cause severe fevers in humans with very high fatality rates [6]. Strikingly, filoviral RNA encodes for only seven structural proteins, which are multifunctional, and diversely hijack, disable or reorganize cellular pathways [7]. For example, EBOV VP24 and VP35 are able to block the cell's interferon system, resulting in the viruses' escape from the innate immune response [8].

A hallmark of the filoviral life cycle is the formation of perinuclear inclusion bodies, where transcription and replication occur and de novo viral nucleocapsids are formed [9]. These nucleocapsids are mainly composed of the nucleoprotein (NP) that encapsidates the viral RNA in highly organized helical structures (approximately 1000 nm long and 50 nm in diameter) [10]. Furthermore, nucleocapsids contain the NP-binding proteins VP24 and VP35, as well as polymerase L and the transcription initiation factor VP30 [10,11,12].

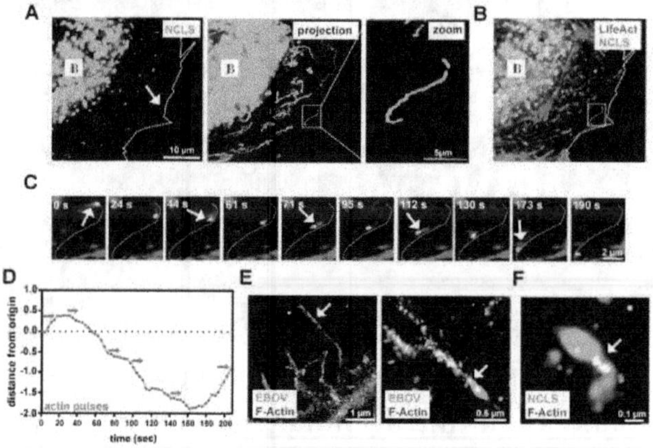

Figure 1. Ebola virus (EBOV) nucleocapsid-like structures induce polar actin tails during transport. (A) Exemplary still image from a movie of a Huh7 cell expressing NP, VP24, VP35 and VP30-GFP (left panel). During imaging, NCLS are tracked for 2 min producing long-distance trajectories as depicted in the maximum intensity projection of the movie (right panel). The white arrow highlights an individual track. IB labels inclusion bodies. (B) Co-imaging of actin using LifeAct-CLIP (stained with Alexa-657 dye) reveals a dense and dynamic actin network. White inset is magnified in (C) showing the time course of an individual NCLS (white arrow, Figure 1B), revealing a pulsative actin tail located on one site of the subviral particle during movement. (D) Graph showing the movement of this NCLS over time. The red arrows indicate actin pulses observed. (E,F) STORM images showing NCLS or EBOV nucleocapsids immunolabelled with NP (green) and stained with Phalloidin (magenta). (D) The left panel shows a filamentous virus particle likely prior to budding (white arrow). Individual intracellular nucleocapsids reveal a preserved actin tail (white arrow). (E) Huh7 cells were transfected with NP, VP24 and VP35, and fixed after 24 h. The samples were stained with anti-NP and Phalloidin, and prepared for STORM microscopy. The zoomed image shows an individual NCLS with a preserved actin tail (white arrow).

Following their assembly in the inclusion bodies, individual nucleocapsids have to be transported over long distances through the cytoplasm to reach the cell periphery [3,5,9]. It has been shown that they accumulate in filopodia structures. Finally, nucleocapsids acquire their envelope at the plasma membrane, containing the viral trans-membrane glycoprotein (GP) and the viral matrix protein VP40, triggering the budding of the filamentous infectious virus particles. [8]. The directed transport is required for the rapid distancing away from the inclusion bodies, and to transfer functional nucleocapsids to the plasma membrane. It has been shown that individual single mutations within the nucleocapsid proteins critically affect long-distance transport, thereby leading to a significant delay and reduction of the release of filamentous nucleocapsids [13,14,15].

While other viruses utilize the microtubule network for intracellular distribution, [16] the transport of nucleocapsids of MARV as well as EBOV is entirely blocked by the pharmacological inhibition of actin polymerization [3,4]. However, the role of actin nucleators and the regulation of their upstream effectors in this process remains elusive. One major obstacle in studying filoviruses is the requirement of high containment laboratories, restricting cell biological experiments that could accelerate our understanding of the critical steps in the virus life cycle. To overcome this problem, we recently established a novel transfection-based system that only requires three viral proteins to produce nucleocapsid-like structures (NCLS), which highly resemble EBOV nucleocapsids in their structure as well as their intracellular dynamics [5]. Through the co-expression of GFP-tagged VP30, which faithfully labels NCLS, it is possible to monitor intracellular transport mechanisms with real time imaging under normal laboratory conditions (biosafety level 1) conditions (Figure 1A,B and Figure 2A).

The host cell contains a highly dynamic actin cytoskeleton that is required for many essential cellular processes, such as cytokinesis, contractility and motility [17,18]. Filamentous actin (called F-actin) is assembled from monomeric globular G-actin, and this polymerization, as well as depolymerization, is highly regulated through a plethora of different cellular factors [19,20,21]. With this high power of regulation, the cell is able to assemble and coordinate diverse structures, such as the strong cortical actin network stabilizing the cell cortex, short and highly dynamic filaments that are involved in vesicular trafficking, and filament networks that form membrane protrusions such as the filopodia and lamellipodia required for cell motility or cell–cell contact formation [22,23,24,25]. The Arp2/3 (actin-related protein) complex efficiently nucleates the actin filaments typically attached to mother filaments to form the highly branched networks that (amongst other things) enable lamellipodia to rapidly adapt during cell migration [26]. This active protein complex has to be tightly regulated, and requires activation for polymerization [27]. So-called nucleation promoting factors (NFPs), such as WAVE1 and WASP proteins, directly induce Arp2/3 complex activity [27,28], and are themselves activated downstream of RhoGTPase signaling in a highly spatial–temporal manner [29].

The Arp2/3 complex is highly conserved, and different pathogens, including Listeria monocytogenes and vaccinia virus, utilize its activity for viral intracellular transport steps [30]. For example, the membrane-integrated Listeria protein ActA mimics the NPF WASP, thereby recruiting and activating the Arp2/3 complex [31,32,33]. The Arp2/3 complex in turn induces local actin polymerization, resulting in so-called actin comet tails that efficiently propel the bacterium through the cytoplasm, pushing it into neighboring cells. Furthermore, actin comet tails have been previously observed at EBOV nucleocapsids [3]; however, the mechanism by which viral nucleocapsids use actin dynamics for their transport has not been described in detail.

In this study, we employ our recently established live cell imaging approach to delineate the cellular pathways by which EBOV exploits host actin signaling, and extend the previously applied manual quantification approach to a semi-automatic high throughput method. Using small inhibitory compounds and siRNA-mediated knockdown, we demonstrate that Arp2/3 complex activity downstream of Rac1 is critically involved in the directed long-distance transport of EBOV nucleocapsid structures. Furthermore, through co-visualization of NCLS transport with the actin marker LifeAct, we detected pulsative actin tails accompanying the movement of NCLS through the cytoplasm, which requires Arp2/3 activity.

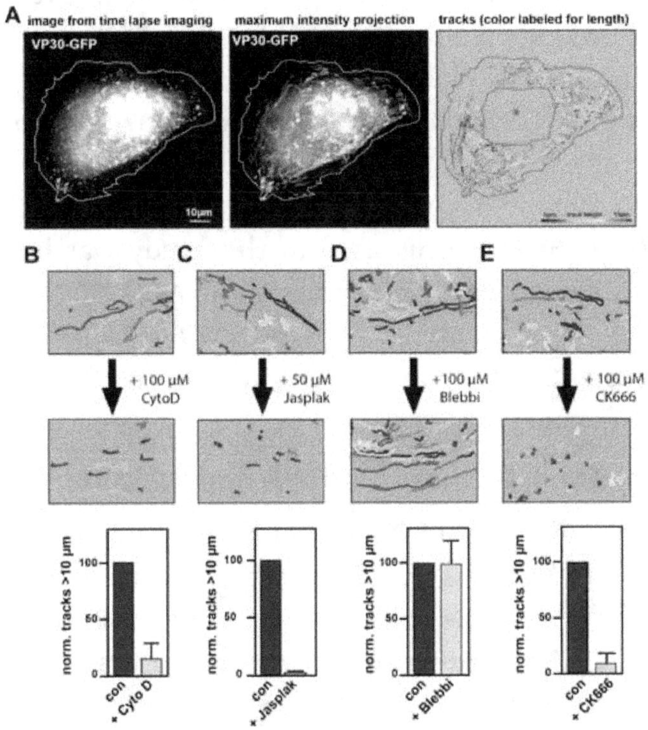

Figure 2. Analysis of NCLS transport using actin-modulating drugs. (A) Exemplary still image from a movie of a Huh7 cell expressing NP, VP24, VP35 and VP30-GFP (left panel). For live cell imaging, NCLS are tracked through the cell over 2 min with images captured every 900 ms. While the middle panel shows the maximum-intensity projection of this movie, the right panel depicts the same movie quantified using the spot algorithm (Imaris). The tracks are color-coded for length. To exclude artefacts, we semi-manually deleted areas with strong accumulation of VP30-GFP, such as the inclusion bodies (asterisk). (B–E) Huh7 cells transfected with NCLS (VP30-GFP) were recorded for 2 min. This was followed by a short incubation with cytoskeletal-modulating drugs after which the cells were re-recorded. Per experiment three cells were recorded and all tracks > 20 s were quantified. The pictures show a magnified area after quantification with Imaris. Cells were incubated with (B) 100 µM Cytochalasin D, (C) 50 µM Jasplakinolide, (D) 100 µM para-nitro-Blebbistatin or (E) 100 µM CK666 following the first imaging. The graphs depict the normalized number of tracks >10 µm, error bars show the SD of n = 3 experiments.

2. MATERIALS AND METHODS

2.1. Cells and Viruses

Huh7 (human hepatoma) cells were cultured in DMEM (Life Technologies, Carlsbad, CA, USA) supplemented with 10% (v/v) fetal calf serum (FCS) (PAN Biotech), 5 mM l-glutamine, 50 U/mL penicillin and 50 µg/mL streptomycin (Life Technologies) and grown at 37 °C with 5% CO2. For live cell imaging experiments, cells were kept in phenol-free Leibovitz's medium (Life Technologies) with PS/Q, non-essential amino acid solution and 20% (v/v) FCS. The virus used in this study was based on EBOV Zaire (Strain Mayinga; GenBank accession no AF27200 (National Center for Biotechnology Information, Bethesda, MD USA)). The experiments with infectious EBOV virus were performed in the BSL-4 facility at the University of Marburg.

2.2. Transfections, Plasmids, siRNA and Inhibitors

The plasmids coding for EBOV proteins (pCAGGS-NP, -VP35, -VP24) and pCAGGS-VP30-GFP were described previously [34]. Transient transfections of Huh7 cells were carried out using TransIT-LT1 (Mirus, Madison, WI, USA) according to the manufacturers' instructions with 3 µL reagent per 1 µg plasmid DNA [5]. Transfection of siRNAs was performed using DharmaFECT (Horizon, Waterbeach, UK) using 25 nM siRNA and 5 µL transfection reagent in Opti-MEM (Thermo Fisher, Waltham, MA, USA). Huh7 cells were transfected with siRNA in a 6-well µ-slide, seeded to a 4-well µ-slide (IBIDI) and then transfected with the plasmids encoding for the viral proteins (200 ng/µL VP30-GFP, 30 ng/µL VP24, 200 ng/µL VP35 and 200 ng/µL NP). Then, NCLS movement was monitored 24 h later through detection of VP30-GFP. The FlexiTube siRNA used were purchased from Qiagen and diluted to 10 µM. The siRNAs used in the study were HS-ACTR3_5 (AAAGTGGGTGATCAAGCTCAA), HS_RAC1_6 (ATGCATTTCCTGGAGAATATA), HS_WASF1_3 (CAAGAACGTGTGGACCGTTTA) and HS-CDC42_16 (5′-CATCAGATTTTGAAAATATTTAA 3′). For treatment with inhibitors, individual transfected cells (NP, VP24, VP35 and VP30-GFP) were monitored, then the inhibitor was added to the cells in the appropriate dilution and after a short incubation the very same cell was imaged. Cytochalasin D, Jasplakinolide and NSC 23766 were obtained from Sigma Aldrich, and CK666 was purchased from EMD Millipore.

2.3. Microscopy Antibodies and Reagents

The following primary antibodies were used in this study: polyclonal anti-chicken EBOV NP [35], polyclonal rabbit anti-Wave1 (Sigma Aldrich, St. Louis, MO, USA), mouse anti-Arp3 (Sigma, A5979), mouse anti-Rac1 (Cytoskeleton) and mouse anti-Tubulin (Sigma). The corresponding secondary antibodies were anti-mouse-HRP and anti-rabbit-HRP (DAKO, Jena, Germany), and anti-chicken-Alexa555 (Invitrogen, Carlsbad, CA, USA). Western Blot analyses were performed as described previously [5]. SNAP/CLIP technology was used to visualize actin by using LifeAct-CLIP. LifeAct-CLIP was cloned using standard cloning procedures and then transfected with the NCLS system. For co-visualization of actin and NCLS transport, transfected cells cells were incubated with the dye CLIP-647 (1:500 diluted in media, NEB) for 30 min prior to the experiment, then washed and replaced with Leibovitz medium for live cell imaging.

2.4. Confocal microscopy, live cell imaging, and STORM microscopy

Huh7 cells were fixed using 4% PFA/DMEM, permeabilized with 0.3%Triton-X100/PBS and blocked with blocking buffer containing 3% glycerol, 2% BSA, 0.2% Tween and 0.05% NaN3. Primary antibodies as well as secondary antibodies were diluted in blocking buffer and both were incubated for 1 h at RT. The actin cytoskeleton was visualized with Phalloidin conjugated with Alexa647 (Cytoskeleton). For STORM analysis of transfected cells, cells were fixed with 4% PFA/PBS, permeabilized with 0.3% Triton-X100 and then stained with Phalloidin-Alexa647 (1:50) in PBS for 48 h at 4 °C, before being immunolabelled for NP using anti-chicken NP (1:1000) in blocking buffer followed by an incubation with anti-chicken-Alexa555 (Invitrogen, 1:250). Every instance of Phalloidin or antibody staining was followed by a post-fixation step using 4% PFA/PBS. Cells infected with EBOV were fixed with 4% PFA/PBS and permeabilized, then stained with Phalloidin for 1 h in the BSL4 lab, followed by 48 h incubation with 4% PFA/DMEM. Cells were then blocked and immunolabeled as described above and re-stained with Phalloidin for 48 h. For STORM analysis, the coverslip was mounted in switching buffer (1 M MEA-HCl in Glucose buffer containing catalase, TCEP and glucose oxidase).

Dual color live-cell imaging was performed on a SP8 confocal laser scanning microscope (Leica) equipped with a 64 × 1.4 oil objective, and movies for single color live cell imaging were recorded using a Nikon ECLIPSE TE2000-E microscope with a 64x-oil objective. The movies for quantification were acquired every 900 ms for 120 s. All live cell imaging was performed at 37 °C. STORM images were acquired using the Oxford Nanoimaging system (ONI). The light was collected by the 100× objective and imaged onto the EM-CCD camera at 30 ms per frame. Images were processed using the NimOS localization software (Oxford Nanoimaging).

2.5. Particle Tracking, Quantification and Statistics

Image processing was performed with Imaris (Bitplane, Oxford Instruments, Abingdon, UK), FIJI (NIH) and Photoshop CS6 (Adobe, San Jose, CA, USA). For figure design, we used Adobe Illustrator. NCLS tracking was performed using Imaris (Bitplane). To this end, VP30-GFP signal was monitored every 900 ms over 2 min and then moving NCLS were analyzed using the "spots feature", a presetting that has been optimized for cell and particle tracking within the software. Within the "spots" feature, we determined an algorithm that automatically classifies objects of over 0.4-μm diameter appearing continuously for at least 20 s as NCLS. The positions of the structures were then calculated in each image of the 2-min movie. From these data, the software is able to calculate trajectories using an autoregressive motion algorithm. After the computational calculation and visualization of all tracks, we semi-manually filtered out particle aggregations, as well as GFP signals derived from inclusion bodies, by deleting spots within certain areas within the cell (see Figure 2A), resulting in approximately 400–800 tracks per cell. We filtered the data (e.g., for tracks >10 μm or straightness > 0.7) which were then used for quantification. This algorithm was then applied to all files of a set of experiments. Each experiment was performed in at least three independent repetitions. Statistical analyses were performed with Prism x (GraphPad, San Diego, CA, USA).

3. RESULTS

3.1. The movement of the nucleocapsid is accompanied by pulsating actin tails

Here, we employ a novel live cell imaging system to monitor the long-distance movement of EBOV NCLS, in which only three viral proteins are expressed, namely, NP, VP24 and VP35 (Figure 1A, left panel) [5]. Using VP30-GFP, we monitor the intracellular transport of NCLS for two minutes, and the corresponding maximum-intensity projection reveals that a robust number of NCLS (6.9% ± 2.7%) are transported in a highly directed fashion over long distances (>10μm–20μm), now referred to as long-distance trajectories (Figure 1A, right panel). However, due to defective assembly or saturation of the cellular transport machinery, we also detect many VP30-GFP-positive NCLS that do not move, or travel only short distances.

Actin tail formation at EBOV nucleocapsids was detected in an earlier study [3]. Therefore, we aimed here to determine whether or not the established transfection system can recapitulate the detection of actin tails at NCLS as well. In cells that co-express the actin marker LifeAct to visualize the actin cytoskeleton (Figure 1B), we observed that NCLS induces actin tails during their movement through the cell, thereby further highlighting the relevance of this transfection system to the study of the transport of EBOV nucleocapsids. Here, we further assessed the movement of an individual NCLS-track over time, starting from its origin and following it through the cell (inset of Figure 1B is magnified in Figure 1C, white arrows). In Figure 1D, the red arrows indicate actin pulses during movement, which appear to correlate with the altered movement of NCLS, such as changes in direction (Figure 1C). While with the live cell imaging setting we used approximately 80% of the tracks that moved in a directed fashion revealed detectable actin tails at their rear end (Figure 3E, control siRNA), we cannot exclude the possibility that smaller NCLS induce actin polymerization below the detection limit, due to the high intracellular background level of LifeAct.

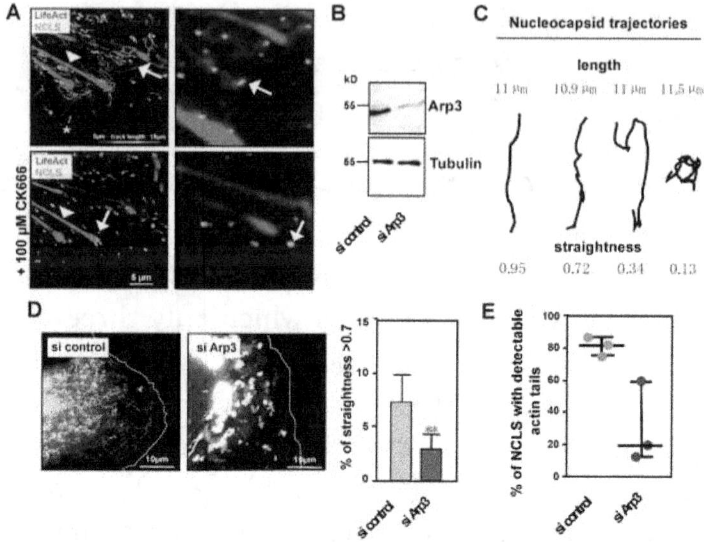

Figure 3. NCLS induction of actin tails depends on Arp2/3 activity. (A) Huh7 cells transfected with NP, VP24 and VP35 and LifeAct-CLIP were monitored over time. The left panels show an overlay image of the tracks as calculated by Imaris (color coded for length) and LifeAct-CLIP. Note, CK666 incubation does not interfere with filopodia (asterisk) or stress fibers (arrow head) . The zoomed image (right panels) show still images of the movie, revealing that CK666 treatment abolishes actin tail formation (white arrows). (B) Western Blot showing the depletion of Arp3 after siRNA treatment. (C) Cartoon showing different movements of trajectories and their corresponding straightness value. (D) Left panel shows maximum intensity projections of time lapse imaging of Huh7 cells transfected either with control or Arp3 siRNA. The right panel shows the graphical analysis of live cell imaging comparing all tracks > 10 µm in cells transfected either with control or Arp3 siRNA. Note, the relative amount of tracks with a straightness > 0.7, representing directed long-distance transport, is reduced in cells transfected with Arp3 siRNA (taken from n = 3 experiments, * p = 0.0027, Mann–Whitney Test). (E) Graph shows the percentage of trajectories with pulsative actin tails in live cell imaging (n = 3, error bars indicate mean with standard deviation).

While live cell imaging was applied to capture rapid and transient phenomena, the image resolution remained low. To precisely determine the localization of actin tails at filoviral nucleocapsids, we decided to use a high resolution, single molecule detection technique (STORM). To this end, we either transfected cells with NP, VP35 and VP24, or infected cells with EBOV and fixed them after 24 h. This was followed by the staining of the actin cytoskeleton and immunolabelling for the viral NP protein. Intracellular NCLS or EBOV nucleocapsids appear as elongated NP-positive structures (Figure 1E,F). However, compared to NCLS in transfected cells, filamentous nucleocapsids in cells infected with EBOV were commonly detected along the plasma membrane and within filopodia, likely representing virus particles prior to or during budding, thereby supporting previous observations from live-cell imaging and electron microscopy (Figure 1D, left panel) [3]. Co-staining with Phalloidin revealed actin tail structures in the vicinity of the nucleocapsids in infected cells and transfected cells (Figure 1D,E), which were less frequent in infected cells. This finding is likely a consequence of the very high spatial-temporal dynamics of actin tail formation, which do not allow a quantifiable detection in fixed cells. Further, in the minimalistic transfection system, NCLS remain in the cytosol and are not released, thus the detection of NCLS with actin tails may occur more frequently. Additionally, it cannot be excluded that in infected cells, the long fixation procedure required for the handling of BSL4 samples might interfere with the fixation of these transient actin structures.

3.2. Analysis of NCLS Transport Dynamics

To measure and characterize the long-distance movement of NCLS in a semi-automated manner, we analyzed our time-lapse movies with a tracking tool within the software Imaris. To this end, we set up an algorithm to follow NCLS, and filtered for trajectories > 10 μm that show mean velocities comparable to those measured in previous reports (here, 167 ± 58 nm/s compared to 187 ± 67 nm/s as in Takamatsu et al., 2018 [5]) (Figure 2A).

These studies demonstrated that the long-distance movement of EBOV nucleocapsids, as well as NCLS, is sensitive to the inhibition of actin polymerization [3,5]. Here, we recapitulated these experiments by monitoring NCLS movement followed by an incubation with an actin-modulating drug, and then re-recorded the very same cell. Using this system, we ensured that the imaged cells were capable of showing long-distance trajectories, and would consequently enable us to analyze alterations in NCLS movement. As observed previously, the application of the inhibitor of actin polymerization, Cytochalasin D, results in a rapid and strong decrease in any long-distance trajectories of NCLS, quantified as a highly reduced number of tracks over 10 μm (Figure 2B) [3,5]. Expanding on previous results, we further incubated the cells with the actin polymerization and stabilization agent Jasplakinolide, resulting in a similar phenotype, thereby indicating that indeed not only actin polymerization, but also depolymerization, is required for the long-distance transport of NCLS (Figure 2C). Treatment with Blebbistatin (here we used para-nitro-Blebbistatin to avoid phototoxicity in live cell experiments) did not interfere with long-distance movement, suggesting that myosin II activity is not required for the long-distance transport of NCLS (Figure 2D). Taken together, these findings reinforce the importance of actin regulation and dynamics in the long-distance transport of NCLS.

3.3. Arp2/3 Complex Activity is necessary for the formation of actin tails and the directed transport across long distances

Previous experiments showed that inhibitors of Arp2/3 affected the long-distance movement of EBOV nucleocapsids [3]. To confirm and further elaborate on these results, we used the minimalistic transfection system to analyze how treatment with an Arp2/3 inhibitor (CK666) affects NCLS movement [36]. Blocking Arp2/3 activity rapidly abolishes any directed long-distance movement of NCLS, and that minimal movement likely derives from Brownian motion (Figure 2E). When dissecting different pools of tracks, we observed that treatment with CK666 strongly decreases tracks >10 μm, subsequently resulting in increases in tracks < 5 μm, which also show altered mean speeds. In cells treated with CK666, we observed a slight reduction in the overall NCLS (−19 ± 7%), which might have resulted from the decreased NCLS transport from the inclusion bodies into the cytoplasm. Furthermore, reveals that the abolishment of NCLS transport, using CK666, coincides with the loss of actin tail formation (Figure 3A, tracks are color-coded for length and actin is visualized with stills from the confocal movie). Time-lapse imaging further demonstrates that Arp2/3-independent structures, such as filopodia or stress fibers, are not compromised during the observed time frame, thereby highlighting a specific role for Arp2/3-dependent actin polymerization in the directed long-distance transport of NCLS.

Next, we continued with the depletion of one protein within the Arp2/3 complex, namely Arp3, using siRNA transfection to further evaluate its role in actin tail-dependent NCLS transport (Figure 3B). The siRNA depletion of Arp3 likely results in the destabilization or reduction of functional Arp2/3 complexes, and does not entirely block all long-distance movement of NCLS, as observed under treatment with CK666. Thus, we assessed whether the remaining tracks showed altered directionality in cells depleted for Arp3, through the calculation of trajectory straightness (ratio of distance to length) (Figure 3C). We determined that tracks with a straightness > 0.7 (Figure 3C), and indeed the depletion of Arp3, interferes with the total number of straight tracks, further highlighting a role for Arp2/3 activity in the long-distance transport of NCLS (Figure 3D,E). Next, we compared the formation of actin tails in control and Arp3-depleted cells. While approximately 80% of the NCLS that travel over long distances showed pulsative actin tails, cells depleted for Arp3 showed a decreased level of detectable actin polymerization at their rear end (Figure 3E,F). Taken together, these results indicate that Arp2/3 activity is required for the efficient directed long-distance transport of NCLS, and that the Arp2/3 complex is directly involved in the formation of actin tails in Huh7 cells.

3.4. Characterization of the Rac1/WAVE1/Arp2/3 signaling pathway implicated in the long-range transportation of NCLS

The Arp2/3 complex resides, in inactive status, within the cell, and requires activation by NFPs, such as N-WASP or Wave proteins, that bind and activate the Arp2/3 complex via their conserved VCA-domain (Figure 4A) [29]. As it was recently demonstrated that the inhibition of N-WASP using Wiskostatin did not interfere with the long-distance transport of EBOV nucleocapsids [3], we focused on Wave1, an alternative activator of the Arp2/3 complex. Interestingly, siRNA directed against WASF1 (Wave1) faithfully recapitulated the phenotype of Arp3 siRNA treatment, leading to the reduced straightness of the NCLS (Figure 4B). This indicates that activation of the Arp2/3 complex downstream of Wave1 is involved in the directionality of long-distance trajectories. Wave1 activity itself can be controlled by the RhoGTPase Rac1 (Figure 4A–C) [37,38]. Again, the treatment of Huh7 cells with siRNA against Rac1 resulted in a decreased level of directed long-distance trajectories (Figure 4D,F,G). In contrast, the depletion of Cdc42, another member of the RhoGTPase family, did not affect the straightness of trajectories, thereby further reinforcing the inference that Rac1 activity is involved in NCLS transport (Figure 4F,H). For further confirmation, we applied the Rac1 inhibitor NSC 23766, which also interfered with long-distance transport, confirming the relevance of Rac1 activity in the regulation of the long-distance transport of EBOV NCLS (Figure 4H).

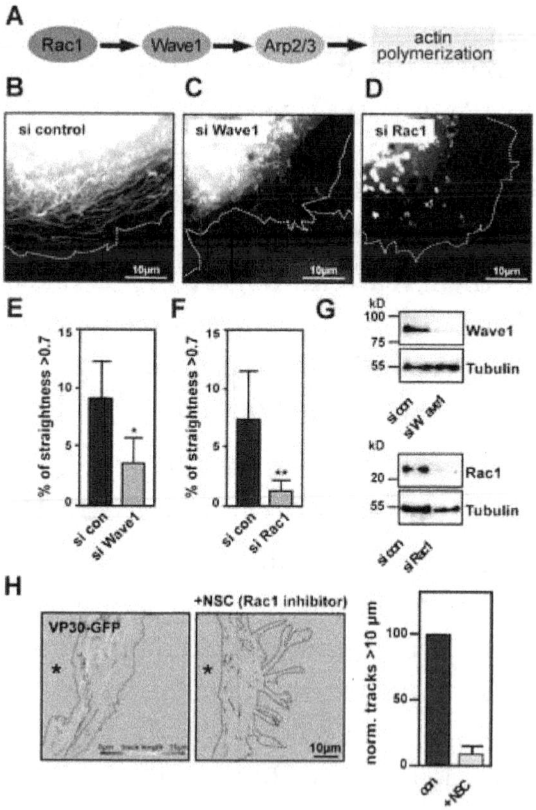

Figure 4. Identification of a Rac1/Wave1/Arp2/3 signaling. (A) Cartoon depicting canonical Arp2/3 signaling downstream of Rac1 [39]. (B–D)

Images show maximum intensity projection of time-lapse images of cells recorded for 2 min; images were captured every 900 ms. Cells were transfected with control siRNA (B), siRNA against (C) Wave1 (WASF1) or (D) Rac1. (E,F) Analysis of live cell imaging comparing relative straightness of tracks > 10 µm and a straightness > 0.7 after either Wave1 siRNA (E) or Rac1 siRNA (F) transfection. Cells were transfected with siRNA against (B) Wave1 (WASF1) or (C) Rac1. Note that siRNA treatment against Wave1 and Rac1 results in a decrease in straight long-distance trajectories (taken from three independent experiments, at least 5 cells per experiments, ns = non-significant, ** p > 0.001, * p > 0.01). (G) Western Blots showing effective siRNA depletion of Wave1 and Rac1. (H) Incubation of Huh7 cells transfected with NCLS (VP30-GFP) was first recorded for 2 min, then they were incubated with 100 nM Rac1 inhibitor (NSC23766) for 1 h and then reimaged. The overview images reveal that NCS23766 also inhibits long-distance transport. Tracks are color-coded for mean speed (6 cells in n = 3, p < 0.0079, t-test).

Taken together, these findings indicate that EBOV nucleocapsids exploit the canonical Rac1/WAVE1/Arp2/3 signaling pathway for long-distance transport of NCLS inside the cell. This signaling pathway classically induces lamellipodial protrusions, in which Rac1 acts coordinately with other upstream signals to activate actin regulators. Future studies shall reveal whether other actin-regulatory mechanisms are also involved in the viral life cycle, and how alterations in long-distance transport affect virus propagation.

4. DISCUSSION

This work has discovered that the Rac1/Wave1/Arp2/3 pathway plays a role in the actin-dependent trafficking of EBOV NCLS in human cell culture. The activity of Arp2/3 is crucial for the formation of propulsive and polar-localized actin tails at the back ends of NCLS. These actin tails can be clearly observed using live cell imaging. In addition, the connection between actin tails and EBOV NCLS and nucleocapsids may also be detected using super resolution in preserved samples. Actin comet tails were first identified in Listeria monocytogenes during the late 1980s [40]. Since then, diverse types of intracellular pathogens have been identified as inducing actin polymerization at their surface. For instance, vaccinia virus protein A36 is able to recruit NCK and GRB2, which in turn recruit N-WASP to stimulate the Arp2/3 complex, or virus protein p78/83 mimics N-WASP and directly activates Arp2/3-induced actin polymerization [30,41]. Here, we showed that EBOV NCLS induce actin tails at one end, and actin tail formation is sensitive to the inhibition of Arp2/3 using CK666, indicating that the hijacking of the highly abundant and conserved Arp2/3 complex for induction of actin tails is a common mechanism in diverse types of pathogens without a common origin [30].

We further show that directed long-distance transport is regulated via Arp2/3 activity, yet the siRNA depletion of Arp3 in cells forming NCLS does not entirely block intracellular transport. Furthermore, the effects of siRNA treatment on track straightness appear more pronounced after Rac1 knockdown, when compared to the depletion of Arp3 or Wave1. These findings also suggest that other actin regulators downstream of Rac1 either compensate Arp2/3 activity, or synergistically regulate actin polymerization. A potential option for consideration is the scaffolding protein IQGAP1, which serves to connect several pathways involved in actin dynamics and has the ability to interact with Rac1. Within infected cells, MARV infection led to the recruitment of IQGAP1 to the inclusion and the trailing end of nucleocapsids. The decrease in IQGAP1 levels caused a hindered release of MARV, indicating that other significant regulators of actin may play a role in this mechanism. [14].

In addition, we also gained unprecedented evidence that Wave1 upstream of Arp2/3 is involved in regulating the long-distance transport of NCLS in Huh7 cells. Given that nucleocapsid transport in EBOV-infected cells does not depend on N-WASP [3], we concluded that other NFPs could be involved in the upstream regulation of Arp2/3. Further supporting this notion, N-WASP is typically involved in endocytotic events, where it regulates actin polymerization in a manner reminiscent of actin tails that propel pathogens through cells [42]. Consequently, enveloped viruses, such as vaccinia and EBOV, might profit from mimicking or recruiting N-WASP during their life cycle.

In recent years, additional NPFs, like WASH, WHAMM and JMY, have been described to promote actin tail formations at intracellular membranes like endosomes and autophagosomes [43,44,45]. In contrast, filoviral nucleocapsids travel in the cytosol without a membrane, and are highly structured protein complexes that probably utilize endocytosis-independent pathways for their transport, thereby likely avoiding membranous structures prior to their arrival at the plasma membrane. Wave1 and Rac1 signaling is considered primarily relevant to actin polymerization in lamellipodial cell protrusions [43], thus it is not surprising that long-distance NCLS transport is best observed in areas with a high activity of this pathway in Huh7 cells. Future studies using high resolution microscopy or electron microscopic approaches shall reveal whether and how actin regulators are actively recruited to EBOV nucleocapsids. It could be that at different stages of the nucleocapsid transport, from inclusion bodies to the budding sites, different actin tail-inducing machineries are exploited by the virus.

As described for other viruses, the long-distance transport of EBOV NCLS is accompanied by the induction of polar actin polymerization, likely resulting in the directionality of movement also observed in other actin tail-inducing pathogens and in in vitro reconstructions [41,46,47]. How this polar induction of actin polymerization is initiated is not understood in detail. One hypothesis derives from studies in Listeria, where it was shown that the surface protein ActA accumulates locally, likely during cell wall growth, which subsequently results in polar interactions with actin regulators [48]. In contrast, filoviral nucleocapsids are not enveloped when they leave inclusion bodies, and only encounter the viral proteins GP and VP40, which are transported independently, when they reach the plasma membrane to form infectious particles [49,50,51,52]. Thus, it remains unclear how this polar actin polymerization is induced in filoviral nucleocapsids, and which viral protein might be relevant for the actin polymerization. One explanation derives from a study investigating the structure of MARV capsids using cryo-electron microscopy, revealing that nucleocapsid assembly itself results in polar structures [10]. Here, it was demonstrated that MARV nucleocapsids are highly oriented towards the plasma membrane, with the pointed end of the nucleocapsid directed towards the plasma membrane prior to budding [11,53]. Therefore, these discoveries provide evidence that the process of polar nucleocapsid assembly may lead to distinct protein structures, which in turn reveal the locations where cellular proteins might trigger actin polymerization. Collectively, our simplified transfection-based method has novel possibilities for investigating the intracellular transportation of filoviral nucleocapsids, as well as accurately measuring the movement of subviral components. Future research needs to more thoroughly analyze how filoviral proteins influence the precise timing and location of RhoGTPase signaling. Additionally, it should aim to determine the specific molecules that directly interact with these proteins and the structural alterations necessary for long-distance transportation in the life cycle of the Ebola virus (EBOV).

REFERENCES

1. Taylor, M.P.; Koyuncu, O.O.; Enquist, L.W. Subversion of the actin cytoskeleton during viral infection. Nat. Rev. Microbiol. 2011, 9, 427–439.
2. Newsome, T.P.; Marzook, N.B. Viruses that ride on the coat-tails of actin nucleation. Semin. Cell. Dev. Biol. 2015, 46, 155–163.
3. Schudt, G.; Dolnik, O.; Kolesnikova, L.; Biedenkopf, N.; Herwig, A.; Becker, S. Transport of Ebolavirus Nucleocapsids Is Dependent on Actin Polymerization: Live-Cell Imaging Analysis of Ebolavirus-Infected Cells. J. Infect. Dis. 2015, 212, 160–166.
4. Schudt, G.; Kolesnikova, L.; Dolnik, O.; Sodeik, B.; Becker, S. Live-cell imaging of Marburg virus-infected cells uncovers actin-dependent transport of nucleocapsids over long distances. Proc. Natl. Acad. Sci. USA 2013, 113, 14402–14407.
5. Takamatsu, Y.; Kolesnikova, L.; Becker, S. Ebola virus proteins NP, VP35, and VP24 are essential and sufficient to mediate nucleocapsid transport. Proc. Natl. Acad. Sci. USA 2018, 115, 1075–1080.
6. Baseler, L.; Chertow, D.S.; Johnson, K.M.; Feldmann, H.; Morens, D.M. The Pathogenesis of Ebola Virus Disease. Annu. Rev. Pathol. 2017, 12, 387–418.

7. Cantoni, D.; Rossman, J.S. Ebolaviruses: New roles for old proteins. PLoS Negl. Trop. Dis. 2018, 12, e0006349.

8. Messaoudi, I.; Amarasinghe, G.K.; Basler, C.F. Filovirus pathogenesis and immune evasion: Insights from Ebola virus and Marburg virus. Nat. Rev. Microbiol. 2015, 13, 663–676.

9. Hoenen, T.; Shabman, R.S.; Groseth, A.; Herwig, A.; Weber, M.; Schudt, G.; Dolnik, O.; Basler, C.F.; Becker, S.; Feldmann, H. Inclusion bodies are a site of ebolavirus replication. J. Virol. 2012, 86, 11779–11788.

10. Wan, W.; Kolesnikova, L.; Clarke, M.; Koehler, A.; Noda, T.; Becker, S.; Briggs, J.A.G. Structure and assembly of the Ebola virus nucleocapsid. Nature 2017, 551, 394–397.

11. Bharat, T.A.; Noda, T.; Riches, J.D.; Kraehling, V.; Kolesnikova, L.; Becker, S.; Kawaoka, Y.; Briggs, J.A. Structural dissection of Ebola virus and its assembly determinants using cryo-electron tomography. Proc. Natl. Acad. Sci. USA 2012, 109, 4275–4280.

12. Biedenkopf, N.; Hartlieb, B.; Hoenen, T.; Becker, S. Phosphorylation of Ebola virus VP30 influences the composition of the viral nucleocapsid complex: Impact on viral transcription and replication. J. Biol. Chem. 2013, 288, 11165–11174.

13. Takamatsu, Y.; Kolesnikova, L.; Schauflinger, M.; Noda, T.; Becker, S. The Integrity of the YxxL Motif of Ebola Virus VP24 Is Important for the Transport of Nucleocapsid-Like Structures and for the Regulation of Viral RNA Synthesis. J. Virol. 2020, 94, e02170–e02219.

14. Dolnik, O.; Kolesnikova, L.; Welsch, S.; Strecker, T.; Schudt, G.; Becker, S. Interaction with Tsg101 is necessary for the efficient transport and release of nucleocapsids in marburg virus-infected cells. PLoS Pathog. 2014, 10, e1004463.

15. Dolnik, O.; Kolesnikova, L.; Stevermann, L.; Becker, S. Tsg101 is recruited by a late domain of the nucleocapsid protein to support budding of Marburg virus-like particles. J. Virol. 2010, 84, 7847–7856.

16. Naghavi, M.H.; Walsh, D. Microtubule Regulation and Function during Virus Infection. J. Virol. 2017, 91, e00538–e00617.
17. Pollard, T.D. Cell Motility and Cytokinesis: From Mysteries to Molecular Mechanisms in Five Decades. Annu. Rev. Cell. Dev. Biol. 2019, 35, 1–28.
18. Murrell, M.; Oakes, P.W.; Lenz, M.; Gardel, M.L. Forcing cells into shape: The mechanics of actomyosin contractility. Nat. Rev. Mol. Cell. Biol. 2015, 16, 486–498.
19. KCampellone, G.; Welch, M.D. A nucleator arms race: Cellular control of actin assembly. Nat. Rev. Mol. Cell. Biol. 2010, 11, 237–251.
20. Pollard, T.D. Actin and Actin-Binding Proteins. Cold Spring Harb. Perspect Biol. 2016, 18, a018226.
21. Lappalainen, P. Actin-binding proteins: The long road to understanding the dynamic landscape of cellular actin networks. Mol. Biol. Cell. 2016, 27, 2519–2522.
22. Grikscheit, K.; Grosse, R. Formins at the Junction. Trends Biochem. Sci. 2016, 41, 148–159.
23. Zech, T.; Calaminus, S.D.; Machesky, L.M. Actin on trafficking: Could actin guide directed receptor transport? Cell. Adh. Migr. 2012, 6, 476–481.
24. Schaks, M.; Giannone, G.; Rottner, K. Actin dynamics in cell migration. Essays Biochem. 2019, 63, 483–495.
25. Schuh, M. An actin-dependent mechanism for long-range vesicle transport. Nat. Cell. Biol. 2011, 13, 1431–1436.
26. Goley, E.D.; Welch, M.D. The ARP2/3 complex: An actin nucleator comes of age. Nat. Rev. Mol. Cell. Biol. 2006, 7, 713–726.
27. Rodal, A.A.; Sokolova, O.; Robins, D.B.; Daugherty, K.M.; Hippenmeyer, S.; Riezman, H.; Grigorieff, N.; Goode, B.L. Conformational changes in the Arp2/3 complex leading to actin nucleation. Nat. Struct. Mol. Biol. 2005, 12, 26–31.

28. Rodnick-Smith, M.; Luan, Q.; Liu, S.L.; Nolen, B.J. Role and structural mechanism of WASP-triggered conformational changes in branched actin filament nucleation by Arp2/3 complex. Proc. Natl. Acad. Sci. USA 2016, 113, E3834–E3843.

29. Rotty, J.D.; Wu, C.; Bear, J.E. New insights into the regulation and cellular functions of the ARP2/3 complex. Nat. Rev. Mol. Cell. Biol. 2013, 14, 7–12.

30. Welch, M.D.; Way, M. Arp2/3-mediated actin-based motility: A tail of pathogen abuse. Cell. Host Microbe 2013, 14, 242–255.

31. Welch, M.D.; Rosenblatt, J.; Skoble, J.; Portnoy, D.A.; Mitchison, T.J. Interaction of human Arp2/3 complex and the Listeria monocytogenes ActA protein in actin filament nucleation. Science 1998, 281, 105–108.

32. Jeng, R.L.; Goley, E.D.; D'Alessio, J.A.; Chaga, O.Y.; Svitkina, T.M.; Borisy, G.G.; Heinzen, R.A.; Welch, M.D. A Rickettsia WASP-like protein activates the Arp2/3 complex and mediates actin-based motility. Cell. Microbiol. 2004, 6, 761–769.

33. Boujemaa–Paterski, R.; Gouin, E.; Hansen, G.; Samarin, S.; le Clainche, C.; Didry, D.; Dehoux, P.; Cossart, P.; Kocks, C.; Carlier, M.F.; et al. Listeria protein ActA mimics WASp family proteins: It activates filament barbed end branching by Arp2/3 complex. Biochemistry 2001, 40, 11390–11404.

34. Hoenen, T.; Groseth, A.; Kolesnikova, L.; Theriault, S.; Ebihara, H.; Hartlieb, B.; Bamberg, S.; Feldmann, H.; Ströher, U.; Becker, S. Infection of naive target cells with virus-like particles: Implications for the function of ebola virus VP24. J. Virol. 2006, 80, 7260–7264.

35. Biedenkopf, N.; Lier, C.; Becker, S. Dynamic Phosphorylation of VP30 Is Essential for Ebola Virus Life Cycle. J. Virol. 2016, 90, 4914–4925.

36. Hetrick, B.; Han, M.S.; Helgeson, L.A.; Nolen, B.J. Small molecules CK-666 and CK-869 inhibit actin-related protein 2/3 complex by blocking an activating conformational change. Chem. Biol. 2013, 20, 701–712.

37. Chen, B.; Chou, H.T.; Brautigam, C.A.; Xing, W.; Yang, S.; Henry, L.; Doolittle, L.K.; Walz, T.; Rosen, M.K. Rac1 GTPase activates the WAVE regulatory complex through two distinct binding sites. Elife 2017, 6, e29795.

38. Eden, S.; Rohatgi, R.; Podtelejnikov, A.V.; Mann, M.; Kirschner, M.W. Mechanism of regulation of WAVE1-induced actin nucleation by Rac1 and Nck. Nature 2002, 418, 790–793.

39. Pollitt, A.Y.; Insall, R.H. WASP and SCAR/WAVE proteins: The drivers of actin assembly. J. Cell. Sci. 2009, 122, 2575–2578.

40. Tilney, L.G.; Portnoy, D.A. Actin filaments and the growth, movement, and spread of the intracellular bacterial parasite, Listeria monocytogenes. J. Cell. Biol. 1989, 109, 1597–1608.

41. Mueller, J.; Pfanzelter, J.; Winkler, C.; Narita, A.; le Clainche, C.; Nemethova, M.; Carlier, M.F.; Maeda, Y.; Welch, M.D.; Ohkawa, T.C.; et al. Electron tomography and simulation of baculovirus actin comet tails support a tethered filament model of pathogen propulsion. PLoS Biol. 2014, 12, e1001765.

42. Rottner, K.; Hänisch, J.; Campellone, K.G. WASH, WHAMM and JMY: Regulation of Arp2/3 complex and beyond. Trends Cell. Biol. 2010, 20, 650–661.

43. Derivery, E.; Helfer, E.; Henriot, V.; Gautreau, A. Actin polymerization controls the organization of WASH domains at the surface of endosomes. PLoS ONE 2012, 7, e39774.

44. Kast, D.J.; Dominguez, R. WHAMM links actin assembly via the Arp2/3 complex to autophagy. Autophagy 2015, 11, 1702–1704.

45. Hu, X.; Mullins, R.D. LC3 and STRAP regulate actin filament assembly by JMY during autophagosome formation. J. Cell Biol. 2019, 218, 251–266.

46. Shenoy, V.B.; Tambe, D.T.; Prasad, A.; Theriot, J.A. A kinematic description of the trajectories of Listeria monocytogenes propelled by actin comet tails. Proc. Natl. Acad. Sci. USA 2007, 104, 8229–8234.

47. Dayel, M.J.; Akin, O.; Landeryou, M.; Risca, V.; Mogilner, A.; Mullins, R.D. In silico reconstitution of actin-based symmetry breaking and motility. PLoS Biol. 2009, 7, e1000201.

48. Lacayo, C.I.; Soneral, P.A.; Zhu, J.; Tsuchida, M.A.; Footer, M.J.; Soo, F.S.; Lu, Y.; Xia, Y.; Mogilner, A.; Theriot, J.A. Choosing orientation: Influence of cargo geometry and ActA polarization on actin comet tails. Mol. Biol. Cell. 2012, 23, 614–629.

49. Becker, S.; Klenk, H.D.; Mühlberger, E. Intracellular transport and processing of the Marburg virus surface protein in vertebrate and insect cells. Virology 1996, 225, 145–155.

50. Mittler, E.; Schudt, G.; Halwe, S.; Rohde, C.; Becker, S. A Fluorescently Labeled Marburg Virus Glycoprotein as a New Tool to Study Viral Transport and Assembly. J. Infect. Dis. 2018, 218, S318–S326.

51. Yamayoshi, S.; Noda, T.; Ebihara, H.; Goto, H.; Morikawa, Y.; Lukashevich, I.S.; Neumann, G.; Feldmann, H.; Kawaoka, Y. Ebola virus matrix protein VP40 uses the COPII transport system for its intracellular transport. Cell. Host Microbe 2008, 3, 168–177.

52. Johnson, K.A.; Taghon, G.J.; Scott, J.L.; Stahelin, R.V. The Ebola Virus matrix protein, VP40, requires phosphatidylinositol 4,5-bisphosphate (PI(4,5)P2) for extensive oligomerization at the plasma membrane and viral egress. Sci. Rep. 2016, 6, 19125.

53. Welsch, S.; Kolesnikova, L.; Krähling, V.; Riches, J.D.; Becker, S.; Briggs, J.A. Electron tomography reveals the steps in filovirus budding. PLoS Pathog 2010, 6, e1000875.

06 THE CELLULAR PROTEIN CAD IS ENLISTED INTO EBOLA VIRUS INCLUSION BODIES BY THE NUCLEOPROTEIN NP TO EXPEDITE GENOME REPLICATION AND TRANSCRIPTION

1. INTRODUCTION: AN OVERVIEW

The Ebola virus (EBOV) is a pathogen that can be transmitted between animals and humans, causing severe fevers with bleeding and a high likelihood of death in both humans and non-human primates. Recently, there has been a rise in the frequency and severity of outbreaks, emphasizing the need for a deeper comprehension of the molecular elements of EBOV infection and the interactions between the virus and host cells in order to enhance the efficiency of virus control. Various viruses, such as EBOV, have been demonstrated to enlist host proteins for various viral activities. Through a comprehensive analysis of the entire genome using siRNA, we have recently shown that the cellular host factor carbamoyl-phosphate synthetase 2, aspartate transcarbamylase, and dihydroorotase (CAD) play a role in the creation of EBOV RNA. Nevertheless, the specific mechanisms by which this host component contributes to the life cycle of the Ebola virus (EBOV) are still unknown. This study focused on examining the functional and molecular interactions that occur between the Ebola virus (EBOV) and CAD. In order to achieve this, we employed siRNA knockdowns along with different reverse genetics-based life cycle modelling systems. Additionally, we conducted co-immunoprecipitation and co-immunofluorescence assays to examine the impact of CAD on specific aspects of the EBOV life cycle and to analyze the interactions between CAD and viral proteins. By adopting this methodology, we could establish that CAD has a direct interaction with the EBOV nucleoprotein NP. Furthermore, we found that NP alone is capable of recruiting CAD into inclusion bodies, which is contingent upon the presence of the glutaminase (GLN) domain of CAD. Moreover, siRNA knockdown tests demonstrated the significance of CAD in both viral genome replication and transcription. Additionally, substrate rescue experiments confirmed the essential role of CAD in pyrimidine synthesis for these activities. Collectively, this indicates that NP utilizes CAD through its GLN domain to incorporate it into inclusion bodies, with the purpose of supplying pyrimidines for the replication and transcription of the EBOV genome. The discoveries establish a new method by which EBOV takes control of the pathways in host cells to aid in the replication and transcription of its genome. This provides additional support for the creation of broad-spectrum antiviral drugs that target the host's cells.

2. MATERIALS AND METHODS

2.1. Cell Lines

Human embryonic kidney cells (HEK 293T, Collection of Cell Lines in Veterinary Medicine CCLV-RIE 1018), African green monkey kidney cells (Vero E6, kindly provided by Stephan Becker, Philipps University Marburg), and human hepatocellular carcinoma cells (Huh7, kindly provided by Stephan Becker, Philipps University Marburg) were cultured in Dulbecco's modified Eagle's medium (DMEM; Thermo Fisher Scientific, Darmstadt, Germany) supplemented with 10% fetal calf serum (FCS), 100 U/mL penicillin, 100 µg/mL streptomycin (PS; Thermo Fisher Scientific), and 1× GlutaMAX (Thermo Fisher Scientific). All cells were incubated at 37°C and 5% CO2.

2.2. Plasmids and Cloning

Minigenome assay components, including expression plasmids coding for the EBOV RNP proteins, T7 polymerase, firefly luciferase, and a classical T7-driven monocistronic minigenome (pT7-1cis-EBOV-vRNA-nLuc) have been previously described [20,32]. A NanoLuc luciferase-expressing T7-driven replication-deficient minigenome was cloned from a classical minigenome expressing NanoLuc luciferase as a reporter by deletion of 55 nucleotides (nt) in the antigenomic replication promoter as previously described [32]. Based on this, a Pol-II-driven replication-deficient minigenome was generated by PCR to amplify a linear version of the replication-deficient minigenome flanked by hammerhead and hepatitis delta virus ribozymes using primers #4571 (5′-AGC TTA CGT GAC TAC TTC CTT CGG ATG CCC AGG TCG GAC CGC G-3′) and #4572 (5′-GAC CGG TAG AAA ACT GAT GAG TCC GTG AGG ACG AAA CGG AGT CTA GAC TCC GTC TTT TCC AGG AAT CCT TTT TGC AAC GTT TAT TCT G-3′). The linearized construct was subsequently inserted into pCAGGS. The CAD gene was cloned from 293T cells into pCAGGS, and deletion mutants and domains of CAD were then generated using PCR-based approaches. All constructs were first cloned into pCAGGS, followed by subcloning into a pCAGGS plasmid encoding an N-terminal FLAG/HA-tag (DYKDDDDKLDGGYPYDVPDYA) immediately upstream of a BsmBI cloning site, allowing a seamless insertion of the open reading frame of interest. The expression plasmid for N-terminally myc-tagged VP35 was constructed by cloning a myc-tag (EQKLISEEDL) immediately before the VP35 ORF. Detailed cloning strategies are available on request.

2.3. Antibodies

The anti-FLAG (clone M2) antibody used for immunofluorescence analyses (IFA), co-immuno precipitation (coIP), and Western blot analyses was purchased from Sigma-Aldrich (Munich, Germany) [F1804], and the anti-c-myc antibody used for IFA analysis was obtained from Thermo Fisher Scientific [A-21281]. Primary antibodies against NP (rabbit anti-EBOV NP polyclonal antibody), GAPDH (mouse anti-GAPDH clone 0411), and CAD (rabbit anti-CAD clone EP710Y) were ordered from IBT Bioservices (San Jose, USA; anti-NP [0301-012]), Santa Cruz (Heidelberg, Germany; anti-GAPDH [sc47724]), or Abcam (Cambridge, UK; anti-CAD [ab40800]). Secondary antibodies used for IFA analysis against mouse (Alexa Fluor 488 anti-mouse [A-11029]), rabbit (Alexa Fluor 568 anti-rabbit [A-11036]), and chicken IgY (Alexa Fluor 647 anti-chicken [A-21449]) were obtained from Thermo Fisher Scientific. For Western blotting, secondary antibodies against mouse (IRDye 680RD anti-mouse [926-68070]) and rabbit IgG (IRDye 800CW anti-rabbit [926-68071]) were purchased from Li-COR (Bad Homburg, Germany), while anti-mouse IgG (Kappa light chain) Alexa Fluor 680 [115-625-174] used for coIP analyses was ordered from Dianova (Hamburg, Germany).

2.4. Viruses

Zaire ebolavirus rec/COD/1976/Mayinga-rgEBOV (GenBank accession number KF827427.1), which is identical in sequence to the EBOV Mayinga isolate with the exception of four silent mutations as genetic markers [33], was used for all infection experiments. rgEBOV was propagated in VeroE6 cells and virus titers were determined by 50% tissue culture infectious dose (TCID50) assay. All work with the infectious virus was performed under BSL-4 conditions at the Friedrich-Loeffler-Institut (Federal Research Institute of Animal Health, Greifswald Insel-Riems, Germany) following approved standard operating procedures.

2.5. Chemical Compounds

100mM uridine or cytidine (both Sigma-Aldrich) stock solutions were prepared in dimethyl sulfoxide (DMSO) and further diluted in cell culture medium. Diluted pyrimidines or DMSO corresponding to 1% of the supernatant volume in 12-well plates was added to the cells at the time of transfection and after medium changes. All concentrations indicated in the figures are final concentrations.

2.6. siRNA Knockdown with EBOV Minigenomes and Pyrimidine Complementation

For siRNA knockdown of endogenous CAD, 293T cells were reverse transfected (i.e., transfected in suspension and subsequently seeded into plates) with 12 pmol pre-designed silencer select siRNAs (CAD-siRNA#1: s2320 [5′-GAG GGU CUC UUC UUA AGU A-3′]; CAD-siRNA#2: 117891 [5′-GCU AGC UGA GAA AAA CUU U-3′]; Negative Control siRNA #2; all Thermo Fisher Scientific) or a self-designed EBOV-anti-L siRNA [5′-UUU AUA UAC AGC UUC GUA CUU-3′] ordered from Eurofins

Genomics (Ebersberg, Germany). Transfection was performed in 12-well plates using Lipofectamine RNAiMax (Thermo Fisher Scientific) following the manufacturer's instructions. 48 h post-siRNA transfection, the cells were transfected using Transit-LT1 (Mirus Bio LLC, Madison, USA) with all minigenome assay components, i.e., pCAGGS-based expression plasmids for NP (62.5 ng), VP35 (62.5 ng), VP30 (37.5 ng), L (500 ng), codon-optimized T7-polymerase (125 ng), firefly luciferase (as a control, 125 ng), and the T7-driven monocistronic minigenome (pT7-EBOV-1cis-vRNA-nLuc; 125 ng). For analyses of vRNA and mRNA levels the control firefly luciferase was replaced with GFP (200 ng), and for the replication-deficient minigenome assay a Pol-II-driven replication-deficient minigenome (pCAGGS-EBOV-1cis-vRNA-nLuc-RdM) was used. Transfections were performed using Transit LT1 as previously described [32]. All samples were harvested 48 h post-transfection for either determination of reporter activity or RNA isolation (see below). For measuring the luciferase activity, cells were lysed for 10 min in 1x Lysis Juice (PJK, Kleinblittersdorf, Germany) at room temperature and lysates were cleared of cell debris by centrifugation for 3 min at 10,000× g. Then, 40 µL of the cleared lysates were added to either 40 µL of Beetle Juice (PJK) or NanoGlo Luciferase Assay Reagent (Promega, Madison, USA) in opaque 96-well plates and luminescence was measured using a Glomax Multi (Promega) microplate reader. NanoLuc luciferase activities were normalized to firefly luciferase activities.

2.7. RNA Isolation and RT-qPCR

RNA isolation from minigenome cell lysates was performed following the manufacturer's instructions using a NucleoSpin RNA kit (Machery-Nagel, Düren, Germany). After RNA purification, all samples were treated with DNase (TURBO DNA-free kit; Thermo Fisher Scientific) following the manufacturer's instructions to avoid plasmid contamination. For cDNA generation, RNA samples were incubated with an oligo(dT)-primer for mRNA quantification or with a strand-specific primer (5′-AGT GTG AGC TTC TAA AGC AAC C-3′) for vRNA quantification using the RevertAid Reverse Transcriptase (Thermo Fisher Scientific) following the manufacturer's instructions. The subsequent qPCR was performed using a PowerUp SYBR Green Master Mix (Thermo Fisher Scientific) with 1 µL of cDNA and primers targeting either the reporter gene (5′-TTC AGA ATC TCG GG GTG TCC-3′, 5′-CGT AAC CCC GTC GAT TAC CA-3′), or GFP as a control (5′-CTT GTA CAG CTC GTC CAT GC-3′, 5′-CGA CAA CCA CTA CCT GAG CAC-3′). Values for vRNA and mRNA levels were normalized to control GFP mRNA levels.

2.8. Immunofluorescence Analysis

Huh7 cells, which are more suitable for IFA than 293T cells, were seeded on coverslips in 12-well plates and transfected 24 h later with 500 ng pCAGGS-EBOV-NP and 500 ng pCAGGS-FLAG-HA-CAD (or CAD mutants) and, in selected experiments, additionally with 500 ng pCAGGS-myc-VP35 as indicated. For a mock control, cells were transfected with pCAGGS. Transfection was performed using polyethylenimine (Sigma-Aldrich) following the manufacturer's instructions. 48 h post-transfection, cells were fixed using 4% paraformaldehyde (Roth, Karlsruhe, Germany) in DMEM for 20 min and then treated with 1 M glycine (in phosphate-buffered saline++ (PBS with 0.9M $Ca2+$ and 0.5M $Mg2+$)) for 10 min. Then, cells were permeabilized with 0.1% Triton X-100 in PBS for another 10 min and incubated with 10% fetal calf serum (FCS) in PBS for 45 min. Primary antibodies (rabbit anti-EBOV-NP 1:500; mouse anti-FLAG 1:2500; chicken anti-myc 1:1200) were diluted in PBS with 10% FCS and cells were incubated for 1 h at room temperature with the prepared antibody solutions. Secondary antibodies (Alexa Fluor 488 anti-mouse 1:1200; Alexa Fluor 568 anti-rabbit 1:500; Alexa Fluor 647 anti-chicken 1:1200) were prepared as described for the primary antibodies. After 45 min of staining, cells were washed with PBS and water before mounting with ProLong Diamond Antifade mountant with 4′,6-diamidino-2-phenylindole (DAPI) (Thermo Fisher Scientific). Slides were analyzed by confocal laser scanning microscopy using a Leica SP5.

2.9. Infection of Transfected Huh7 Cells

To investigate the localization of CAD during EBOV infection, Huh7 cells were seeded in 8-well chambered slides (ibidi, Martinsried, Germany) and transfected as described above (immunofluorescence analysis) with

500 ng pCAGGS-FLAG/HA-CAD. At 48 h post-transfection, the transfected cells were infected with EBOV at an MOI of 1, and the samples were fixed 16 h post-infection in 10% formalin twice overnight prior to removal from the BSL4 facility and immunofluorescence analysis.

2.10. Co-Immunoprecipitation of Viral Proteins

CoIPs were performed as previously described [19]. Briefly, 293T cells were seeded in 6-well plates and transfected with expression plasmids encoding FLAG/HA-tagged CAD and EBOV-NP using Transit LT-1 (Mirus Bio LLC) following the manufacturer's instructions. The medium was changed after 24 h and the cells were harvested 48 h post-transfection. For coIP, cells were lysed in 1 mL coIP lysis buffer (1% NP-40; 50 mM Tris pH 7.4; 167 mM NaCl in water) with protease inhibitors (cOmplete; Roche, Mannheim, Germany). To investigate a possible RNA dependency of the interaction between CAD and NP, 100 µg/mL RNase A (Machery-Nagel) were added to the samples. Subsequently, the samples were incubated rotating at 15 RPM for 2 h at 4 °C. Then, 150 µL of the cleared lysates were taken as an input control (representing a sixth of the complete pre-immune lysate and 20% of the sample used for immunoprecipitation) and subjected to acetone precipitation. The remaining 750 µL of cell lysate were mixed with the prepared bead-antibody solution (Dynabeads Protein G, Thermo Fisher Scientific; 1 µL anti-FLAG M2 antibody per 10 µL beads). Immunoprecipitation was performed for 10 min, as recommended by the manufacturer, at room temperature and rotation at 15 RPM. Then, samples were transferred to new tubes and boiled for 10 min at 99 °C. Input and coIP samples were analyzed by SDS-PAGE and Western blotting.

2.11. Western Blotting

For validation of CAD knockdown efficiency and analyses of coIP input and lysates, samples were subjected to SDS-PAGE and Western blotting as previously described [34]. FLAG-tagged CAD was detected using a monoclonal anti-FLAG antibody (1:2000), while NP, wild type CAD, and GAPDH were detected using anti-NP (1:1000), anti-CAD (1:250), and anti-GAPDH (1:1000) antibodies. As secondary antibodies, 680RD-coupled goat-anti-mouse, goat-anti-mouse-Alexa Fluor 680, and 800CW-coupled goat-anti-rabbit antibodies (1:14000) were used. Fluorescent signals were detected and quantified using an Odyssey CLx infrared imaging system (Li-Cor Biosciences). For knockdown quantification, CAD signals were normalized to GAPDH signals.

2.12. Statistical Analyses

One-way ANOVA with a Dunnett's multiple comparisons test was performed using the GraphPad Prism 8.1.0 software.

3. RESULTS

3.1. CAD Knockdown Affects Both EBOV Genome Replication and Transcription

Using a genome-wide siRNA screen, we previously identified CAD to be important for EBOV RNA synthesis and/or viral protein expression [20]. However, since only the effect of CAD knockdown on the sum of these processes had been tested, we now analyzed the role of CAD on individual aspects of the EBOV life cycle. As a first step, we assessed the efficiency of endogenous CAD knockdown using two different siRNAs via quantitative Western blotting, which revealed a 60% to 80% reduction in endogenous CAD expression levels for the two siRNAs (Figure 1A,B).

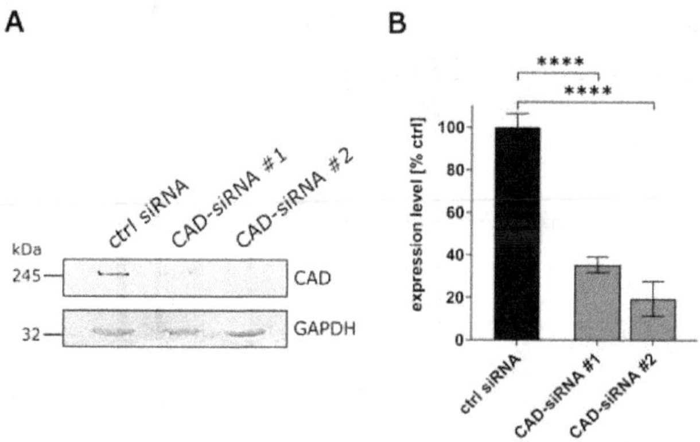

Figure 1. Quantification of CAD knockdown. (A) Analysis of CAD knockdown. 293T cells were transfected with siRNAs targeting CAD (CAD-siRNA), or a negative control (ctrl siRNA). The cells were harvested 48 h post-transfection and the lysates were subjected to SDS-PAGE and Western blotting. (B) Quantification of CAD knockdown. The Western blot

signals for CAD knockdown (as shown in Figure 1A) were measured and normalized to the GAPDH signals. The negative control (ctrl siRNA) was set to 100% and the efficiency of CAD knockdown was calculated (** $p \leq 0.0001$).**

Next, we performed a classical minigenome assay (Figure 2A) in connection with an siRNA knockdown of CAD. As previously shown, knockdown of CAD led to a 40 to 53-fold reduction in reporter activity, verifying an influence of CAD on EBOV viral RNA synthesis and protein expression (Figure 2B) [20]. In order to identify whether CAD knockdown affects transcription and/or protein expression independent of replication, we next used a replication-deficient minigenome system [32]. In contrast to a replication-competent minigenome, the replication-deficient minigenome lacks 55 nt in the antigenomic replication promoter leading to a block of minigenome vRNA replication, while minigenome transcription still takes place [32]. However, when using this system, which is based on T7-driven initial transcription of minigenomes, we observed a very low dynamic range between our controls, which made it difficult to evaluate a possible influence of CAD knockdown. Therefore, in order to increase the dynamic range of this system, we generated a Pol-II-driven replication-deficient minigenome that resulted in a ~10-fold higher dynamic range. Using this system, CAD knockdown resulted in a clear reduction in reporter activity, indicating that CAD is important for EBOV transcription and/or protein expression independent of viral genome replication (Figure 2C).

To further dissect the influences of CAD on viral genome replication, mRNA transcription, and later steps of viral protein expression, we performed classical minigenome assays in the context of an siRNA knockdown of CAD and measured vRNA and mRNA levels in cell lysates using RT-qPCR. For this, we used either an oligo-dT primer for reverse transcription of mRNAs, or a strand-specific primer for reverse transcription of vRNA, followed by qPCR against the reporter gene. CAD siRNA-treated cells showed a strong reduction in both vRNA and mRNA levels in comparison to the control cells, demonstrating that CAD is important for both EBOV transcription and viral genome replication (Figure 2D,E).

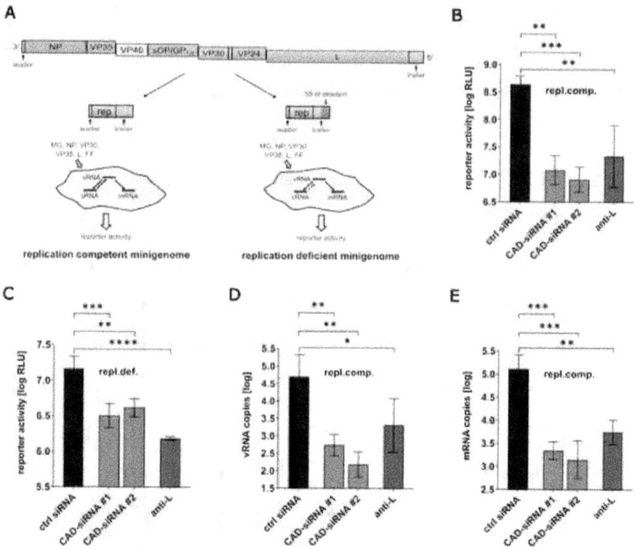

Figure 2. Influence of CAD knockdown on the Ebola virus life cycle. (A) Replication-competent and -deficient minigenome systems. The full-length genome structure of EBOV, as well as replication-competent and -deficient minigenomes derived from this full-length genome, are shown. Abbrevations: MG: minigenome, rep: reporter; FF: Firefly luciferase. Figure modified from [35] under CC BY 4.0 license. (B) Influence of CAD knockdown on EBOV RNA synthesis. 293T cells were transfected with siRNAs targeting either CAD (CAD-siRNA), EBOV-L (anti-L), or a negative control (ctrl siRNA). 48 h post-transfection, cells were transfected with all the components required for a replication-competent minigenome assay (repl.comp.). Another 48 h later, cells were harvested and the reporter activity was measured. (C) Analysis of CAD knockdown on EBOV transcription and gene expression. 293T cells were transfected with siRNAs targeting either CAD (CAD-siRNA), EBOV-L (anti-L), or a negative control (ctrl siRNA). 48 h post-transfection, cells were transfected with all the components required for a replication-deficient minigenome assay (repl. def.). Another 48 h later, cells were harvested and the reporter activity was measured. (D) Impact of CAD knockdown on EBOV replication. Cells were treated as described in 2B. After cell harvesting, RNA was extracted from the cell lysates and RT-qPCR for vRNA was performed. (E) Influence of CAD knockdown on EBOV mRNA levels. Cells were treated as described in 2B. After cell harvesting, RNA was extracted from cell lysates and RT-qPCR for mRNA was performed. The means and standard deviations of 3 independent experiments are shown for each panel. Asterisks indicate p-values from a one-way ANOVA (* $p \leq 0.05$; ** $p \leq 0.01$; *** $p \leq 0.001$; **** $p \leq 0.0001$; ns: $p > 0.05$).

3.2. The Effect of CAD Knockdown Can Be Compensated for by Exogenous Pyrimidines

As CAD is an important component for pyrimidine synthesis [23], we wanted to investigate the effect of providing exogenous pyrimidines on EBOV transcription and replication during siRNA knockdown of CAD. To this end, we performed an siRNA-mediated knockdown of CAD with EBOV minigenomes and treated the cells with 1 mM of either uridine or cytidine. Complementation of uridine resulted in reporter activities similar to the positive controls, indicating that the effect of CAD knockdown on EBOV genome replication and transcription is due to a lack of pyrimidines (Figure 3). When providing cytidine, a similar rescue effect was seen, albeit less pronounced, possibly because cytidine is not metabolized into uridine, whereas exogenous uridine can be metabolized into cytidine during natural pyrimidine synthesis.

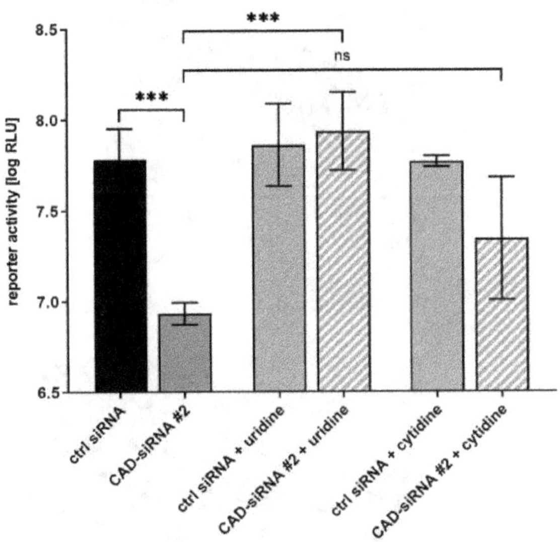

Figure 3. Supplementation of pyrimidines compensates for the effect of CAD knockdown. 293T cells were transfected with siRNAs targeting CAD (CAD-siRNA) or a negative control (ctrl siRNA). 48 h post-transfection, the cells were transfected with all the components required for a replication-competent minigenome assay and treated with 1 mM pyrimidines, either uridine or cytidine. Another 48 h later, the cells were harvested and the reporter activity was measured. The means and standard deviations of 3 independent experiments are shown. Asterisks indicate p-values from a one-way ANOVA (* $0.0001 < p \leq 0.001$; ns: $p > 0.05$).**

3.3. CAD Colocalizes with NP-Induced Inclusion Bodies

Similar to other negative-sense RNA viruses, EBOV and in particular its nucleoprotein NP is known to induce the formation of cytoplasmic inclusion bodies, which are sites of viral genome replication and transcription [8,9]. Since we had shown that CAD is important for EBOV replication and transcription, we wanted to investigate whether the presence of inclusion bodies has an influence on the intracellular distribution of CAD, and in particular whether recruitment of CAD into NP-induced inclusion bodies can be detected. As previously reported, expression of only NP resulted in the formation of inclusion bodies, predominantly in the perinuclear region [5,10,11], while sole expression of CAD led to an even distribution throughout the cytoplasm, with small amounts of CAD present in the nucleus [30] (Figure 4A).

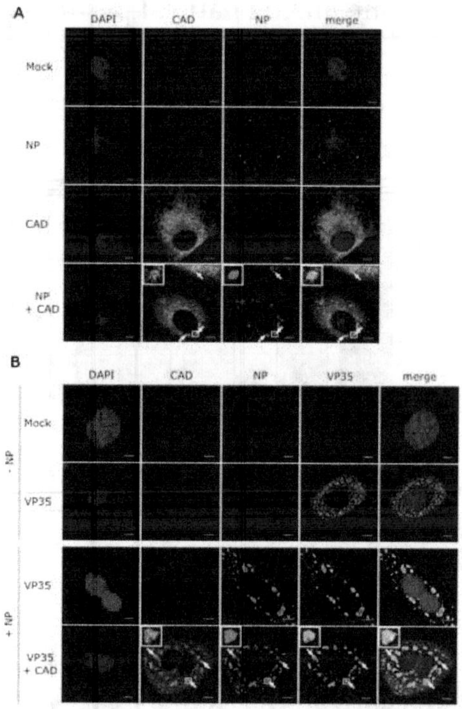

Figure 4. Recruitment of CAD into NP-induced inclusion bodies. (A) Colocalization between CAD and NP-induced inclusion bodies. Huh7 cells were transfected with plasmids encoding FLAG/HA-CAD and EBOV-NP as indicated. 48 h post-transfection, the cells were fixed with 4% paraformaldehyde and permeabilized with 0.1% Triton X-100. FLAG-tagged

CAD (shown in green) was detected using an anti-FLAG antibody and NP (shown in red) was stained with anti-EBOV NP antibodies. (B) Recruitment of CAD into inclusion bodies occurs in the presence of VP35. Huh7 cells were transfected with plasmids encoding FLAG/HA-CAD, EBOV-NP, and myc-EBOV-VP35 as indicated. 48 h post-transfection, the cells were fixed with 4% PFA and permeabilized with 0.1% Triton X-100. FLAG-tagged CAD (shown in green) was detected using an anti-FLAG antibody, NP (shown in red) was stained with anti-EBOV NP antibodies, and myc-tagged VP35 (shown in turquoise) with an anti-myc antibody. The nuclei were stained with DAPI (shown in blue), and the cells were visualized by confocal laser scanning microscopy. The scale bars indicate 10 μm. The arrows point out colocalization, and the insets show magnifications of the indicated areas. Merge shows an overlay of all three channels.

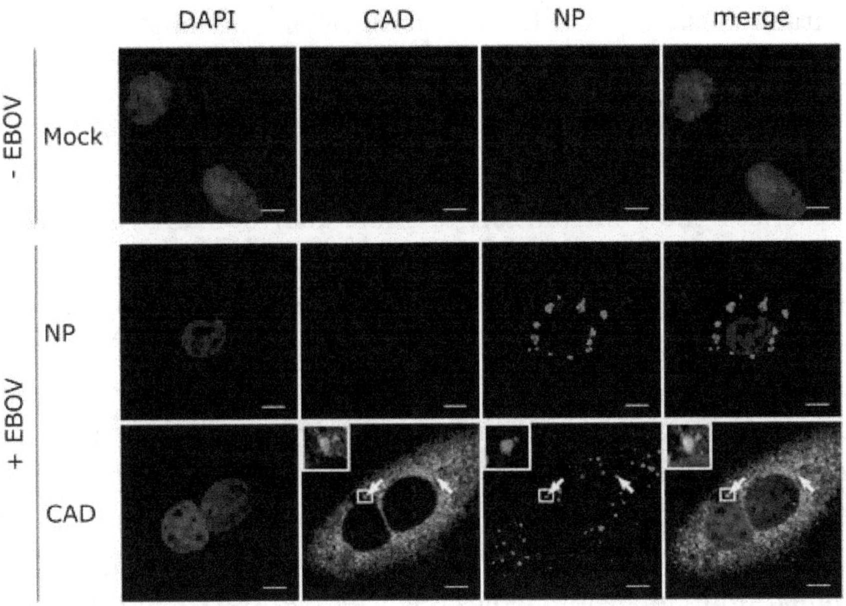

Figure 5. CAD localizes in EBOV inclusion bodies. Huh7 cells were transfected with a plasmid encoding FLAG/HA-CAD. 48 h post-transfection, the cells were infected with rgEBOV at an MOI of 1. After incubation for 16 h, the cells were fixed with 10% formalin and permeabilized with Triton X-100. CAD (shown in green) was detected with an anti-FLAG antibody and NP (shown in red) with an anti-NP antibody. The nuclei were stained with DAPI (shown in blue), and the cells were visualized by confocal laser scanning microscopy. Scale bars indicate 10 μm. The arrows point out colocalization, and the insets show magnifications of the indicated areas. Merge shows an overlay of all three channels.

During coexpression of NP and CAD we observed relocalization of CAD into NP-induced inclusion bodies (with clear accumulation in inclusion bodies in 70% of the cells, clear exclusion in 0%, and an unclear phenotype in 30%). When we additionally coexpressed VP35, which is involved in nucleocapsid formation during EBOV infection, together with NP [36], we observed a similar relocalization (Figure 4B). To confirm these results, we also performed experiments with infectious EBOV and stained the samples for NP as an inclusion body marker and CAD (Figure 5). Colocalization of CAD and inclusion bodies was still detectable, albeit not as apparent as under conditions of recombinant overexpression of NP and VP35. Taken together, these results suggest that CAD is recruited into viral inclusion bodies to provide sufficient amounts of pyrimidines for EBOV genome replication and transcription.

3.4. The GLN Domain of CAD Is Required for its Accumulation in Inclusion Bodies

To assess the contribution of individual domains of CAD in its recruitment into NP-induced inclusion bodies, we focused on the GLN and the CPS domains. When we expressed deletion mutants lacking these domains, they showed a similar intracellular distribution compared to wild-type CAD when expressed alone in cells. During coexpression of NP and CAD-ΔCPS, we observed recruitment of this mutant into NP-driven inclusion bodies (with clear accumulation in inclusion bodies in 50% of the cells, clear exclusion in 0%, and an unclear phenotype in 50%), indicating that the CPS domain of CAD is not required for its accumulation in inclusion bodies (Figure 6). In stark contrast, when NP was expressed together with CAD-ΔGLN, colocalization with inclusion bodies was abolished (with clear accumulation in inclusion bodies in 0% of the cells, clear exclusion in 68%, and unclear phenotype in 32%), suggesting that the GLN domain is required for recruitment and accumulation in NP-induced inclusion bodies.

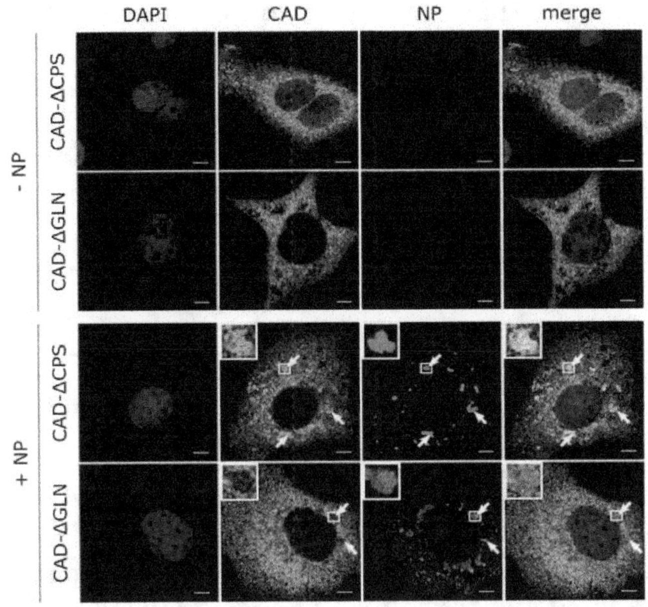

Figure 6. Recruitment of CAD deletion mutants into inclusion bodies. Huh7 cells overexpressing FLAG/HA-CAD-ΔGLN, FLAG/HA-CAD-ΔCPS and EBOV-NP, as indicated, were fixed with 4% PFA and permeabilized with 0.1% Triton X-100 48 h post-transfection. FLAG-tagged CAD (shown in green) was detected using an anti-FLAG antibody and NP (shown in red) was stained with EBOV anti-NP antibodies. The nuclei were stained with DAPI (shown in blue), and the cells were visualized by confocal laser scanning microscopy. Scale bars indicate 10 μm. The arrows point out inclusion bodies, and the insets show magnifications of the indicated areas. Merge shows an overlay of all three channels.

3.5. CAD Interacts with NP in an RNA-Independent Manner

As NP recruits CAD into EBOV inclusion bodies, we next assessed whether CAD interacts with NP. To this end, we performed coIP assays using FLAG-CAD expressed in the presence of NP by precipitating CAD with an anti-FLAG antibody and then detecting NP by Western blotting. We could readily co-precipitate NP with CAD, indicating that CAD is able to interact with NP (Figure 6). Because NP is an RNA-binding protein [37], we also tested whether this interaction between CAD and NP is RNA-dependent by treating the samples prior to coIP with RNase A. Under these conditions, we were still able to co-precipitate NP with

CAD, demonstrating that the interaction between CAD and NP is not dependent on the presence of RNA (Figure 7).

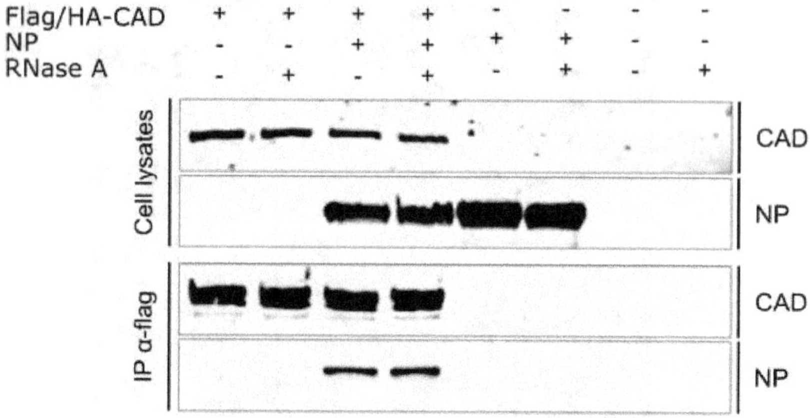

Figure 7. Interaction of CAD with NP. 293T cells were transfected with plasmids encoding FLAG/HA-CAD and EBOV-NP. 48 h post-transfection, the cells were lysed and treated with RNase A (100 μg/mL) or remained untreated. FLAG/HA-CAD was precipitated using anti-FLAG antibodies, and input and precipitates were analyzed via SDS-PAGE and Western blotting using anti-FLAG and anti-NP antibodies. In the CAD IP sample, several bands for CAD are visible, possibly due to posttranslational modifications that are not visible in the lysates because of the overall lower CAD signals in those samples.

4. DISCUSSION

In this study, we have shown that CAD, a crucial element of the de novo pyrimidine synthesis pathway, plays a significant role in both the replication and transcription of the EBOV genome. Furthermore, we have proven that the impact of CAD on pyrimidine synthesis is responsible for this effect. Additionally, the inhibition of CAD has been demonstrated to impact the reproduction and transcription processes of various other viruses, such as hepatitis C viruses. [38]. Furthermore, inhibitors of CAD, e.g., the antinucleoside N-phosphonacetyl-l-aspartate (PALA), which transiently inhibits the aspartate transcarbamylase activity of CAD, were effective in vitro against various viruses, including vaccinia virus and arenaviruses [39,40]. The fact that these compounds exhibit antiviral activity against a broad range of viruses qualifies CAD as a promising indirect antiviral target. However, whether PALA shows antiviral efficiency against EBOV remains to be investigated. Further, whether targeting viral RNA synthesis by inhibition of CAD will be synergistic with other inhibitors of EBOV RNA synthesis, such as remdesivir [41], will have to be addressed in future studies.

Our results are consistent with the fact that several pyrimidine synthesis inhibitors are effective against EBOV in vitro, underlining the importance of the pyrimidine pathway for these viruses [20,42]. Examples are the FDA-approved drug leflunomide and its active metabolite teriflunomide, as well as SW835, a racemic version of GSK983, which has been described to exhibit a broad-spectrum antiviral activity [20,42,43]. These compounds all impair de novo pyrimidine biosynthesis through inhibition of dihydroorotate dehydrogenase (DHODH), an enzyme downstream of CAD in the pyrimidine pathway. Interestingly, treatment with these inhibitors seems to have similar inhibitory effects on EBOV minigenome assays compared to the effect we observe for CAD knockdown, although CAD activity is not directly affected [20,42]. Provision of pyrimidines or upstream metabolites, e.g., orotic acid, reversed antiviral activity of all pyrimidine pathway inhibitors in EBOV minigenome assays, which is consistent with our observation that supplementation with pyrimidines restores reporter activity after CAD knockdown. Interestingly, inhibition of DHODH by using SW835 not only showed pyrimidine depletion, but also stimulated ISG (interferon-stimulated gene) expression, which contributes to the innate immune response [42]. However, currently, the mechanism behind this stimulation of the innate immune response by DHODH inhibitors remains incompletely understood and needs to be further analyzed. Further supporting the importance of CAD for the EBOV lifecycle is the fact that the de novo pyrimidine synthesis activity of CAD is a prerequisite for cell division, which has been suggested to be necessary for productive infection of cells with EBOV [44].

We were further able to show that CAD is recruited to EBOV inclusion bodies, which represent the site of EBOV replication and transcription [8,9]. Since we observed CAD recruitment into NP-induced inclusion bodies during expression of NP alone and detected an interaction of CAD with NP using CoIP studies, we suggest that this recruitment is mediated via an interaction of CAD with NP. So far, knowledge regarding direct interactions between CAD and the proteins of other viruses is limited, but Angeletti et al., showed that CAD recruits the preterminal protein (pTP) of adenoviruses to the site of adenovirus replication in the nuclear matrix via direct interaction. This interaction is believed to be required for anchorage of the adenovirus replication complex at the nuclear matrix in close proximity of the cellular factors required to segregate replicated and genomic viral DNA [45,46].

In the context of its cellular function, CAD has been shown to localize primarily in the cytoplasm, although small amounts can also be detected in the nucleus of dividing cells. Redistribution of CAD into nuclear compartments during cell growth and proliferation is believed to be in response to phosphorylation by MAP kinases at position Thr-456, which results in upregulation of the enzymatic activity of CAD [30]. Since NP is known to recruit a number of factors, including kinases and phosphatases, into inclusion bodies [16,17,18], Recruited CAD may be activated within inclusion bodies to supply pyrimidines for EBOV replication and transcription. Nevertheless, CAD without the CPS domain, which includes Thr-456, was still enlisted in NP-induced inclusion bodies, so ruling out the specific enlistment of Thr-456-phosphorylated and consequently activated CAD into inclusion bodies.

In summary, our research has demonstrated that CAD is utilized in NP-induced and virus-induced inclusion bodies to supply ample quantities of pyrimidines for EBOV genome replication and transcription. In addition, we have shown that the GLN domain of CAD is necessary for being recruited into inclusion bodies. These discoveries enhance our comprehension of EBOV and its interactions with host cells, and lay the groundwork for identifying molecular targets in the future to create new treatments against this virus.

REFERENCES

1. Feldmann, H.; Geisbert, T.W. Ebola haemorrhagic fever. Lancet 2011, 377, 849–862.

2. Burk, R.; Bollinger, L.; Johnson, J.C.; Wada, J.; Radoshitzky, S.R.; Palacios, G.; Bavari, S.; Jahrling, P.B.; Kuhn, J.H. Neglected filoviruses. FEMS Microbiol. Rev. 2016, 40, 494–519.
3. Bharat, T.A.; Noda, T.; Riches, J.D.; Kraehling, V.; Kolesnikova, L.; Becker, S.; Kawaoka, Y.; Briggs, J.A. Structural dissection of Ebola virus and its assembly determinants using cryo-electron tomography. Proc. Natl. Acad. Sci. USA 2012, 109, 4275–4280.
4. Ruigrok, R.W.; Crepin, T.; Kolakofsky, D. Nucleoproteins and nucleocapsids of negative-strand RNA viruses. Curr. Opin. Microbiol. 2011, 14, 504–510.
5. Becker, S.; Rinne, C.; Hofsass, U.; Klenk, H.D.; Muhlberger, E. Interactions of Marburg virus nucleocapsid proteins. Virology 1998, 249, 406–417.
6. Muhlberger, E.; Weik, M.; Volchkov, V.E.; Klenk, H.D.; Becker, S. Comparison of the transcription and replication strategies of marburg virus and Ebola virus by using artificial replication systems. J. Virol. 1999, 73, 2333–2342.
7. Weik, M.; Modrof, J.; Klenk, H.D.; Becker, S.; Muhlberger, E. Ebola virus VP30-mediated transcription is regulated by RNA secondary structure formation. J. Virol. 2002, 76, 8532–8539.
8. Hoenen, T.; Shabman, R.S.; Groseth, A.; Herwig, A.; Weber, M.; Schudt, G.; Dolnik, O.; Basler, C.F.; Becker, S.; Feldmann, H. Inclusion bodies are a site of ebolavirus replication. J. Virol. 2012, 86, 11779–11788.
9. Lier, C.; Becker, S.; Biedenkopf, N. Dynamic phosphorylation of Ebola virus VP30 in NP-induced inclusion bodies. Virology 2017, 512, 39–47.
10. Boehmann, Y.; Enterlein, S.; Randolf, A.; Muhlberger, E. A reconstituted replication and transcription system for Ebola virus Reston and comparison with Ebola virus Zaire. Virology 2005, 332, 406–417.

11. Groseth, A.; Charton, J.E.; Sauerborn, M.; Feldmann, F.; Jones, S.M.; Hoenen, T.; Feldmann, H. The Ebola virus ribonucleoprotein complex: A novel VP30-L interaction identified. Virus Res. 2009, 140, 8–14.
12. Fang, J.; Pietzsch, C.; Ramanathan, P.; Santos, R.I.; Ilinykh, P.A.; Garcia-Blanco, M.A.; Bukreyev, A.; Bradrick, S.S. Staufen1 Interacts with Multiple Components of the Ebola Virus Ribonucleoprotein and Enhances Viral RNA Synthesis. mBio 2018, 9.
13. Chen, J.; He, Z.; Yuan, Y.; Huang, F.; Luo, B.; Zhang, J.; Pan, T.; Zhang, H.; Zhang, J. Host factor SMYD3 is recruited by Ebola virus nucleoprotein to facilitate viral mRNA transcription. Emerg. Microbes Infect. 2019, 8, 1347–1360.
14. Batra, J.; Hultquist, J.F.; Liu, D.; Shtanko, O.; Von Dollen, J.; Satkamp, L.; Jang, G.M.; Luthra, P.; Schwarz, T.M.; Small, G.I.; et al. Protein Interaction Mapping Identifies RBBP6 as a Negative Regulator of Ebola Virus Replication. Cell 2018, 175, 1917–1930.
15. Gabriel, G.; Feldmann, F.; Reimer, R.; Thiele, S.; Fischer, M.; Hartmann, E.; Bader, M.; Ebihara, H.; Hoenen, T.; Feldmann, H. Importin-alpha7 Is Involved in the Formation of Ebola Virus Inclusion Bodies but Is Not Essential for Pathogenicity in Mice. J. Infect. Dis. 2015, 212 Suppl 2, S316–S321.
16. Morwitzer, M.J.; Tritsch, S.R.; Cazares, L.H.; Ward, M.D.; Nuss, J.E.; Bavari, S.; Reid, S.P. Identification of RUVBL1 and RUVBL2 as Novel Cellular Interactors of the Ebola Virus Nucleoprotein. Viruses 2019, 11, 372.
17. Kruse, T.; Biedenkopf, N.; Hertz, E.P.T.; Dietzel, E.; Stalmann, G.; Lopez-Mendez, B.; Davey, N.E.; Nilsson, J.; Becker, S. The Ebola Virus Nucleoprotein Recruits the Host PP2A-B56 Phosphatase to Activate Transcriptional Support Activity of VP30. Mol. Cell 2018, 69, 136–145.
18. Takamatsu, Y.; Krahling, V.; Kolesnikova, L.; Halwe, S.; Lier, C.; Baumeister, S.; Noda, T.; Biedenkopf, N.; Becker, S. Serine-Arginine

Protein Kinase 1 Regulates Ebola Virus Transcription. mBio 2020, 11.

19. Wendt, L.; Brandt, J.; Bodmer, B.S.; Reiche, S.; Schmidt, M.L.; Traeger, S.; Hoenen, T. The Ebola Virus Nucleoprotein Recruits the Nuclear RNA Export Factor NXF1 into Inclusion Bodies to Facilitate Viral Protein Expression. Cells 2020, 9, 187.
20. Martin, S.; Chiramel, A.I.; Schmidt, M.L.; Chen, Y.C.; Whitt, N.; Watt, A.; Dunham, E.C.; Shifflett, K.; Traeger, S.; Leske, A.; et al. A genome-wide siRNA screen identifies a druggable host pathway essential for the Ebola virus life cycle. Genome Med. 2018, 10, 58.
21. Uebelhoer, L.S.; Albarino, C.G.; McMullan, L.K.; Chakrabarti, A.K.; Vincent, J.P.; Nichol, S.T.; Towner, J.S. High-throughput, luciferase-based reverse genetics systems for identifying inhibitors of Marburg and Ebola viruses. Antiviral Res. 2014, 106, 86–94.
22. Nelson, E.V.; Pacheco, J.R.; Hume, A.J.; Cressey, T.N.; Deflube, L.R.; Ruedas, J.B.; Connor, J.H.; Ebihara, H.; Muhlberger, E. An RNA polymerase II-driven Ebola virus minigenome system as an advanced tool for antiviral drug screening. Antiviral Res. 2017, 146, 21–27.
23. Coleman, P.F.; Suttle, D.P.; Stark, G.R. Purification from hamster cells of the multifunctional protein that initiates de novo synthesis of pyrimidine nucleotides. J. Biol. Chem. 1977, 252, 6379–6385.
24. Jones, M.E. Pyrimidine nucleotide biosynthesis in animals: Genes, enzymes, and regulation of UMP biosynthesis. Annu. Rev. Biochem. 1980, 49, 253–279.
25. Lee, L.; Kelly, R.E.; Pastra-Landis, S.C.; Evans, D.R. Oligomeric structure of the multifunctional protein CAD that initiates pyrimidine biosynthesis in mammalian cells. Proc. Natl. Acad. Sci. USA 1985, 82, 6802–6806.
26. Christopherson, R.I.; Jones, M.E. The overall synthesis of L-5,6-dihydroorotate by multienzymatic protein pyr1-3 from hamster cells. Kinetic studies, substrate channeling, and the effects of inhibitors. J. Biol. Chem. 1980, 255, 11381–11395.

27. Irvine, H.S.; Shaw, S.M.; Paton, A.; Carrey, E.A. A reciprocal allosteric mechanism for efficient transfer of labile intermediates between active sites in CAD, the mammalian pyrimidine-biosynthetic multienzyme polypeptide. Eur. J. Biochem. 1997, 247, 1063–1073.

28. Evans, D.R.; Guy, H.I. Mammalian pyrimidine biosynthesis: Fresh insights into an ancient pathway. J. Biol. Chem. 2004, 279, 33035–33038.

29. Sigoillot, F.D.; Berkowski, J.A.; Sigoillot, S.M.; Kotsis, D.H.; Guy, H.I. Cell cycle-dependent regulation of pyrimidine biosynthesis. J. Biol. Chem. 2003, 278, 3403–3409.

30. Sigoillot, F.D.; Kotsis, D.H.; Serre, V.; Sigoillot, S.M.; Evans, D.R.; Guy, H.I. Nuclear localization and mitogen-activated protein kinase phosphorylation of the multifunctional protein CAD. J. Biol. Chem. 2005, 280, 25611–25620.

31. Chaparian, M.G.; Evans, D.R. Intracellular location of the multidomain protein CAD in mammalian cells. FASEB J. 1988, 2, 2982–2989.

32. Hoenen, T.; Jung, S.; Herwig, A.; Groseth, A.; Becker, S. Both matrix proteins of Ebola virus contribute to the regulation of viral genome replication and transcription. Virology 2010, 403, 56–66.

33. Shabman, R.S.; Hoenen, T.; Groseth, A.; Jabado, O.; Binning, J.M.; Amarasinghe, G.K.; Feldmann, H.; Basler, C.F. An upstream open reading frame modulates ebola virus polymerase translation and virus replication. PLoS Pathog. 2013, 9, e1003147.

34. Kamper, L.; Zierke, L.; Schmidt, M.L.; Muller, A.; Wendt, L.; Brandt, J.; Hartmann, E.; Braun, S.; Holzerland, J.; Groseth, A.; et al. Assessment of the function and intergenus-compatibility of Ebola and Lloviu virus proteins. J. Gen. Virol. 2019, 100, 760–772.

35. Schmidt, M.L.; Hoenen, T. Characterization of the catalytic center of the Ebola virus L polymerase. PLoS Negl. Trop. Dis. 2017, 11, e0005996.

36. Huang, Y.; Xu, L.; Sun, Y.; Nabel, G.J. The assembly of Ebola virus nucleocapsid requires virion-associated proteins 35 and 24 and posttranslational modification of nucleoprotein. Mol. Cell. 2002, 10, 307–316.

37. Noda, T.; Hagiwara, K.; Sagara, H.; Kawaoka, Y. Characterization of the Ebola virus nucleoprotein-RNA complex. J. Gen. Virol. 2010, 91, 1478–1483.

38. Borawski, J.; Troke, P.; Puyang, X.; Gibaja, V.; Zhao, S.; Mickanin, C.; Leighton-Davies, J.; Wilson, C.J.; Myer, V.; Cornellataracido, I.; et al. Class III phosphatidylinositol 4-kinase alpha and beta are novel host factor regulators of hepatitis C virus replication. J. Virol. 2009, 83, 10058–10074.

39. Ortiz-Riano, E.; Ngo, N.; Devito, S.; Eggink, D.; Munger, J.; Shaw, M.L.; de la Torre, J.C.; Martinez-Sobrido, L. Inhibition of arenavirus by A3, a pyrimidine biosynthesis inhibitor. J. Virol. 2014, 88, 878–889.

40. Katsafanas, G.C.; Grem, J.L.; Blough, H.A.; Moss, B. Inhibition of vaccinia virus replication by N-(phosphonoacetyl)-L-aspartate: Differential effects on viral gene expression result from a reduced pyrimidine nucleotide pool. Virology 1997, 236, 177–187.

41. Warren, T.K.; Jordan, R.; Lo, M.K.; Ray, A.S.; Mackman, R.L.; Soloveva, V.; Siegel, D.; Perron, M.; Bannister, R.; Hui, H.C.; et al. Therapeutic efficacy of the small molecule GS-5734 against Ebola virus in rhesus monkeys. Nature 2016, 531, 381–385.

42. Luthra, P.; Naidoo, J.; Pietzsch, C.A.; De, S.; Khadka, S.; Anantpadma, M.; Williams, C.G.; Edwards, M.R.; Davey, R.A.; Bukreyev, A.; et al. Inhibiting pyrimidine biosynthesis impairs Ebola virus replication through depletion of nucleoside pools and activation of innate immune responses. Antiviral Res. 2018, 158, 288–302.

43. Deans, R.M.; Morgens, D.W.; Okesli, A.; Pillay, S.; Horlbeck, M.A.; Kampmann, M.; Gilbert, L.A.; Li, A.; Mateo, R.; Smith, M.; et al. Parallel shRNA and CRISPR-Cas9 screens enable antiviral drug target identification. Nat. Chem. Biol. 2016, 12, 361–366.

44. Kota, K.P.; Benko, J.G.; Mudhasani, R.; Retterer, C.; Tran, J.P.; Bavari, S.; Panchal, R.G. High content image based analysis identifies cell cycle inhibitors as regulators of Ebola virus infection. Viruses 2012, 4, 1865–1877.

45. Angeletti, P.C.; Engler, J.A. Adenovirus preterminal protein binds to the CAD enzyme at active sites of viral DNA replication on the nuclear matrix. J. Virol. 1998, 72, 2896–2904.

46. Fredman, J.N.; Engler, J.A. Adenovirus precursor to terminal protein interacts with the nuclear matrix in vivo and in vitro. J. Virol. 1993, 67, 3384–3395.

07 THE EBOLA VIRUS NUCLEOPROTEIN RECRUITS THE NUCLEAR RNA EXPORT FACTOR NXF1 INTO INCLUSION BODIES TO ENHANCE VIRAL PROTEIN EXPRESSION

1. INTRODUCTION: AN OVERVIEW

The Ebola virus (EBOV) is a member of the Ebolavirus genus in the Filoviridae family. It is responsible for causing a very lethal hemorrhagic fever known as Ebola virus illness in humans, with case fatality rates ranging from 40% to 60% [1,2]. Ongoing and past outbreaks of Ebola virus disease in Africa highlight the importance of a better understanding of the EBOV life cycle in order to develop new therapeutic approaches. During the viral life cycle the EBOV nucleoprotein (NP) encapsidates the negative-stranded RNA genome and is essential for viral replication and transcription [3]. NP interacts with the transcriptional activator viral protein 30 (VP30), which bridges NP and the RNA-dependent RNA polymerase L [4,5,6]. Furthermore, NP interacts with the polymerase cofactor VP35 [5,6]. This interaction regulates the oligomerization and RNA-binding of NP, and also bridges NP to L [5,6,7,8,9]. NP, VP35, VP30, and L, together with the RNA genome, form the ribonucleoprotein complex (RNP) and are sufficient to mediate viral replication and transcription [3], which takes place in cytoplasmic inclusion bodies [10]. The formation of these inclusion bodies is driven by expression of NP, which is localized in these structures not only during infection, but also after sole expression of this protein [5,6]. However, only limited knowledge exists regarding host factors that interact with the viral proteins and RNAs found in these structures. One such host factor that has been identified is importin-α7, which seems to be involved in inclusion body formation [11]. Marburg virus, a close relative of EBOV, was shown to recruit components of the endosomal sorting complex required for transport (ESCRT) to inclusion bodies to facilitate the trafficking of nucleocapsids to the plasma membrane for viral assembly and budding [12,13]. Kinases and phosphatases such as PP2A-B56 are also known to be recruited to inclusion bodies, and are important in regulating the activity of VP30 in viral RNA synthesis, which is dependent on its phosphorylation status [14,15]. Similarly, RBBP6 appears to regulate the balance of replication and transcription by binding to VP30, and also Staufen1 was described to influence viral RNA synthesis [16,17]. Finally, EBOV VP35 appears to sequester cellular stress granule proteins within inclusion bodies in order to prevent stress granule formation [18].

To obtain a comprehensive picture of the pro- and anti-viral factors that are important for EBOV RNA synthesis (i.e., genome replication and transcription) and/or protein expression, we recently performed a genome-wide siRNA screen [19]. As primary readout we used a minigenome assay (reviewed in [20]). In this assay RNA minigenomes, i.e., miniature versions of the EBOV genome with all viral genes removed and replaced with a reporter gene, are expressed in mammalian cells together with the viral RNP proteins. Since the minigenomes still contain the regulatory terminal leader and trailer regions of the EBOV genome that carry the replication and transcription promoters, the RNP proteins recognize these minigenomes as authentic templates. This results in their replication and transcription, and ultimately reporter protein expression reflecting these aspects of the viral life cycle. In our siRNA screen we showed that the nuclear RNA export factor 1 (NXF1) is necessary for the EBOV life cycle, and also for the life cycle of the highly pathogenic New World arenavirus Junín virus. These data suggest a mechanism of action that may be conserved among several cytoplasmically replicating negative-stranded RNA viruses (NSVs) that share commonalties in their replication cycles, such as replication in cytoplasmic inclusion bodies [10,21,22,23,24]. Importantly, we also could show that NXF1 is important for the life cycle of EBOV in context of infectious virus [19]. However, the precise function of NXF1 in the EBOV life cycle remained unclear.

NXF1 is a crucial component of the nuclear mRNA export pathway, where it exports cellular mRNAs from the nucleus by interacting with nucleoporins [25,26]. For this interaction NXF1 forms a stable heterodimer with p15 (also called NXT1) via its NTF2-like domain (NTF2) [27,28]. The NTF2-like domain is one of the five functional domains found in NXF1, with the other ones being the RNA-binding domain (RBD), the pseudo RNA recognition motif (RRM), a region of leucine-rich repeats (LRR), and the ubiquitin-associated domain (UBA) [27,29,30]. NTF2 and UBA promote interaction with nucleoporins, while the three amino-terminal domains (RBD, RRM, and LRR) are important for mRNA binding [26,27,31]. Recently, NXF1 was shown to be loaded onto cellular mRNAs co-transcriptionally, a process that depends on several other export adapters that promote the interaction between NXF1 and mRNA, in order to enable mRNA export from the nucleus [29,31,32]. However, the NXF1-driven nuclear mRNA export pathway is also utilized by many viruses that replicate in the nucleus [33,34,35,36]. For instance, herpesviruses and influenza viruses usurp this host pathway to export their viral RNAs, and particularly their mRNAs, from the nucleus [33,34,37,38,39]. Furthermore, NXF1 is known to export retroviral RNAs containing constitutive transport elements and Hepatitis B virus pregenomic RNA [35,36]. So far, the sole known role of NXF1, apart from its involvement in nuclear functions, is its function in gammaretroviruses. In these viruses, NXF1 is responsible for loading gag RNA into polysomes, rather than exporting mRNA [40]. However, due to the replication of EBOV in cytoplasmic inclusion bodies, the role of NXF1 during an EBOV infection must be different from what has been previously reported [10]. Thus, we thoroughly examined the biochemical and functional aspects of the interaction between NXF1 and EBOV. Our findings suggest a paradigm in which NXF1 is brought to inclusion bodies, where it interacts with NP and later with viral mRNAs, facilitating their exportation from the inclusion bodies.

2. MATERIALS AND METHODS

2.1. Cells

HEK 293T (human embryonic kidney) cells (Collection of Cell Lines in Veterinary Medicine CCLV-RIE 1018) and Huh7 (human hepatocarcinoma) cells (kindly provided by Stephan Becker, Philipps University Marburg, Marburg, Germany) were maintained in Dulbecco's modified Eagle's medium (DMEM; ThermoFisher Scientific, Waltham, MA, USA) supplemented with 10% fetal bovine serum (FBS), 100 U/mL penicillin and 100 μg/mL streptomycin (PS; ThermoFisher Scientific) and 1× GlutaMAX (ThermoFisher Scientific). All cells were grown at 37 °C with 5% CO2.

2.2. Plasmids and Antibodies

Expression plasmids for the RNP proteins, the T7 polymerase and the T7-driven monocistronic minigenomes (pT7-1cis-vRNA-rLuc, pT7-1cis-vRNA-nLuc) have been previously described [19,41]. Plasmids for the expression of N-terminally flag/HA-tagged constructs were generated by insertion of the respective genes (NXF1, p15, NP) into a pCAGGS-flag/HA vector. All mutant versions and individual domains of NXF1 were first inserted into pCAGGS and subsequently subcloned into the pCAGGS-flag/HA plasmid. EBOV VP40 and VP24 were also cloned into pCAGGS. The PolII-driven replication-deficient minigenome expressing NanoLuc luciferase as its reporter was generated by flanking the minigenome in vRNA-orientation with hammerhead and Hepatitis Delta Virus ribozymes and insertion of the minigenome into pCAGGS. Deletion of 55 nt in the antigenomic replication promotor to generate a replication-deficient version of the PolII-driven minigenome was performed as previously described [41]. Detailed cloning strategies are available on request.

The anti-flag (mouse anti-flag mAb, clone M2, cat. no. F1804) antibody used for co-immunoprecipitation assays (coIPs) and Western blot analyses was purchased from Sigma-Aldrich. Primary antibodies against NP (rabbit anti-EBOV NP pAb, cat. no. 0301-012), VP35 (rabbit anti-ZEBOV VP35 affinity purified polyclonal antibody, cat. no. 0301-

040) and VP40 (mouse anti-EBOV VP40 mAb, clone 3G5, cat. no. 0201-016) were obtained from IBT Bioservices, and the antibody against NXF1 (mouse anti-NXF1 mAb, clone 53H8, cat. no. ab50609) was obtained from Abcam. For VP30, a mouse monoclonal antibody (clone 5F11 A1) was produced by the FLI biobank as previously described [42] and following approved protocols using affinity-purified flag/HA-tagged VP30 expressed in mammalian cells as the immunogen. Secondary antibodies against mouse (Peroxidase AffiniPure Goat Anti-Mouse IgG, light chain specific, cat. no. 115-035-174, used for Western blots with exception of that shown in Figure 4) and rabbit (Peroxidase AffiniPure Goat Anti-Rabbit IgG H+L, cat. no. 111-035-003) were obtained from Jackson Immunoresearch or Abcam (anti-mouse IgG for IP, HRP, cat. no. ab131368, used for the Western blot shown in Figure 4).

2.3. Co-Immunoprecipitation of Viral Proteins

For coIPs, 293T cells were seeded into 6-well plates and transfected at 30% confluency with expression plasmids encoding the flag/HA-tagged host proteins (NXF1, p15, mutated versions or individual domains of NXF1) and for their putative interaction partners (p15, NXF1 or EBOV proteins) using Transit LT-1 (Mirus Bio LLC, Madison, WI, USA) following the manufacturer's instructions. Supernatant was exchanged after 24 h and coIP was performed 48 h post transfection (p.t.). To this end, cells were washed in PBS and pelleted before being lysed in 1 mL coIP lysis buffer (1% NP-40; 50 mM Tris; 150 mM NaCl; pH 7.4) with protease inhibitor (cOmplete; Roche, Basel, Switzerland) and incubated with rotation at 15 rpm for 2 h at 4 °C. Initial experiments investigating an interaction between viral proteins and NXF1 (Figure 1A) where performed in the absence of RNase treatment. However, since RNase treatment increases the interaction between NXF1 and NP (see Figure 2), in all further experiments 100 µg/mL RNase A (Machery-Nagel, Düren, Germany) was added to the samples prior to incubation. Lysates were separated from cell debris by centrifugation (11,000× g, 10 min, 4 °C) and 150 µL was used for the input control (representing a sixth of the whole lysate and 20% of the sample subjected to IP) and subjected to acetone precipitation. To this end 750 µL acetone pre-chilled to −20 °C was added to the samples, and they were incubated for at least 30 min at -20 °C. Proteins were then pelleted by centrifugation (20,000× g, 15 min, 4 °C), resuspended in 60 µL 1× SDS sample buffer and boiled for 10 min at 99 °C.

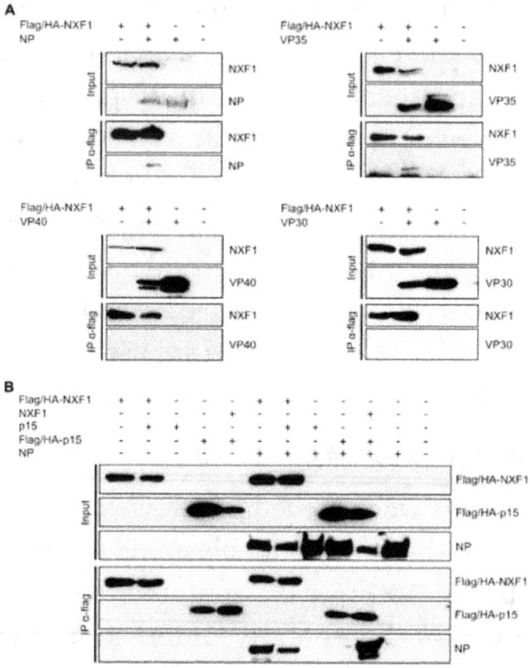

Figure 1. Interaction of NXF1 with nucleoprotein (NP) and VP35. (A) Immunoprecipitation of NXF1 and coIP of viral proteins. 293T cells were transfected with the plasmids encoding for flag/HA-NXF1 and one of the viral proteins NP, VP35, VP40 or VP30. Forty-eight hours p.t. flag/ HA-NXF1 was precipitated using anti-flag antibodies, and input and precipitates were analyzed via SDS-PAGE and subsequent Western blot. NXF1 was detected with anti-flag antibodies, and anti-NP, anti-VP35, anti-VP40 or anti-VP30 antibodies were used for the detection of the respective viral protein. In case of the VP35 blot, the band at the lower edge of the blot that is partially cut off is caused by the light chain of the IP antibody. (B) Influence of p15 on the interaction between NXF1 and NP. Flag/HA-tagged NXF1 or p15 was precipitated from transfected 293T cells using anti-flag antibodies. Input and precipitate samples were subjected to SDS-PAGE and analyzed by Western blot using anti-flag antibodies for the detection of p15 and NXF1, whereas an anti-NP antibody was used for the detection of NP.

The remaining 750 µL of lysate was added to the prepared bead-antibody solution (Dynabeads Protein G, ThermoFisher Scientific;

1 μL anti-flag M2 antibody per 10 μL beads), resuspended, and the IP was performed for 10 min (as recommended by the manufacturer to minimize non-specific binding) at room temperature with rotation at 15 rpm. Beads were washed three times with PBS containing 0.02% Tween-20 and transferred to new tubes. The beads were resuspended in 60 μL 1× SDS sample buffer and boiled for 10 min at 99 °C. Input and coIP samples were analyzed via SDS-PAGE and semi-dry Western blot, using an ethanol-containing blotting buffer (5.8 g Tris(hydroxymethyl) aminomethan; 2.9 g Glycin; 200 mL Ethanol; ad 1 mL H2O) together with Nitrocellulose membranes and a blotting current of 0.8 mA/cm2 for 1 to 2 h, and 7% skim milk powder in PBS for blocking.

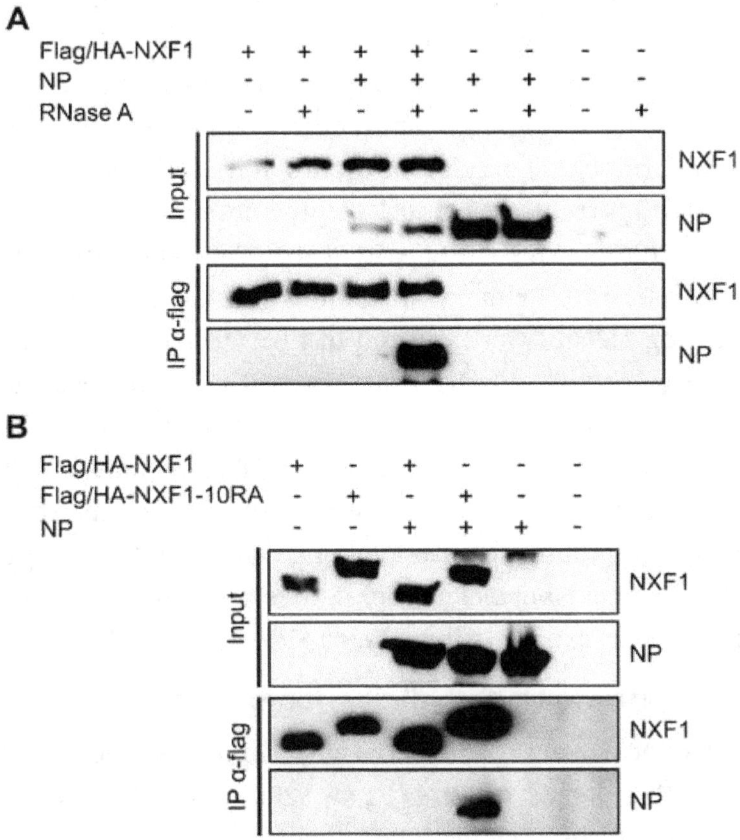

Figure 2. RNA dependence of the interaction between NXF1 and NP. (A) Absence of single-stranded RNA enhances interaction between NP and NXF1. 293T cell lysates containing overexpressed flag/HA-NXF1 and NP

were treated with RNase A (100 µg/mL) or remained untreated before flag/HA-NXF1 was precipitated using anti-flag antibodies. Input and precipitates were analyzed via SDS-PAGE and Western blot using anti-flag and anti-NP antibodies. (B) RNA binding-deficient NXF1-10RA displays a stronger interaction with NP than wildtype NXF1. 293T cells were transfected with plasmids encoding for either flag/HA-NXF1 or flag/HA-NXF1-10RA together with NP, and 48 h p.t. cells were lysed and subjected to coIP analyses. Pulldowns were performed using anti-flag antibodies, and precipitates as well as input samples were analyzed via SDS-PAGE and Western blot using anti-flag and anti-NP antibodies.

2.4. Immunofluorescence Analysis

Huh7 cells were seeded onto coverslips in 12-well plates and transfected the next day with 500 ng pCAGGS-EBOV-NP, 500 ng pCAGGS-p15 and 500 ng pCAGGS-flag/HA-NXF1 (or pCAGGS-flag/HA-NXF1-10RA or pCAGGS-flag/HA-NXF1-ΔRBD) using Polyethylenimine (Sigma-Aldrich) following the manufacturer's instructions, or remained untransfected (mock). Forty-eight hours p.t., cells were fixed using 4% PFA in DMEM, treated with 0.1 M glycine and permeabilized with 0.1% Triton X-100 (in PBS++ (PBS with 1 mM Ca2+ and 1 mM Mg2+) with 10% fetal calf serum). Primary antibodies (rabbit anti-EBOV-NP 1:500; mouse anti-flag 1:2000) were diluted in PBS with 10% fetal calf serum, and cells were incubated for 1 h at 4 °C with the antibody solutions. The same procedure was used for the secondary antibodies (Alexa Fluor-568 anti-rabbit 1:1500, Alexa Fluor-488 anti-mouse 1:1200; both ThermoFisher Scientific). After staining, cells were washed with PBS and water before mounting with ProLong Diamond Antifade mountant with DAPI (ThermoFisher Scientific). Slides were analyzed by confocal laser scanning microscopy using a Leica SP5.

2.5. siRNA Knockdown and Minigenome Assay

293T cells were reverse transfected with 12 pmol pre-designed silencer select siRNAs (NXF1 #1: s20532; NXF1 #2: s20533; Negative Control siRNA No. 2; all ThermoFisher Scientific) or an anti-L siRNA (5'-UUU AUA UAC AGC UUC GUA CUU-3') using Lipofectamine RNAiMax (ThermoFisher Scientific) following the manufacturer's instructions in 12-well plates. Forty-eight hours p.t., cells were transfected with all plasmids for the respective minigenome assay using Transit LT-1 as described before [41]. For the classical T7-driven replication-competent minigenome system (including the analysis of mRNA levels), the cells were transfected with the plasmids encoding the T7-polymerase, the T7-driven monocistronic minigenome (1cis-vRNA-nLuc), the RNP proteins L, NP, VP30 and VP35 and firefly luciferase or green fluorescent protein (GFP) as a control. For the replication-deficient minigenome system, cells were transfected with the plasmids encoding the PolII-driven replication-deficient minigenome (pCAGGS-1cis-vRNA-nLuc-RdM), the RNP proteins L, NP, VP30 and VP35 and firefly luciferase as a control. A further 48 h later cells were harvested for either determination of reporter activity or RNA isolation (see below). For measuring the reporter activity, cells were lysed for 10 min in 1× Lysis Juice (PJK, Kleinblittersdorf, Germany) and lysates were cleared by centrifugation (3 min at 10,000× g, room temperature). Forty µL of cleared lysate was added to either 40 µL of Beetle Juice (PJK) or Nano-Glo Luciferase Assay Reagent (Promega, Madison, WI, USA) in opaque 96-well plates. Luminescence was measured using a Glomax Multi (Promega) microplate reader. Minigenome reporter activity was normalized to control firefly luciferase activity, which acts as a measure for plasmid-driven gene expression.

2.6. Infection Experiments

siRNA-transfected 293T cells in 12-well format were infected with 5000 TCID50 rgEBOV-luc2 [43] 24–48 h p.t. Twenty-four hours post infection (p.i.) cells were lysed in 300 µL GloLysis buffer (Promega) for 10 min and 50 µL samples were mixed with either 50 µL BrightGlo (Promega) or 50 µL CellTiterGlo (Promega) in white opaque 96-well plates and measured on a Tecan F200Pro. Firefly luciferase activity reflecting viral gene expression was normalized to luciferase activity in the CellTiterGlo assay (which quenches firefly luciferase activity and instead measures intracellular ATP levels as proxy for cell viability using an UltraGlo luciferase). For mRNA quantification, identically transfected and infected cells were harvested in 700 µL RLT buffer (Qiagen, Hilden, Germany), and after addition of 700 µL 70% EtOH and removal from the BSL4-laboratory RNA was extracted using the RNEasy kit (Qiagen) following the manufacturer's instructions. mRNA quantification was performed as described below.

For immunofluorescence analysis, Huh7 cells in ibidi 8-well chamber slides were infected with rgEBOV-wt [44] at an MOI of 1. Sixteen hours p.i., cells were washed with PBS and fixed using 10% formalin. The subsequent procedure for immunofluorescence staining was performed as described above. All experiments involving infectious EBOV were done in the BSL4 laboratory of the Friedrich-Loeffler-Institut following approved SOPs.

2.7. RNA Co-Immunoprecipitation

For mRNA-coIPs 293T cells were transfected using Transit LT-1 (Mirus Bio LLC) following the manufacturer's instructions with the plasmids encoding for flag/HA-NXF1 or its mutants (1500 ng; amounts per well), GFP (250 ng) as a transfection control, as well as all T7 monocistronic minigenome components (T7—250 ng, L—1000 ng, NP—125 ng, VP30—65 ng, VP35—125 ng, monocistronic EBOV minigenome encoding Renilla luciferase—250 ng), while as a control either L (-L) or flag/HA-NXF1 (-NXF1) was omitted. For vRNA-coIPs 293T cells were transfected with the plasmids encoding for the T7 polymerase (250 ng) and the T7-driven monocistronic minigenome encoding Renilla luciferase (250 ng) together with either flag/HA-NXF1 (1500 ng) or a combination of flag/HA-NP (1500 ng), VP35 (1250 ng) and VP24 (1250 ng). As controls, either the minigenome (-1cis) or the flag/HA-tagged protein (1cis w/o flag-protein) was omitted. The coIP procedure was then carried out 48 h p.t. as described above for the protein coIPs, but the beads were resuspended in 140 μL PBS instead of 1× SDS sample buffer and not boiled before they were subjected to RNA isolation and RT-qPCR.

2.8. RNA Isolation and RT-qPCR

RNA isolation from minigenome-transfected cell lysates was performed using the NucleoSpin RNA kit (Machery-Nagel), and RNA isolation from coIP precipitates was performed using the QIAmp Viral RNA Mini kit (Qiagen), both according to the manufacturers' instructions. The beads were separated from the precipitate-buffer solution with the use of a magnetic rack prior to loading of the spin columns to avoid clogging of the columns. All RNA samples were treated with DNase (TURBO DNA-free kit; ThermoFisher Scientific) following the manufacturer's instructions to avoid plasmid contamination. The resulting RNA samples were then used for the generation of cDNA with either oligo(dT)-primers for total mRNA or strand-specific primers for the reporter gene in vRNA-orientation (Renilla: luc(-) 5'-CCA AAC AAG CAC CCC AAT CA-3') using RevertAid Reverse Transcriptase (ThermoFisher Scientific) following the manufacturer's instructions. The subsequent qPCR was performed using the PowerUp SYBR Green Master Mix (ThermoFisher Scientific) with 1 µL of cDNA and primers targeting either the reporter genes (NanoLuc: nluc(-) 5'-TTC AGA ATC TCG GG GTG TCC-3', nluc(+): 5'-CGT AAC CCC GTC GAT TAC CA-3'; Renilla: luc(-), luc(+) 5'-TGT GCC ACA TAT TGA GCC AG-3'; firefly: luc2(-) 5'-CAG TCG TCG TGC TGG AAC AC-3', luc2(+) 5'-GTC CAA CTT GCC GGT CAG TC-3'), or as a control for the minigenome experiments GFP (GFP(+) 5'-CTT GTA CAG CTC GTC CAT GC-3', GFP(-) 5'-CGA CAA CCA CTA CCT GAG CAC-3') and for the infection experiments GAPDH (GAPDH(-) 5'-CCA GGT GGT CTC CTC TGA CTT CAA-3', GAPDH(+) 5'-ATA CCA GGA AAT GAG CTT GAC A-3'). Values for luciferase reporter mRNA levels were normalized to GFP/GAPDH mRNA levels and values for the mRNA coIP samples were normalized to their respective -L controls.

3. RESULTS & DISCUSSION

3.1. NXF1 Interacts with NP and VP35

To assess whether NXF1 interacts with EBOV proteins, we performed co-immunoprecipitation (coIP) assays using overexpressed flag/HA-NXF1 in the presence of either NP, VP35, VP30, or the EBOV matrix protein VP40, to find a possible interaction partner (Figure 1A). Using this approach, we could co-precipitate NP as well as VP35 with NXF1 (Figure 1A, second lanes), indicating that NXF1 interacts with both viral proteins. In contrast, VP40 and VP30 could not be co-precipitated with NXF1, indicating that most likely there is no interaction between NXF1 and these viral factors. Since NXF1 is known to form stable heterodimers with p15 in order to promote the interaction of NXF1 with nucleoporins and thereby fulfill its cellular function in nuclear export, we also performed coIPs of NP with NXF1 in presence and absence of overexpressed p15 [25,28]. The amount of NP co-precipitated with NXF1 was not increased in presence of p15 (Figure 1B, sixth and seventh lane). Further, p15 alone failed to co-precipitate NP, while co-precipitation could be recovered by NXF1 overexpression (Figure 1B, ninth and 10th lane). These data show that p15 is dispensable for the interaction between NXF1 and NP, and that the formation of the NXF1:p15 heterodimer is not necessary to mediate an interaction with NP. This is in contrast to other viruses interacting with NXF1 such as Hepatitis B virus, where HBc was shown to interact with both NXF1 and p15 [35].

3.2. NP and Cellular RNAs Compete for Interaction with NXF1

Since NP, VP35, and NXF1 are all known to be RNA-binding proteins, we next analyzed the dependence of their interaction on RNA by treating samples with RNase A to digest single-stranded RNAs before they were used for coIPs [25,45,46]. In case of VP35 the interaction with NXF1 was clearly dependent on the presence of RNA, raising the possibility that the observed interaction is merely due to unspecific binding of the same RNA, so that we did not further investigate this interaction, but rather focused on the interaction between NP and NXF1. Here, to our surprise, the experiments revealed that the interaction between NP and NXF1 was stronger in the presence of RNase A (Figure 2A, third and fourth lane), indicating a negative impact of single-stranded RNA on their interaction. To confirm this result, we performed coIPs with a mutant version of NXF1 that has been described to have lost its RNA-binding capability due to an exchange of ten arginines to alanines within its RNA-binding domain [29,31]. Using this mutant (NXF1-10RA) we observed that much more NP was co-precipitated than when using wildtype NXF1 (Figure 2B, third and fourth lane), similar to the results seen after RNase A treatment. Indeed, it was not possible to find conditions where both the very strong band for NP co-precipitated by NXF1-10RA and the much weaker band for NP co-precipitated by wildtype NXF1 could be visualized together with reasonable band intensities. This suggests that the observed differences in NXF1:NP interaction in the presence/absence of RNA are linked to the RNA-binding activity of NXF1. The more prominent interaction between NXF1 and NP in the absence of RNA-binding by NXF1 further suggests that there is competition between NP and single-stranded RNA for the binding of NXF1. Finally, the strong interaction between NXF1-10RA and NP also shows that the amino acids within NXF1 that are important for its RNA-binding capability are dispensable for the interaction with NP.

3.3. Interaction with NP Is Mediated by the RNA-Binding Domain of NXF1

To identify the domain of NXF1 that facilitates the interaction with NP, we generated different mutants of NXF1, each lacking one of its five described domains (Figure 3A) [29]. Using these mutants in coIP studies, we could co-precipitate NP with four of the five deletion mutants (ΔRRM, ΔLRR, ΔNTF2, ΔUBA), demonstrating that none of these domains is essential for the interaction (Figure 3B, fourth lanes). In contrast, NXF1-ΔRBD failed to co-precipitate NP, indicating that it is the RBD that is required for the interaction between these proteins.

To further analyze whether the RBD is also sufficient to mediate an interaction with NP, we also performed coIPs of NP with only the RBD of NXF1, as well as with NXF1-RRM and NXF1-RBD-RRM as controls. In these experiments, NP could be co-precipitated with the RBD and the RBD-RRM, but not with the RRM alone (Figure 4, sixth, eighth and seventh lane). This demonstrates that the RBD is not only required, but also sufficient for the interaction between NP and NXF1, while the RRM is neither required or sufficient for this interaction, nor does it interfere with it.

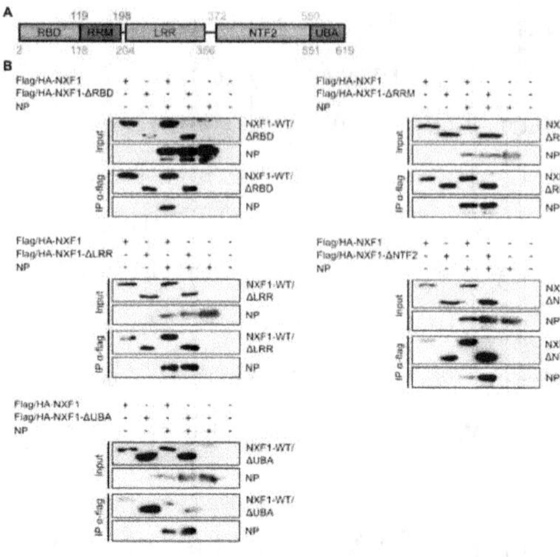

Figure 3. Identification of NXF1 domains necessary for the interaction with NP. (A) Domain organization of NXF1. Numbers indicate amino acid positions. (B) The RNA-binding domain of NXF1 is essential for its interaction with NP. 293T cells were transfected with plasmids encoding flag/HA-NXF1 or flag/HA-tagged NXF1 mutants together with NP. Forty-eight hours p.t., immunoprecipitation was performed using anti-flag antibodies. Subsequent SDS-PAGE and Western blot analyses with input and precipitates were performed using anti-flag and anti-NP antibodies.

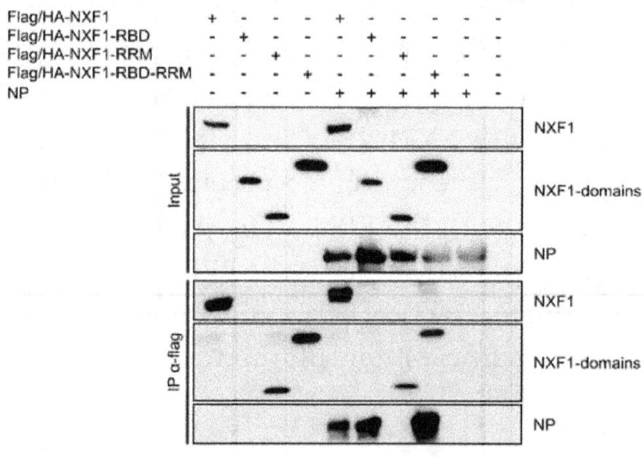

Figure 4. Role of the RNA-binding domain (RBD) and RNA recognition motif (RRM) in the interaction between NXF1 and NP. Immunoprecipitations of lysates from 293T cells transfected with flag/HA tagged NXF1, NXF1-RBD, NXF1-RRM or NXF1-RBD-RRM together with NP using anti-flag antibodies were performed 48 h p.t. Input and precipitate samples were subjected to SDS-PAGE and Western blot, and NXF1 and NP were detected using anti-flag and anti-NP antibodies, respectively.

3.4. NXF1 Is Recruited into NP-Derived Inclusion Bodies

Genome replication and transcription of EBOV is known to take place in inclusion bodies within the cytoplasm [10,14], and overexpression of NP alone leads to formation of such inclusion bodies [6]. In contrast, NXF1 is mainly localized in and around the nucleus [25,26]. Therefore, we wanted to analyze the influence of NP overexpression (and thus inclusion body formation) on the intracellular localization of NXF1. As previously shown [5,6], NP formed the typical inclusion bodies in the cytoplasm when expressed alone (Figure 5), and thus was used as a marker for these structures, while overexpression of NXF1 alone showed a localization in the nucleus and in the perinuclear region. This is in line with previous publications, which have described both a nuclear localization, and particularly a nuclear rim staining mediated by the NTF2 and UBA domains of NXF1 [27], as well as a cytoplasmic localization of NXF1, e.g., inside stress granules [47]. Co-expression of NP and wildtype NXF1 led to no dramatic changes in the intracellular distribution of NXF1, although small amounts of NXF1 could be detected in the NP-derived inclusion bodies. In contrast, the intracellular localization of NXF1-10RA changed upon co-expression of NXF1-10RA, and it clearly accumulated in NP-derived inclusion bodies. Similarly, NXF1-ΔRBD also accumulated in inclusion bodies, indicating that the accumulation of RNA binding-deficient NXF1 in inclusion bodies is not due to the identified direct interaction of NP with the RBD of NXF1, but that NXF1 is recruited via another mechanism. Importantly, while in the coIP experiments NXF1-10RA showed a stronger interaction with NP than wildtype NXF1, this was not the case for NXF1-ΔRBD, which is no longer able to interact with NP. Therefore, the fact that both of these NXF1 mutants accumulate in inclusion bodies suggests that the loss of their ability to bind RNA, but not changes to their ability to bind NP, dictates this phenotype. Together, these data suggest that NXF1 can enter inclusion bodies, but that only RNA binding-deficient NXF1 is retained and, therefore, accumulates in these structures, whereas RNA binding-competent NXF1 rapidly leaves them again.

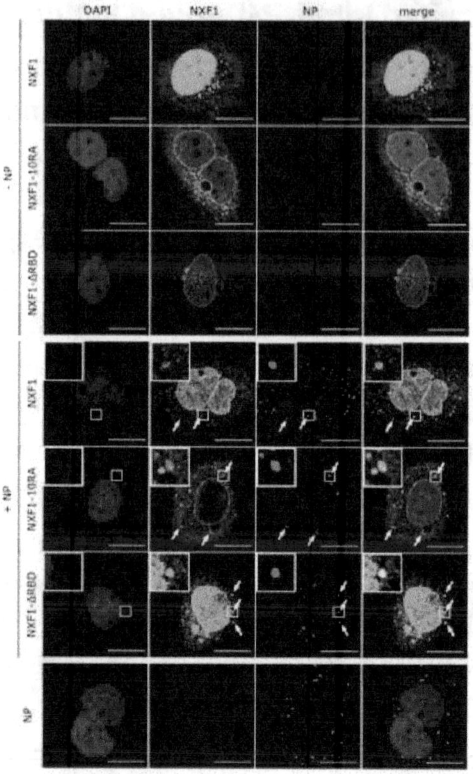

Figure 5. Recruitment of NXF1 into NP-derived inclusion bodies. Huh7 cells were transfected with plasmids encoding flag/HA-NXF1, flag/HA-NXF1-10RA or flag/HA-NXF1-ΔRBD together with NP as indicated. Forty-eight hours p.t., cells were fixed with 4% PFA and permeabilized with Triton X-100. NXF1 (shown in green) was detected with anti-flag antibodies and NP (shown in red) with anti-NP antibodies. Nuclei were stained with DAPI (shown in blue), and cells were visualized by confocal laser scanning microscopy. Scale bars indicate 20 µm. Arrows point out colocalization, and insets show magnifications of indicated areas.

To further substantiate these results, we analyzed the localization of endogenous NXF1 in context of infectious EBOV (Figure 6). As shown above with overexpressed NXF1, endogenous NXF1 was mostly localized in the nucleus. In contrast, NP was only detectable in the cytoplasmic inclusion bodies. Staining of both NXF1 and NP in the context of an EBOV infection showed a subtle but clearly visible relocalization of NXF1 into inclusion bodies, supporting the results obtained using overexpression of NP and NXF1.

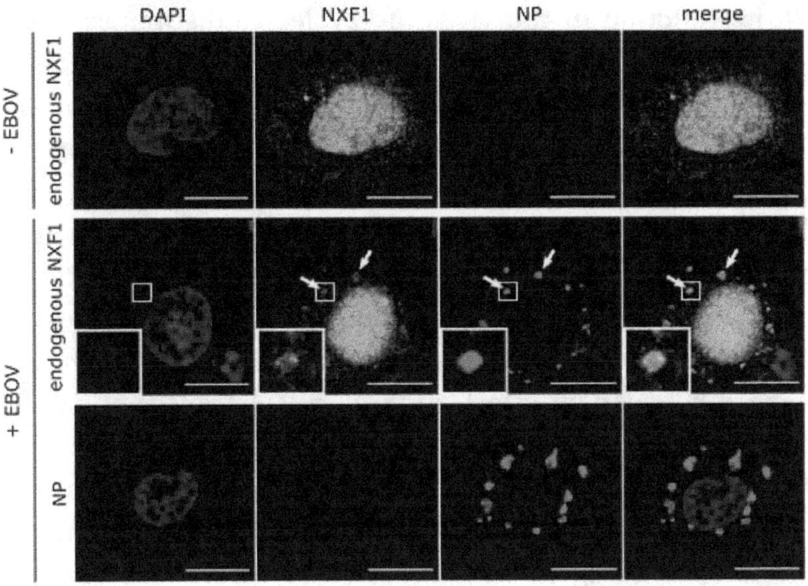

Figure 6. Recruitment of endogenous NXF1 into EBOV inclusion bodies. Huh7 cells were infected with rgEBOV. Sixteen hours p.i., cells were fixed with 10% formalin and permeabilized with Triton X-100. NXF1 (shown in green) was detected with anti-NXF1 antibodies and NP (shown in red) with anti-NP antibodies. Nuclei were stained with DAPI (shown in blue), and cells were visualized by confocal laser scanning microscopy. Scale bars indicate 20 μm. Arrows point out colocalization, and insets show magnifications of indicated areas.

3.5. NXF1 Influences Protein Expression but Not Viral mRNA Transcription

As we had previously identified NXF1 in a genome-wide siRNA screen as being necessary for viral RNA synthesis and/or viral protein expression [19], we wanted to further dissect its impact on viral RNA synthesis and protein expression by assessing which of the different steps of these processes are influenced by NXF1. Therefore, we first performed classical EBOV minigenome assays in the context of an siRNA-mediated knockdown of endogenous NXF1. As shown before, the knockdown of NXF1 led to a reduction of reporter activity, confirming a role of NXF1 in viral RNA synthesis and/or gene expression (Figure 7A). In contrast to this, overexpression of NXF1 did not influence reporter activity, suggesting that the endogenous levels of NXF1 are sufficient to fulfill its function in the virus life cycle. To distinguish between a function of NXF1 in viral genome replication or in transcription/protein expression, we next used a replication-deficient minigenome, which lacks 55 nt in the trailer region, thus abolishing viral genome replication while still allowing transcription of the minigenome as previously described [41]. Using this minigenome system, NXF1 knockdown resulted in an 11.2 to 33.5-fold reduction in reporter activity, demonstrating that NXF1 is important for either viral transcription or protein expression, independent of viral genome replication (Figure 7B). In order to discriminate between an effect on viral mRNA transcription and later steps of viral protein expression such as mRNA transport from inclusion bodies or mRNA translation we next analyzed the influence of NXF1 knockdown on minigenome-derived viral mRNA levels. To this end, we again performed minigenome assays in NXF1 siRNA-treated cells, and analyzed the levels of viral mRNA in the cell lysates via RT-qPCR (Figure 7C). As the mRNA levels in the NXF1 knockdown cells were comparable to the control cells, NXF1 seems not to be important for viral transcription, but rather for either mRNA transport out of inclusion bodies towards ribosomes or efficient mRNA translation.

To confirm these findings in context of infectious EBOV, we repeated these experiments under BSL4 conditions using a recombinant EBOV that expresses firefly luciferase from an additional transcriptional unit, and thus allows precise quantification of viral gene expression [43]. In case of NXF1 knockdown we observed a significant reduction in luciferase activity (Figure 7D), and thus in viral gene expression compared to the negative control. In contrast, when we determined the amount of luciferase mRNAs (Figure 7E), there was no reduction but rather an increase in the mRNA levels upon NXF1 knockdown compared to the negative control, although this increase was not statistically significant. In contrast, knockdown of the viral polymerase L resulted in reductions in both reporter activity and mRNA levels by 62%. This indicates that also in context of a viral infection NXF1 is important for a late step in viral protein expression, i.e., either mRNA transport or efficient mRNA translation, but not for mRNA synthesis.

To further determine whether the influence of NXF1 on reporter protein expression can be explained by binding of viral mRNAs by NXF1, coIPs of viral mRNA in the context of a minigenome assay were performed (Figure 7F). To this end, NXF1 was precipitated and mRNA was isolated from the precipitate. When viral mRNA was produced (i.e., in the sample where the viral polymerase L was present; NXF1 +L in Figure 7F), we detected viral mRNA that could be co-precipitated together with NXF1. This was in contrast to the background signals observed in the control sample without polymerase (NXF1 -L) or in the negative control samples (-NXF1, mock IP, -RT). In contrast, the mutant lacking the RNA-binding domain (NXF1-ΔRBD) failed to co-precipitate the viral mRNA. However, coIP with NXF1-ΔRRM led to similar amounts of viral mRNA in the precipitate compared to wildtype NXF1. These results indicate that NXF1 interacts with the minigenome-derived viral mRNA, and that this interaction is mediated by the RBD of NXF1.

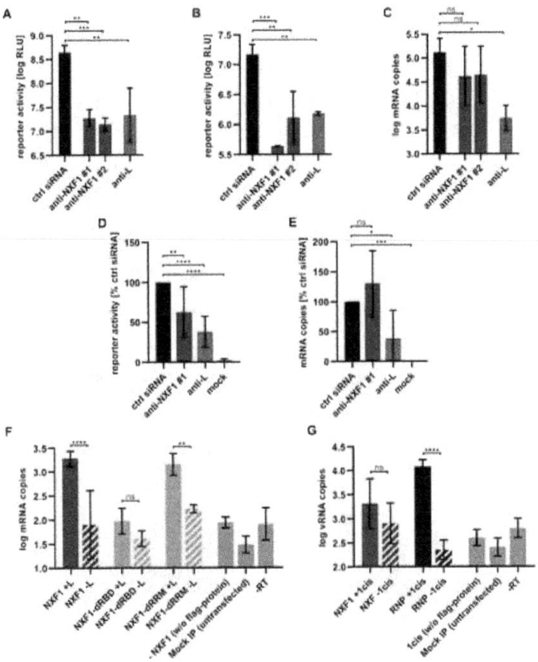

Figure 7. Influence of NXF1 on viral RNA synthesis and protein expression and interaction of NXF1 with viral RNAs. (A) Analysis of viral RNA synthesis and protein expression in NXF1 knockdown cells. 293T cells were transfected with siRNAs targeting either NXF1 (anti-NXF1) or L

(anti-L) or with a negative control (ctrl) siRNA. Forty-eight hours p.t., cells were transfected with the plasmids required for a replication-competent minigenome assay. Forty-eight hours later, reporter activity was measured. (B) Analysis of viral transcription and protein expression in NXF1 knockdown cells in the absence of genome replication. 293T cells were transfected with siRNAs and 48 h p.t. with the plasmids required for a replication-deficient minigenome assay. After a further 48 h, reporter activity was analyzed. (C) Analysis of viral mRNA levels in NXF1 knockdown cells. Forty-eight hours after siRNA transfection of 293T cells, the cells were transfected with all components for the replication-competent minigenome assay. Forty-eight hours later RNA was isolated and RT-qPCR for viral mRNA was performed. (D) Analysis of viral gene expression after infection of NXF1 knockdown cells infected with recombinant EBOV expressing firefly luciferase. Reporter activity was determined one day p.i. and is shown relative to the reporter activity of cells treated with negative control siRNA. (E) Influence of NXF1 knockdown on viral mRNA synthesis. RNA from cells from panel D was isolated and luciferase mRNA was quantified using RT-qPCR. (F) Interaction of NXF1 with viral mRNA. 293T cells were transfected with plasmids encoding flag/HA-NXF1 or flag/HA-NXF1-mutants as indicated, in addition to the replication-competent minigenome, T7-polymerase, and the RNP proteins. As a control L was omitted (-L). Further negative controls (shown in grey) were as indicated. Forty-eight hours p.t., pulldowns were performed using anti-flag antibodies and RNA was isolated from the precipitates. mRNA levels in the precipitates were analyzed by subsequent RT-qPCR. (G) Interaction of NXF1 with viral genomic RNA. IP with anti-flag antibodies was performed with cell lysates from 293T cells transfected with plasmids encoding the replication-competent minigenome, T7-polymerase, and either flag/HA-NXF1, or flag/HA-NP, VP35, and VP24 (RNP). In controls the minigenome was omitted (-1cis). Further negative controls (shown in grey) were as indicated. RNA was isolated from the precipitates and analyzed via RT-qPCR. Means and standard deviations for two independent experiments (six biological replicates) are shown in panels D and E, and for three independent experiments in all other panels. Asterisks indicate p values from one-way ANOVA (*: $p \leq 0.05$; **: $p \leq 0.01$; ***: $p \leq 0.001$; ****: $p \leq 0.0001$; ns: $p > 0.05$).

Finally, to check whether the interaction of NXF1 with the viral mRNA is specific for mRNA, or whether NXF1 binds viral RNA (vRNA) as well, we performed coIPs of minigenome vRNA together with NXF1 (Figure 7G). For this experiment the vRNA was generated by ectopically expressed T7 bacteriophage RNA polymerase in the absence of any viral proteins. As a positive control, vRNA was co-expressed with flag/HA-NP, VP35 and VP24, since these three proteins have been described as being minimally required for the formation of vRNA-containing nucleocapsids, and as such are necessary for co-precipitation of vRNA together with NP [7,48]. RNA was isolated from the precipitates and subsequent RT-qPCR showed that vRNA could indeed be co-precipitated with NP under these conditions (compare RNP +1cis vs. RNP -1cis in Figure 7G). In contrast, NXF1 failed to efficiently co-precipitate minigenome vRNA, suggesting a specific interaction with viral mRNAs.

3.6. A Model for the Function of NXF1 in the EBOV Life Cycle

Our results show that NXF1 is recruited to inclusion bodies, which we and others have shown to be sites of EBOV replication and transcription [10,14]. Furthermore, it is able to interact with NP, with the RBD of NXF1 being both necessary and sufficient for this interaction. In the context of its endogenous function, NXF1 has been shown to interact with many cellular proteins, including export adaptor proteins such as Aly and THOC5, to promote efficient mRNA binding (reviewed in [49]). Binding of these adaptor proteins, which are both components of the transcription and export complex (TREX), leads to a structural rearrangement of NXF1 [29,31]. This conformational change leads to exposure of the RBD resulting in an enhanced RNA-binding activity and subsequently to a handover of mRNA from Aly to NXF1. Since the interaction of NP with NXF1 resembles that of Aly and NXF1, as it involves the RBD, but not the amino acids that are important for the interaction with RNA [31], we propose a model in which a binding of NP to NXF1 results in a handover of viral mRNA from NP to NXF1 (Figure 8). Indeed, we were able to show that NXF1 binds mRNAs that were produced in inclusion bodies. We further propose that NXF1 then rapidly exports the bound RNA from inclusion bodies. This is in line with our observations that only RNA binding-deficient NXF1 accumulates in these structures, while wildtype NXF1 can only be observed in inclusion bodies in minute amounts, as well as the fact that the interaction of NP and NXF1 is weakened when NXF1 binds to RNA.

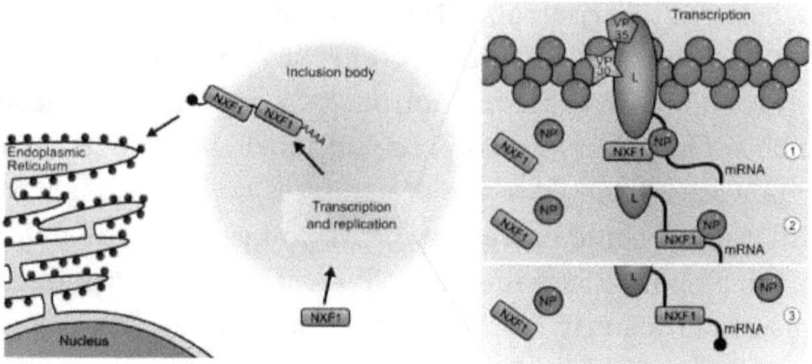

Figure 8. Model for the role of NXF1 and its interaction with NP and the viral mRNA during the EBOV life cycle. NXF1 enters inclusion bodies, which are the site of viral genome replication and transcription. Once in the inclusion bodies, NP recruits NXF1 to nascent mRNA during viral transcription (1). NXF1 then takes over the viral mRNA from NP (2), and thereby avoids encapsidation of the mRNA (3). After the completion of transcription, NXF1 together with the viral mRNA is exported from the inclusion bodies and transports it to ribosomes for translation.

This model could explain how EBOV is able to overcome a number of fundamental challenges it faces during its life cycle, and which it shares with other cytoplasmically replicating NSVs. First, due to the RNA nature of its genome EBOV has to perform both genome replication and transcription, which exhibit fundamental differences: while both processes involve the generation of RNAs from a viral RNA template, during genome replication an accurate and complete copy of the RNA template has to be produced that is then encapsidated by NP to form a new RNP. This is in contrast to the subgenomic RNA fragments produced during transcription, which are processed by the addition of a cap structure and of a poly(A)-tail, and are not encapsidated by NP. While the production of subgenomic RNAs as well as their capping and poly(A)-tailing are all carried out by the viral polymerase L (reviewed in [50]), it is not clear how viral mRNAs evade encapsidation by NP, which is present in abundant quantities in inclusion bodies. However, co-transcriptional binding of viral mRNAs by NXF1 could shield the mRNA from such encapsidation by NP and/or displace already bound NP.

A second challenge during the EBOV life cycle is the fact that viral RNA synthesis occurs in inclusion bodies. These are increasingly appreciated for their importance in the genome replication of cytoplasmically replicating NSVs [12,21,22,23,51], and have for some of these viruses been described to show properties of liquid organelles [24,52]. Similar properties have been observed for EBOV inclusion bodies [10], suggesting that inclusion bodies of NSVs might generally represent liquid organelles. Such liquid organelles consist in most cases of RNA and RNA-binding proteins, and are formed by liquid-liquid phase separation, with multivalent low affinity interactions and intrinsically disordered protein regions being important drivers of their formation (reviewed in [53,54]). RNAs are important constituents of these liquid organelles and sometimes even the seed for their formation; however, in the case of mRNAs they have to be exported in order to reach ribosomes for translation. It is not clear how naked viral RNAs such as mRNAs would be able to escape this liquid organelle environment. To this end, their decoration with NXF1 might help viral mRNA to be exported from inclusion bodies by phase separation between the RNA:NXF1 complexes and other inclusion body constituents.

Finally, in contrast to most cellular mRNAs, the mRNAs of cytoplasmically replicating NSVs do not undergo splicing. Besides the removal of introns, splicing also leads to the association of mRNAs with a multitude of cellular proteins including NXF1, and has been shown to greatly improve translation from both cellular and viral mRNAs, independent of any effects on nuclear mRNA export [55,56,57]. Remarkably, gammaretroviruses use the NXF1 pathway to enhance the efficacy of translating their unspliced mRNAs [40]. Consequently, it is plausible that the enlistment of NXF1 to EBOV mRNAs could have a comparable function.

However, we cannot completely rule out the possibility that NXF1 serves an as-yet-undescribed role in the later phases of viral protein expression. Further research will be necessary to confirm or refute the model suggested by us. Moreover, it would be intriguing to analyze the intricate mechanisms involved in the suggested transportation of viral mRNAs from inclusion bodies. [19]Additionally, it is important to determine the exact positioning of NXF1 within inclusion bodies, and specifically investigate whether there is a segregation of various stages of the virus life cycle within inclusion bodies that aligns with the localization of NXF1 and other host factors.[21]. It is interesting to note that when we initially discovered NXF1 as a factor in viral RNA synthesis and protein expression, we found that it also plays a crucial role in these processes for Junín virus. Junín virus is an arenavirus that causes viral hemorrhagic fever, similar to EBOV. Both viruses replicate their genomes in inclusion bodies. It will be intriguing to determine if NXF1 plays a similar role in the life cycle of arenaviruses as it does in the life cycle of filoviruses. Additionally, since the challenges mentioned above are inherent to most, if not all, negative-sense RNA viruses that replicate in the cytoplasm, it is worth investigating whether other viruses of this type also rely on NXF1.

CONCISE SUMMARY

The Ebola virus (EBOV) is responsible for causing intense outbreaks of viral hemorrhagic fever in people. Although virus-host interactions show potential as targets for antiviral drugs, our understanding of the interplay between EBOV and cellular host components is currently inadequate. Our recent study involved conducting a comprehensive siRNA screen of the entire genome, which led to the identification of the nuclear RNA export factor 1 (NXF1) as a crucial host factor in the life cycle of the Ebola virus (EBOV). NXF1 plays a crucial role in the nuclear mRNA export route, which is used by numerous viruses that have nuclear phases in their life cycles. Nevertheless, the specific function of NXF1 in the life cycle of EBOV, a virus that reproduces within cytoplasmic inclusion bodies, has yet to be determined. To gain a deeper understanding of the role of NXF1 in the life cycle of EBOV, we conducted a series of experiments involving co-immunoprecipitation and double immunofluorescence assays. These experiments were aimed at studying the interactions between NXF1 and viral proteins as well as RNAs. In addition, we employed siRNA-mediated knockdown of NXF1 and functional tests to examine the specific contributions of NXF1 to different stages of the virus life cycle. Using this methodology, we have determined that the EBOV nucleoprotein (NP) is a viral counterpart that interacts with NXF1. Subsequent investigations have demonstrated that NP engages with the RNA-binding domain of NXF1 and vies with RNA for this contact. Co-localization investigations revealed that NXF1 with impaired RNA binding ability, but not the normal NXF1, gathers in inclusion bodies formed from NP. Knockdown tests confirmed that NXF1 is essential for viral protein production, but not for viral RNA synthesis. Ultimately, our findings demonstrated that NXF1 engages with viral mRNAs, but not with viral genomic RNAs. These data indicate that NXF1 is brought into inclusion bodies to facilitate the export of viral mRNA:NXF1 complexes from these locations. This discovery demonstrates a novel role for NXF1 in the life cycle of cytoplasmically replicating viruses. It has the potential to serve as a foundation for developing new therapeutic strategies against EBOV and other emerging viruses.

REFERENCES

1. Burk, R.; Bollinger, L.; Johnson, J.C.; Wada, J.; Radoshitzky, S.R.; Palacios, G.; Bavari, S.; Jahrling, P.B.; Kuhn, J.H. Neglected filoviruses. FEMS Microbiol. Rev. 2016, 40, 494–519.
2. Feldmann, H.; Sanchez, A.; Geisbert, T.W. Filoviridae: Marburg and Ebola Viruses. In Fields Virology, 6th ed.; Knipe, D.M., Howley, P.M., Eds.; Lippincott Williams & Wilkins: Philadelphia, PA, USA, 2013; pp. 923–956.
3. Mühlberger, E.; Weik, M.; Volchkov, V.E.; Klenk, H.D.; Becker, S. Comparison of the transcription and replication strategies of marburg virus and Ebola virus by using artificial replication systems. J. Virol. 1999, 73, 2333–2342.
4. Modrof, J.; Mühlberger, E.; Klenk, H.D.; Becker, S. Phosphorylation of VP30 impairs ebola virus transcription. J. Biol. Chem. 2002, 277, 33099–33104.
5. Groseth, A.; Charton, J.E.; Sauerborn, M.; Feldmann, F.; Jones, S.M.; Hoenen, T.; Feldmann, H. The Ebola virus ribonucleoprotein

complex: A novel VP30-L interaction identified. Virus Res. 2009, 140, 8–14.

6. Becker, S.; Rinne, C.; Hofsass, U.; Klenk, H.D.; Muhlberger, E. Interactions of Marburg virus nucleocapsid proteins. Virology 1998, 249, 406–417.
7. Huang, Y.; Xu, L.; Sun, Y.; Nabel, G.J. The assembly of Ebola virus nucleocapsid requires virion-associated proteins 35 and 24 and posttranslational modification of nucleoprotein. Mol. Cell 2002, 10, 307–316.
8. Kirchdoerfer, R.N.; Abelson, D.M.; Li, S.; Wood, M.R.; Saphire, E.O. Assembly of the Ebola Virus Nucleoprotein from a Chaperoned VP35 Complex. Cell Rep. 2015, 12, 140–149.
9. Leung, D.W.; Borek, D.; Luthra, P.; Binning, J.M.; Anantpadma, M.; Liu, G.; Harvey, I.B.; Su, Z.; Endlich-Frazier, A.; Pan, J.; et al. An Intrinsically Disordered Peptide from Ebola Virus VP35 Controls Viral RNA Synthesis by Modulating Nucleoprotein-RNA Interactions. Cell Rep. 2015, 11, 376–389.
10. Hoenen, T.; Shabman, R.S.; Groseth, A.; Herwig, A.; Weber, M.; Schudt, G.; Dolnik, O.; Basler, C.F.; Becker, S.; Feldmann, H. Inclusion bodies are a site of ebolavirus replication. J. Virol. 2012, 86, 11779–11788.
11. Gabriel, G.; Feldmann, F.; Reimer, R.; Thiele, S.; Fischer, M.; Hartmann, E.; Bader, M.; Ebihara, H.; Hoenen, T.; Feldmann, H. Importin-alpha7 Is Involved in the Formation of Ebola Virus Inclusion Bodies but Is Not Essential for Pathogenicity in Mice. J. Infect. Dis. 2015, 212 (Suppl. 2), S316–S321.
12. Dolnik, O.; Stevermann, L.; Kolesnikova, L.; Becker, S. Marburg virus inclusions: A virus-induced microcompartment and interface to multivesicular bodies and the late endosomal compartment. Eur. J. Cell Biol. 2015, 94, 323–331.
13. Jasenosky, L.D.; Neumann, G.; Lukashevich, I.; Kawaoka, Y. Ebola virus VP40-induced particle formation and association with the lipid bilayer. J. Virol. 2001, 75, 5205–5214.

14. Lier, C.; Becker, S.; Biedenkopf, N. Dynamic phosphorylation of Ebola virus VP30 in NP-induced inclusion bodies. Virology 2017, 512, 39–47.

15. Kruse, T.; Biedenkopf, N.; Hertz, E.P.T.; Dietzel, E.; Stalmann, G.; Lopez-Mendez, B.; Davey, N.E.; Nilsson, J.; Becker, S. The Ebola Virus Nucleoprotein Recruits the Host PP2A-B56 Phosphatase to Activate Transcriptional Support Activity of VP30. Mol. Cell 2018, 69, 136–145.e136.

16. Batra, J.; Hultquist, J.F.; Liu, D.; Shtanko, O.; Von Dollen, J.; Satkamp, L.; Jang, G.M.; Luthra, P.; Schwarz, T.M.; Small, G.I.; et al. Protein Interaction Mapping Identifies RBBP6 as a Negative Regulator of Ebola Virus Replication. Cell 2018, 175, 1917–1930.e1913.

17. Fang, J.; Pietzsch, C.; Ramanathan, P.; Santos, R.I.; Ilinykh, P.A.; Garcia-Blanco, M.A.; Bukreyev, A.; Bradrick, S.S. Staufen1 Interacts with Multiple Components of the Ebola Virus Ribonucleoprotein and Enhances Viral RNA Synthesis. MBio 2018, 9.

18. Nelson, E.V.; Schmidt, K.M.; Deflube, L.R.; Doganay, S.; Banadyga, L.; Olejnik, J.; Hume, A.J.; Ryabchikova, E.; Ebihara, H.; Kedersha, N.; et al. Ebola Virus Does Not Induce Stress Granule Formation during Infection and Sequesters Stress Granule Proteins within Viral Inclusions. J. Virol. 2016, 90, 7268–7284.

19. Martin, S.; Chiramel, A.I.; Schmidt, M.L.; Chen, Y.C.; Whitt, N.; Watt, A.; Dunham, E.C.; Shifflett, K.; Traeger, S.; Leske, A.; et al. A genome-wide siRNA screen identifies a druggable host pathway essential for the Ebola virus life cycle. Genome Med. 2018, 10, 58.

20. Wendt, L.; Bostedt, L.; Hoenen, T.; Groseth, A. High-throughput screening for negative-stranded hemorrhagic fever viruses using reverse genetics. Antiviral. Res. 2019, 170, 104569.

21. Baird, N.L.; York, J.; Nunberg, J.H. Arenavirus infection induces discrete cytosolic structures for RNA replication. J. Virol. 2012, 86, 11301–11310.

22. Lahaye, X.; Vidy, A.; Pomier, C.; Obiang, L.; Harper, F.; Gaudin, Y.; Blondel, D. Functional characterization of Negri bodies (NBs) in rabies virus-infected cells: Evidence that NBs are sites of viral transcription and replication. J. Virol. 2009, 83, 7948–7958.

23. Rincheval, V.; Lelek, M.; Gault, E.; Bouillier, C.; Sitterlin, D.; Blouquit-Laye, S.; Galloux, M.; Zimmer, C.; Eleouet, J.F.; Rameix-Welti, M.A. Functional organization of cytoplasmic inclusion bodies in cells infected by respiratory syncytial virus. Nat. Commun. 2017, 8, 563.

24. Zhou, Y.; Su, J.M.; Samuel, C.E.; Ma, D. Measles Virus Forms Inclusion Bodies with Properties of Liquid Organelles. J. Virol. 2019.

25. Katahira, J.; Strasser, K.; Podtelejnikov, A.; Mann, M.; Jung, J.U.; Hurt, E. The Mex67p-mediated nuclear mRNA export pathway is conserved from yeast to human. EMBO J. 1999, 18, 2593–2609.

26. Bachi, A.; Braun, I.C.; Rodrigues, J.P.; Pante, N.; Ribbeck, K.; von Kobbe, C.; Kutay, U.; Wilm, M.; Gorlich, D.; Carmo-Fonseca, M.; et al. The C-terminal domain of TAP interacts with the nuclear pore complex and promotes export of specific CTE-bearing RNA substrates. RNA 2000, 6, 136–158.

27. Fribourg, S.; Braun, I.C.; Izaurralde, E.; Conti, E. Structural basis for the recognition of a nucleoporin FG repeat by the NTF2-like domain of the TAP/p15 mRNA nuclear export factor. Mol. Cell 2001, 8, 645–656.

28. Guzik, B.W.; Levesque, L.; Prasad, S.; Bor, Y.C.; Black, B.E.; Paschal, B.M.; Rekosh, D.; Hammarskjold, M.L. NXT1 (p15) is a crucial cellular cofactor in TAP-dependent export of intron-containing RNA in mammalian cells. Mol. Cell Biol. 2001, 21, 2545–2554.

29. Viphakone, N.; Hautbergue, G.M.; Walsh, M.; Chang, C.T.; Holland, A.; Folco, E.G.; Reed, R.; Wilson, S.A. TREX exposes the RNA-binding domain of Nxf1 to enable mRNA export. Nat. Commun. 2012, 3, 1006.

30. Herold, A.; Suyama, M.; Rodrigues, J.P.; Braun, I.C.; Kutay, U.; Carmo-Fonseca, M.; Bork, P.; Izaurralde, E. TAP (NXF1) belongs to a

multigene family of putative RNA export factors with a conserved modular architecture. Mol. Cell Biol. 2000, 20, 8996–9008.

31. Hautbergue, G.M.; Hung, M.L.; Golovanov, A.P.; Lian, L.Y.; Wilson, S.A. Mutually exclusive interactions drive handover of mRNA from export adaptors to TAP. Proc. Natl. Acad. Sci. USA 2008, 105, 5154–5159.

32. Viphakone, N.; Sudbery, I.; Griffith, L.; Heath, C.G.; Sims, D.; Wilson, S.A. Co-transcriptional Loading of RNA Export Factors Shapes the Human Transcriptome. Mol. Cell 2019.

33. Larsen, S.; Bui, S.; Perez, V.; Mohammad, A.; Medina-Ramirez, H.; Newcomb, L.L. Influenza polymerase encoding mRNAs utilize atypical mRNA nuclear export. Virol. J. 2014, 11, 154.

34. Tunnicliffe, R.B.; Hautbergue, G.M.; Kalra, P.; Jackson, B.R.; Whitehouse, A.; Wilson, S.A.; Golovanov, A.P. Structural basis for the recognition of cellular mRNA export factor REF by herpes viral proteins HSV-1 ICP27 and HVS ORF57. PLoS Pathog. 2011, 7, e1001244.

35. Yang, C.C.; Huang, E.Y.; Li, H.C.; Su, P.Y.; Shih, C. Nuclear export of human hepatitis B virus core protein and pregenomic RNA depends on the cellular NXF1-p15 machinery. PLoS ONE 2014, 9, e106683.

36. Grüter, P.; Tabernero, C.; von Kobbe, C.; Schmitt, C.; Saavedra, C.; Bachi, A.; Wilm, M.; Felber, B.K.; Izaurralde, E. TAP, the human homolog of Mex67p, mediates CTE-dependent RNA export from the nucleus. Mol. Cell 1998, 1, 649–659.

37. Chen, I.H.B.; Sciabica, K.S.; Sandri-Goldin, R.M. ICP27 Interacts with the RNA Export Factor Aly/REF To Direct Herpes Simplex Virus Type 1 Intronless mRNAs to the TAP Export Pathway. J. Virol. 2002, 76, 12877–12889.

38. Satterly, N.; Tsai, P.L.; van Deursen, J.; Nussenzveig, D.R.; Wang, Y.; Faria, P.A.; Levay, A.; Levy, D.E.; Fontoura, B.M. Influenza virus targets the mRNA export machinery and the nuclear pore complex. Proc. Natl. Acad. Sci. USA 2007, 104, 1853–1858.

39. Tunnicliffe, R.B.; Hautbergue, G.M.; Wilson, S.A.; Kalra, P.; Golovanov, A.P. Competitive and cooperative interactions mediate RNA transfer from herpesvirus saimiri ORF57 to the mammalian export adaptor ALYREF. PLoS Pathog. 2014, 10, e1003907.
40. Bartels, H.; Luban, J. Gammaretroviral pol sequences act in cis to direct polysome loading and NXF1/NXT-dependent protein production by gag-encoded RNA. Retrovirology 2014, 11, 73.
41. Hoenen, T.; Jung, S.; Herwig, A.; Groseth, A.; Becker, S. Both matrix proteins of Ebola virus contribute to the regulation of viral genome replication and transcription. Virology 2010, 403, 56–66.
42. Bussmann, B.M.; Reiche, S.; Jacob, L.H.; Braun, J.M.; Jassoy, C. Antigenic and cellular localisation analysis of the severe acute respiratory syndrome coronavirus nucleocapsid protein using monoclonal antibodies. Virus. Res. 2006, 122, 119–126.
43. Hoenen, T.; Groseth, A.; Callison, J.; Takada, A.; Feldmann, H. A novel Ebola virus expressing luciferase allows for rapid and quantitative testing of antivirals. Antivir. Res. 2013, 99, 207–213.
44. Shabman, R.S.; Hoenen, T.; Groseth, A.; Jabado, O.; Binning, J.M.; Amarasinghe, G.K.; Feldmann, H.; Basler, C.F. An upstream open reading frame modulates ebola virus polymerase translation and virus replication. PLoS Pathog. 2013, 9, e1003147.
45. Noda, T.; Hagiwara, K.; Sagara, H.; Kawaoka, Y. Characterization of the Ebola virus nucleoprotein-RNA complex. J. Gen. Virol. 2010, 91, 1478–1483.
46. Basler, C.F.; Wang, X.; Muhlberger, E.; Volchkov, V.; Paragas, J.; Klenk, H.D.; Garcia-Sastre, A.; Palese, P. The Ebola virus VP35 protein functions as a type I IFN antagonist. Proc. Natl. Acad. Sci. USA 2000, 97, 12289–12294.
47. Hochberg-Laufer, H.; Schwed-Gross, A.; Neugebauer, K.M.; Shav-Tal, Y. Uncoupling of nucleo-cytoplasmic RNA export and localization during stress. Nucleic Acids Res. 2019, 47, 4778–4797.

48. Banadyga, L.; Hoenen, T.; Ambroggio, X.; Dunham, E.; Groseth, A.; Ebihara, H. Ebola virus VP24 interacts with NP to facilitate nucleocapsid assembly and genome packaging. Sci. Rep. 2017, 7, 7698.

49. Heath, C.G.; Viphakone, N.; Wilson, S.A. The role of TREX in gene expression and disease. Biochem. J. 2016, 473, 2911–2935.

50. Hume, A.J.; Muhlberger, E. Distinct Genome Replication and Transcription Strategies within the Growing Filovirus Family. J. Mol. Biol. 2019.

51. Zhang, S.; Chen, L.; Zhang, G.; Yan, Q.; Yang, X.; Ding, B.; Tang, Q.; Sun, S.; Hu, Z.; Chen, M. An amino acid of human parainfluenza virus type 3 nucleoprotein is critical for template function and cytoplasmic inclusion body formation. J. Virol. 2013, 87, 12457–12470.

52. Nikolic, J.; Le Bars, R.; Lama, Z.; Scrima, N.; Lagaudriere-Gesbert, C.; Gaudin, Y.; Blondel, D. Negri bodies are viral factories with properties of liquid organelles. Nat. Commun. 2017, 8, 58.

53. Weber, S.C.; Brangwynne, C.P. Getting RNA and protein in phase. Cell 2012, 149, 1188–1191.

54. Gomes, E.; Shorter, J. The molecular language of membraneless organelles. J. Biol. Chem. 2019, 294, 7115–7127.

55. Sadek, J.; Read, G.S. The Splicing History of an mRNA Affects Its Level of Translation and Sensitivity to Cleavage by the Virion Host Shutoff Endonuclease during Herpes Simplex Virus Infections. J. Virol. 2016, 90, 10844–10856.

56. Nott, A.; Meislin, S.H.; Moore, M.J. A quantitative analysis of intron effects on mammalian gene expression. RNA 2003, 9, 607–617.

57. Nott, A.; Le Hir, H.; Moore, M.J. Splicing enhances translation in mammalian cells: An additional function of the exon junction complex. Genes Dev. 2004, 18, 210–222.

08 OLD WORLD ALPHAVIRUSES ACTIVATE THE PI3K-AKT PATHWAY

1. INTRODUCTION: AN OVERVIEW

1.1. PI3K-AKT Pathway

Viruses stimulate metabolic pathways to fulfill their requirements for synthesizing certain macromolecules. One example of such a pathway is the multifunctional phosphatidylinositol 3-kinases (PI3K)-AKT pathway. Given its crucial role in metabolism and several physiological activities, this route is frequently targeted by viruses [1,2,3,4]. In this review, we describe the activation of the PI3K-AKT pathway by alphaviruses and the consequent cellular effects.

PI3Ks are a large family of kinases that is divided in three classes: class 1 (1A and 1B), class II, and class III. The PI3K-AKT pathway involves class 1A PI3Ks. These PI3Ks are heterodimers with a catalytic subunit (p110) and a regulatory subunit (p85). Class 1A PI3Ks are activated by receptor tyrosine kinases (RTK) or G protein-coupled receptors (GPCR) after binding of growth factors, either directly or indirectly via activation of the small GTPase RAS [5] (Figure 1). Once the regulatory action of p85 has been relieved, active class 1A PI3Ks phosphorylate phosphatidylinositol-4,5-biphosphate (PIP2, or PtdIns(4,5)P2) to phosphatidylinositol-3,4,5-triphosphate (PIP3, or PtdIns(3,4,5)P3) at the plasma membrane (PM). PIP3-enriched membranes are a docking site for the PI3K-dependent kinase-1 (PDK1) and mTORC2 (or PDK2). Furthermore, the serine/threonine kinase AKT (or protein kinase B, PKB) is relocated to membrane sites with PIP3, where it can be activated via phosphorylation by PDK1 and mTORC2 [6,7]. A number of downstream targets of AKT have been identified, several of which are multifunctional nodes, integrating AKT signalling with signalling through other pathways [7]. One of these is the serine/threonine kinase mammalian target of rapamycin (mTOR) in the multi-protein complex mTORC1. AKT signalling leads to the activation of mTORC1, which promotes cell growth by inducing lipid biogenesis through activation of the transcription factors SREBP1 and PPARγ and by promoting protein synthesis by activating the S6 kinase (S6K) and by inactivating the translational inhibitor 4E-BP1. mTORC1 also inhibits autophagy by blocking ULK1 [6]. The serine/threonine kinase glycogen synthase kinase 3 (GSK3) is another multifunctional target of AKT. Through phosphorylation, active GSK3 inhibits most of its substrates. Upon phosphorylation by AKT, GSK3 itself is inhibited and thereby the downstream targets are positively regulated. These targets include the prosurvival BCL-2 family member MCL-1 and the transcription factor c-Myc, which is required for expression of many genes involved in proliferation.

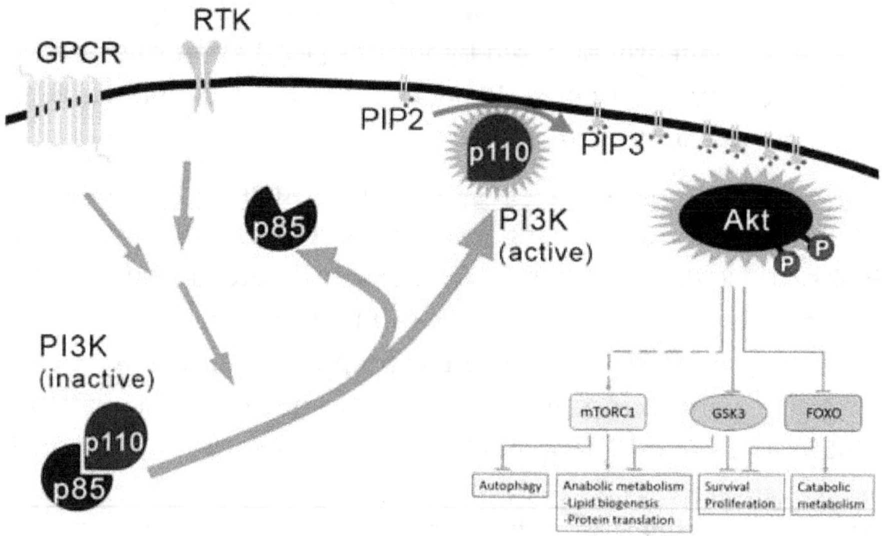

Figure 1. Activation of PI3K-AKT pathway by external signals via G-protein coupled receptors (GPCR) or receptor tyrosine kinases (RTK) leads to dissociation of the p85 regulatory subunit from the active p110 PI3K subunit. The active subunit catalyses the conversion of PIP2 to PIP3 at the plasma membrane, leading to the recruitment and activation of the AKT kinase. Via multiple downstream effector pathways, cellular states of growth, proliferation, heightened metabolic activity and survival are promoted. For more details, see text.

Other GSK3-targets, such as glycogen synthase, are involved in (regulation of) cellular metabolism [7]. The forkhead box O (FoxO) transcription factors are the third multifunctional target of AKT. The process of phosphorylation, carried out by AKT, causes the FoxO transcription factors to promptly move out of the nucleus. Therefore, AKT signaling inhibits the expression of FoxO targets. These targets encompass processes related to the initiation of programmed cell death, halting of cell division, breakdown of molecules for energy production, and suppression of cellular growth. Therefore, the PI3K-AKT signaling pathway, through its multifunctional targets, enhances cell survival, growth, and proliferation, and directs cellular metabolism towards anabolism.

The PI3K-AKT pathway is regulated in many ways. The RAF-MEK-ERK pathway also promotes cell survival and growth and the two pathways have co-regulated proteins and negatively regulate each other. For example, MEK promotes membrane localisation of the phosphatase and tensin homologue (PTEN), where it dephosphorylates PIP3 and inhibits AKT activation [6]. Also, post-translational modifications (including (de)phosphorylation and acetylation) of AKT play important roles in regulation. An important immediate negative feedback loop is provided by mTORC1. Through a variety of downstream targets of mTORC1, AKT signalling is inhibited [7].

1.2. Alphaviruses

Alphaviruses, belonging to the family Togaviridae, are enveloped, icosahedral, positive-sense single-stranded RNA viruses. They are arthropod-borne, typically establishing persistent, asymptomatic infections in their vectors, which are usually mosquitoes, and acute infections in a broad range of vertebrate hosts, including humans, primates, horses and birds. They are globally distributed and can be distinguished into Old World and New World alphaviruses by geographical distribution and by clinical symptoms. Much of what is known of the biology of the Old World alphaviruses stems from decades of research on model viruses Semliki Forest (SFV) and Sindbis viruses (SINV), but the group also contains important human pathogens such as chikungunya virus (CHIKV) and Ross River virus (RRV). Currently, no vaccines or antiviral therapies have been approved for any alphavirus.

The viral genome is a single strand of positive-sense RNA with a 5′ 7-methyl-GpppA cap and a 3′ poly(A) tail. It contains two open reading frames (ORFs). The first ORF encodes the non-structural polyprotein (nsP) P1234 and is translated immediately after infection and the second, encoding the structural proteins, is translated later and to very high levels. All nsPs fulfil a role in viral RNA synthesis [8]. Those roles are quite well understood although the function of nsP3 remained obscure for a long time but is starting to be more understood [9,10]. nsP3 has three domains: the macrodomain, the alphavirus unique or zinc-binding domain (AUD/ZBD) and the hypervariable region (HVD). The macrodomain and AUD/ZBD are conserved among alphaviruses, while the HVD shows high variability in sequence and length even between closely related alphaviruses. The nsP3 HVD is a hub for interactions with multiple cellular proteins and can modulate multiple cellular pathways during infection [11].

1.3. Effect of Alphaviruses on PI3K-AKT Pathway During Infection

Several alphaviruses influence the PI3K-AKT(-mTOR) pathway during infection. For four of the Old World alphaviruses (SFV, RRV, CHIKV and SINV) PI3K-AKT activation has been studied and will be discussed in the following sections. Briefly, activation appears to manifest itself as either extremely strong or weak, depending on the virus. The complexity of PI3K-AKT pathway activation, cross-talk with other pathways and the numerous effectors, make it very difficult to appreciate what outcomes are important for replication of the different members of the alphavirus genus.

1.4. SFV and RRV Hyperactivate the PI3K-AKT-mTOR Pathway and Downstream Effectors

Both SFV and RRV have been shown to very strongly activate the PI3K-AKT-mTOR pathway, via the presence of YXXM motifs in the C-terminal HVD of nsP3 (Y369-E370-P371-M372 for SFV, and Y356-E357-T358-M359 for RRV) [12] (Figure 2). YXXM motifs are known activators of class 1A PI3Ks [13,14] and, after localisation to the PM in nascent viral RNA replication complexes, the YXXM motifs of SFV and RRV nsP3 bind to one or both of the SH2 domains of p85, thereby releasing the enzymatic p110 subunit, leading to activation of AKT [12]. Strikingly, the hyperactivation of the pathway even occurs when cells are starved of amino acids and growth factors prior to infection. The presence of PM-associated nsP3 is sufficient to induce AKT phosphorylation [15], suggesting interaction with a PM-localised cellular factor important for this activation. By comparison with other YXXM-containing proteins, we propose that the tyrosine residue is phosphorylated by a PM-localised cellular tyrosine kinase, but the identity of such a kinase(s) is not known.

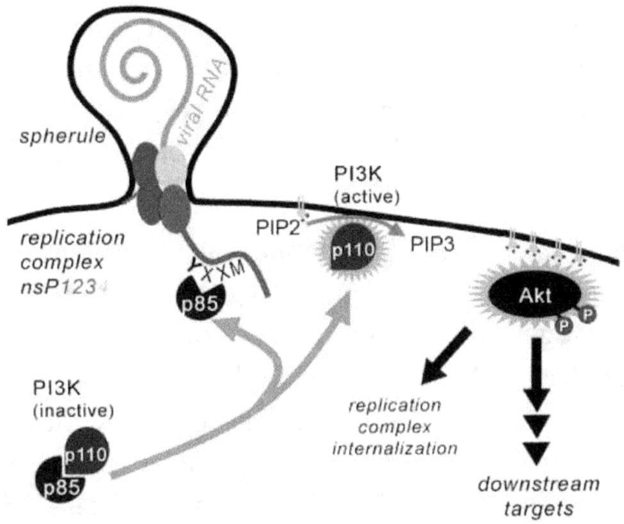

Figure 2. In Semliki Forest virus (SFV) and Ross River virus (RRV)-infected cells, viral replication complexes form at the PM. Via recruitment of p85 to a YXXM motif in nsP3, p110 is freed from regulation and can catalyse the conversion of PIP2 to PIP3 and recruitment and activation of AKT to the PM. Via multiple downstream effector pathways, the internalization of viral replication complexes as well as a heightened cellular metabolic activity and likely other yet undescribed states are promoted. For more details, see text.

All alphaviruses encode nsP3, but the HVD greatly varies in length and sequence between alphaviruses. We analysed the nsP3 sequence of all alphaviruses and found that YXXM motifs are only present in SFV, RRV and the equine alphaviruses Getah virus and Sagiyama virus [12]. There are no studies showing whether the latter viruses also hyperactivate

AKT in their respective host cells. Interestingly, some viruses from other families also activate AKT via YXX(X)M motifs, including influenza virus [16] and herpes simplex virus type 1 [17].

SFV infection leads to the induction of strong phosphorylation of AKT and two downstream targets of mTORC1, S6 and 4E-BP1 [15]. Furthermore, it induces phosphorylation of the AKT substrates phosphofructokinase 2 (PFK2), Rab GTPase-activating protein AS160 (or TBC1D4) and ATP citrate lyase (ACL) [12]. These latter substrates all play a role in cell metabolism. Treatment of infected cells with wortmannin, a specific PI3K inhibitor, reduces but does not eliminate phosphorylation of AKT, PFK2, AS160 and ACL [12,15]. This suggests that SFV-induced AKT hyperactivation is mediated by PI3K-dependent and independent mechanisms. S6 phosphorylation is not reduced upon wortmannin treatment, probably due to the fact that mTORC1 is also stimulated via other pathways [15].

For both SFV and RRV, wortmannin treatment inhibits viral replication at late stages only, suggesting that PI3K-AKT activation and consequent changes in cellular metabolism are more important for later stages of viral replication. For RRV however, there is no difference in release of new virions when cells are infected with an RRV mutant that cannot hyperactivate AKT. However, in a murine model of RRV infection, wildtype RRV causes a more severe disease than the non-hyperactivating mutant [12]. Probably, AKT activation is not essential for in vitro replication, but does contribute to infection in vivo. Effects of the pathway on SFV pathogenesis in vivo have not been studied.

1.5. CHIKV Activates PI3K-AKT Moderately

CHIKV infection of a number of different cell types and species also leads to activation of PI3K-AKT [18,19,20,21], and PI3K expression was found to be upregulated in total white blood cells isolated from CHIKV infected patients with both mild and severe disease [22]. As the PI3K inhibitor wortmannin completely blocks CHIKV-induced activation of AKT [15], the activation is likely mediated entirely via PI3K. However, when directly compared to SFV, the PI3K-AKT activation by CHIKV is significantly slower and less extensive [15]. In line with that, CHIKV nsP3 does not have a YXXM motif and likely does not hyperactivate by binding to SH2 domains of p85. The mechanism of this moderate activation of PI3K-AKT is not known. Src family kinases (SFK), which have been shown to be important for CHIKV-replication and with which CHIKV might interact [19], directly interact with AKT and are required for its activation [23,24]. It has been shown that microRNAs that target proteins in the PI3K-AKT pathway are modulated by CHIKV [21] and may play some role in the activation. Nevertheless, despite less extensive activation than SFV or RRV, the pathway appears to be important for replication and pathogenesis of CHIKV. Inhibition of PI3K has been shown to inhibit CHIKV replication, translation of CHIKV proteins and the amount of new infectious viruses [19,25].

In contrast, another study showed that inhibition of PI3K did not affect CHIKV replication, whereas direct inhibition of AKT or inhibition of protein kinase A (PKA), a kinase which regulates AKT-activation independent of PI3K, did inhibit CHIKV replication [18]. This might suggest that CHIKV could activate AKT independently of PI3K, through PKA. Another AKT inhibitor, GSK 690693, did not influence CHIKV replication [19]. CHIKV infection transiently inhibits activation of mTORC1 by increasing reactive oxygen species (ROS) levels [25,26]. Inhibition of mTORC1 with either rapamycin, TORISEL or RAD001 is reported to not affect CHIKV replication [18], and might even favour it [19,25]. Treatment of CHIKV infected human fibroblasts with dasitinib strongly reduced viral titres by blocking translation of subgenomic RNA [19]. This might be through PI3K-AKT activation, but also through MAPK activation, as MAPK signalling pathways were shown to be required for optimal replication of CHIKV [27]. However, inhibition of both mTORC1 and mTORC2 by Torin 1 decreases CHIKV replication, suggesting that CHIKV needs mTORC2, though it does not depend on mTORC1 [19]. mTORC2 activates AKT, but also other kinases, such as protein kinase C isotypes and serum and glucocorticoid-induced kinase 1 (GSK1) [28]. SFK and mTORC1/mTORC2 seem to be important for more alphaviruses, as Mayaro virus (MAYV), o'nyong'nyong virus (ONNV) and RRV titres are also decreased upon inhibition with dasitinib or Torin 1 [19]. The contradicting results of different inhibitors on CHIKV replication might be due to the relatively weak activation of the PI3K-AKT pathway, which may be more sensitive to variation.

1.6. SINV Differentially Activates PI3K-AKT in Different Species

Lacking a YXXM motif, SINV also does not hyperactivate the PI3K-AKT pathway [12], but rather appears to induce low or transient activation in cell lines from different species [12,29,30]. LY294002, which inhibits class I PI3Ks [31], inhibits SINV RNA replication, but not virus titres in human cells [30]. Specific mTORC1 inhibition does not decrease SINV replication [30], and might even increase it [25]. Taken together, these reports do not suggest a strong requirement for the PI3K-AKT-mTOR activation for efficient SINV infection and replication.

1.7. Benefits of Activation of PI3K Pathway for Alphaviruses

As described above, activation of the PI3K-AKT pathway can have many effects on diverse cellular processes, including cell survival, growth and proliferation and metabolism, and many more functions specific for certain cell types [7]. In the subsequent sections, we discuss what is known about the effects of alphavirus-induced PI3K-AKT activation on vertebrate cells.

2. METABOLIC CHANGE

Alphaviruses Influence Cellular Metabolism

PI3K-AKT activation by SFV leads to a change in cellular metabolism, probably through increased phosphorylation of AKT targets. It increases glycolysis and causes an increased glucose flux through the tricarboxylic acid cycle (TCA). Consequently, more citrate is exported to the cytoplasm and more fatty acids are de novo synthesised. Furthermore, there is increased activation of the pentose phosphate pathway (PPP), which results in increased synthesis of nucleotides, such as UMP [12]. Treating cells for 16h with the glucose analogue 2-deoxyglucose (2-DG), which inhibits glycolysis and prevents entrance of glucose into the PPP, followed by SFV infection, strongly reduces production of new virions [12,32]. This effect is still seen when 2-DG is administered at 3 hpi, which indicates that glycolysis is not required for early events in replication [32]. Inhibition of the PPP specifically through the glucose-6-phosphate dehydrogenase inhibitor (DHEA), reduces production of new virions almost 10-fold. Furthermore, treatment of infected cells with wortmannin or the AKT-inhibitor MK-2206 dramatically inhibits glycolysis and reduces the release of new virions. This indicates the importance of the metabolic shift for SFV replication [12].

RRV infection increases glycolysis moderately in a manner that appears to be independent of the hyperactivation of PI3K-AKT [12]. In contrast, wildtype RRV increases fatty acid levels significantly, whereas RRV-YF (which does not hyperactivate PI3K-AKT) does not. It is not shown whether inhibition of fatty acid synthesis hampers RRV replication. RRV-YF is attenuated in an in vivo mouse model, but this attenuation is not directly attributed to effects on metabolism [12].

Conversely, although SINV does not induce sustained AKT phosphorylation in human cells, it does activate glycolysis at 8 hpi [12]. SINV also increases glycolytic flux in mouse neuroblastoma cells (Neuro 2a), perhaps to compensate for virus-induced mitochondrial dysfunction [33]. Inhibition of glycolysis or PPP in SINV-infected cells strongly decreases the release of progeny virus in both human and African green monkey kidney (Vero) cells [12,32]. Similar to SFV, glycolysis is not required for virus entry, but is important later during the viral life cycle [32]. It does not seem that SINV infection increases fatty acid synthesis. Thus, SINV induces a slightly different metabolic profile, but needs glycolysis and PPP for replication in cells of various species. SINV does not induce sustained AKT-activation in human cells, and probably has evolved other mechanisms to stimulate those metabolic pathways [12].

Metabolic pathways have not been well studied in the context of CHIKV infection. There are some indications that CHIKV activates glycolysis. PKM2, an isoenzyme of the glycolytic enzyme pyruvate kinase, is overexpressed in CHIKV-infected mouse muscles [34]. Furthermore, in human hepatic cells, CHIKV induces upregulation of PDHA1, a subunit of the pyruvate dehydrogenase complex involved in transforming pyruvate to acetyl-CoA in the TCA. However, other glycolytic and TCA-enzymes, such as alpha-enolase and isocitrate dehydrogenase, are downregulated in CHIKV-infected cells [35]. Regarding the other alphaviruses, only MAYV is known to increase glycolytic flux strongly, through activation of the enzyme 6-phosphofructo 1-kinase (PFK1) [36]. However, it is not shown whether MAYV activates AKT.

To conclude, all abovementioned alphaviruses influence cellular metabolism. However, this is not always entirely AKT-dependent, and all alphaviruses modulate cellular metabolism slightly differently. Processes such as increased glycolysis are essential for some viruses, but dispensable for others, at least in vitro. Further study will be necessary to gain a more complete understanding of the importance of metabolic alterations to alphavirus infections.

Many viruses influence cellular metabolism in order to create optimal circumstances for replication. Most viruses stimulate glycolysis and the PPP and fatty acid synthesis. Through these mechanisms, viruses can increase the availability of energy and building blocks for replication and progeny production [37]. During acute infection, viruses need much energy and building blocks for viral replication and production of viruses. Thus, it is very plausible that alphaviruses as well as other viruses causing acute infection, activate PI3K-AKT to influence cellular mechanism and ensure this. However, PI3K-AKT-dependent increased glucose uptake and glycolysis was also shown to promote an interferon-induced antiviral response [38]. Increased glycolysis is beneficial for the virus but comes at a cost.

3. INTERACTION WITH AUTOPHAGY

3.1. Various Viruses Influence Autophagy

Macroautophagy, the process by which intracellular components, such as long-lived proteins and organelles are degraded and recycled is also regulated at least in part by the PI3K-AKT pathway [39]. It is a constitutive process but can be further induced through activation of the kinase Ulk1 by AMP activated protein kinase (AMPK). Activation of the PI3K-AKT-mTORC1 pathway inhibits autophagy through inhibition of the interaction between Ulk1 and AMPK by mTORC1 [40].

Autophagy in the context of virus infection has been well studied, although it is not always clear if the effects of the pathway are pro- or antiviral [41,42]. The benefits of autophagy seem to be different for different viruses, and dependent on the time after infection. Some viruses, such as ZIKV and rotavirus require the process early in infection [43,44,45,46], whereas this is detrimental for others. Conversely, some viruses, such as influenza A virus induce autophagy late after infection to increase replication [47]. Finally, there are viruses in which autophagy does not appear to influence replication. Autophagy has been studied in the context of CHIKV, SFV and SINV infection, though most of these studies did not investigate the direct role of PI3K-AKT activation. These alphaviruses appear to have different effects on autophagy, though there also might be variation due to different cell types used and different experimental conditions.

3.2. CHIKV Induce Autophagy, Whereas SFV Blocks It

CHIKV infection increases the amount of autophagosomes in various human cell lines: human embryonic kidney (HEK) 293 cells [48], HeLa cells [49,50], human glioblastoma cells (U-87) [51], and murine embryonic fibroblasts (MEF) [26] from early in infection and continuously increasing during the course of infection. CHIKV induces oxidative stress, leading to increased intracellular ROS and NO levels. This inhibits mTORC1, which results in increased autophagy [26]. CHIKV-induced autophagy has the effect to reduce or delay CHIKV-induced cell death [26]. The correlation between increased autophagy and efficient CHIKV replication might be species dependent [26,51,52]. An explanation for the difference between species came with the discovery that in human cells, CHIKV nsP2 interacts with the autophagy receptor NDP52, but not in murine cells [50]. The situation appears more complicated in vivo, where one study reported that induction of autophagy attenuates disease, when compared between wildtype and autophagy-deficient mice [26], while another showed that inhibition of autophagy with Tat-beclin 1 peptide reduces viral titres and improves clinical outcomes in CHIKV-infected mice [52].

SFV also increases the amount of autophagosomes in the infected cell [53]. However, in contrast to CHIKV, SFV infection does not appear to strongly affect the production of autophagosomes but rather blocks their degradation in a manner dependent on expression of the viral glycoproteins. Since SFV replication was unaffected by ATG5 depletion, it is concluded that the pathway plays no role in replication in vitro [53].

In conclusion, many acute viruses influence AKT and mTORC1 activation to influence autophagy. For most alphaviruses it is not clear how PI3K-AKT activation relates to autophagy. SFV strongly activates PI3K-AKT-mTORC1, but specifically blocks autophagosome degradation, which seems contradictory as active mTORC1 blocks both autophagosome formation and degradation. CHIKV activates PI3K-AKT moderately, but inhibits mTORC1 via ROS and NO, resulting in increased autophagy.

4. PROMOTION OF CELL SURVIVAL

Many AKT substrates play a role in promoting cell survival and proliferation, but there are currently no studies addressing the effects of AKT on cell death in alphavirus infection. The New World alphavirus Venezuelan equine encephalitis virus stimulates early growth factor response genes in infected astrocytes, but this was shown not to be dependent on MAPK or PI3K signaling [54]. Rubella virus (RV) which also belongs to the Togaviridae family, induces caspase-dependent apoptosis, leading to a cytopathic effect, but delays and reduces this process through PI3K-AKT activation [55]. However, in our preliminary experiments, we did not detect any significant delay in cell death in wildtype SFV infection compared to mutants, which do not hyperactivate the pathway (unpublished).

5. THE STRANGE CASE OF THE TRAFFICKING OF REPLICATION COMPLEXES

Some Alphaviruses Stimulate Internalisation of Replication Complexes Early in infection, alphavirus replication complexes (RC) assemble in membrane invaginations/spherules at the PM in which viral replication takes place [56,57]. A striking effect of PI3K-AKT hyperactivation by SFV and RRV is the trafficking of RC from the PM. First, RC localise in small and scattered cytoplasmic vesicles and then in large acidic perinuclear vacuoles (called cytopathic vacuoles of type I (CPV-I)). In SFV infection, RCs are relocalised from the PM to CPV-I at 8 hpi [56]. When PI3K is inhibited or when the YXXM motif in nsP3 is mutated, trafficking of RC does not take place [12,15,56]. It is not known which downstream targets of AKT mediate RC trafficking. Remarkably, viral RNA synthesis is not hindered when PI3K is inhibited, indicating that RC at the PM alone can sustain RNA replication [15]. Therefore, it is difficult to determine whether internalisation of RCs has benefits for the virus.

In other alphaviruses, the effect is less clear. CHIKV activates PI3K-AKT, although to a lesser extent than SFV, and CHIKV RCs mostly remain at the PM [15]. RC of SINV, which does not activate PI3K-AKT in human cells, but does activate it in murine cells, partially relocalises into the cytoplasm in baby hamster kidney cells (BHK-21), though the majority of RCs remain PM membrane associated [57]. Altogether, it seems that only strong AKT activation is required to induce trafficking of alphavirus RCs and replication happens to a similar extent in RC at the PM as in CPV-I.

6. REMARKS IN CONCLUSION

Alphaviruses are pathogens of growing importance, causing increasing numbers of cases throughout the world but for which there are no vaccines or antiviral therapies. We have reviewed the literature showing that several alphaviruses activate the PI3K-AKT pathway in vertebrate cells. The extent of activation differs (Table 1), as SFV and RRV induce hyperactivation and CHIKV causes moderate activation. SINV does not induce sustained AKT activation in humans but activates AKT in mice. The mechanism of hyperactivaton by SFV and RRV is known [12], and future work will reveal the mechanism of the moderate activation by CHIKV, SINV and mutants of SFV and RRV which do not hyperactivate. Inhibition of PI3K-AKT most clearly inhibits replication of SFV and RRV, while for CHIKV there are contradicting results and SINV does not seem to rely on PI3K-AKT for replication.

Table 1. Mechanisms and Effects of PI3K-AKT activation.

	Mechanism of PI3K-AKT Activation	Effects of PI3K-AKT Activation on			
		Metabolism	Autophagy	Apoptosis	Trafficking RC
SFV	Strong activation via YXXM motif in nsP3	Increases glycolysis and fatty acid synthesis	Blocks degradation of autophagosomes	Small, not significant delay	RCs traffic from PM to CPV-I
RRV	Strong activation via YXXM motif in nsP3	Increases fatty acid synthesis	Unknown	Small, not significant delay	RCs traffic from PM to CPV-I
CHIKV	Moderate activation by unknown mechanism	Unknown	Increases production of autophagosomes	Unknown	RC mostly remain at PM
SINV	Weak or transient activation by unknown mechanism	Unknown	Unknown	Unknown	RC mostly remain at PM

Therefore, the activation of PI3K-AKT has multiple potential advantages for alphaviruses, with the alteration of cellular metabolism appearing to be the most significant. PI3K-AKT-mTORC1-S6K is responsible for facilitating an antiviral response mediated by interferon [58]. In addition, AKT plays a role in activating NF-κB, resulting in the generation of inflammatory cytokines [2]. Although the study of PI3K-AKT in alphavirus-infected cells is lacking, it is crucial to ascertain the degree to which PI3K-AKT is necessary for antiviral immunity and when PI3K-AKT activation is advantageous for the virus but harmful to the host. Proteins from the PI3K-AKT pathway could serve as promising targets in the development of antiviral treatments. Excessive suppression of PI3K-AKT is likely to harm the host by increasing the susceptibility of cells to apoptosis and potentially activating innate immune pathways. Hence, it is crucial to investigate further the proteins involved in the processes activated by AKT and that are advantageous for alphaviruses, particularly in terms of anabolic metabolism.

CONCISE SUMMARY

Alphaviruses have the ability to infect a wide variety of vertebrate hosts, such as birds, horses, monkeys, and humans. Infection in humans can result in symptoms such as rash, fever, encephalitis, and joint pain or inflammation. These diseases are mostly spread by mosquitoes, which carry and maintain long-lasting, symptomless infections. At present, there are no vaccinations or antiviral treatments available for any alphavirus. Several Old World alphaviruses, such as Semliki Forest virus, Ross River virus, and chikungunya virus, stimulate or intensify the activity of the phosphatidylinositol-3-kinase (PI3K)-AKT pathway in vertebrate cells. This has the potential to affect various cellular processes, including cell survival, proliferation, metabolism, and autophagy. The replication of several alphaviruses can be effectively suppressed either in laboratory settings or in living organisms by inhibiting the activity of PI3K or AKT. This highlights the crucial role of these enzymes in viral replication. This review focuses on elucidating the mechanisms responsible for activating the pathway during infection and examining the advantageous effects of PI3K-AKT activation on alphaviruses. This knowledge can be valuable for identifying and creating treatments.

REFERENCES

1. Diehl, N.; Schaal, H. Make yourself at home: Viral hijacking of the PI3K/Akt signaling pathway. Viruses 2013, 5, 3192–3212.
2. Dunn, E.F.; Connor, J.H. HijAkt: The PI3K/Akt pathway in virus replication and pathogenesis. Prog. Mol. Biol. Transl. Sci. 2012, 106, 223–250.
3. Buchkovich, N.J.; Yu, Y.; Zampieri, C.A.; Alwine, J.C. The TORrid affairs of viruses: Effects of mammalian DNA viruses on the PI3K-Akt-mTOR signalling pathway. Nat. Rev. Microbiol. 2008, 6, 266–275.

4. Cooray, S. The pivotal role of phosphatidylinositol 3-kinase-Akt signal transduction in virus survival. J. Gen. Virol. 2004, 85, 1065–1076.

5. Vanhaesebroeck, B.; Guillermet-Guibert, J.; Graupera, M.; Bilanges, B. The emerging mechanisms of isoform-specific PI3K signalling. Nat. Rev. Mol. Cell Biol. 2010, 11, 329–341.

6. Ersahin, T.; Tuncbag, N.; Cetin-Atalay, R. The PI3K/AKT/mTOR interactive pathway. Mol. Biosyst. 2015, 11, 1946–1954.

7. Manning, B.D.; Toker, A. AKT/PKB Signaling: Navigating the Network. Cell 2017, 169, 381–405.

8. Rupp, J.C.; Sokoloski, K.J.; Gebhart, N.N.; Hardy, R.W. Alphavirus RNA synthesis and non-structural protein functions. J. Gen. Virol. 2015, 96, 2483–2500.

9. Gotte, B.; Panas, M.D.; Hellstrom, K.; Liu, L.; Samreen, B.; Larsson, O.; Ahola, T.; McInerney, G.M. Separate domains of G3BP promote efficient clustering of alphavirus replication complexes and recruitment of the translation initiation machinery. PLoS Pathog. 2019, 15, e1007842.

10. Lark, T.; Keck, F.; Narayanan, A. Interactions of Alphavirus nsP3 Protein with Host Proteins. Front. Microbiol. 2017, 8, 2652.

11. Gotte, B.; Liu, L.; McInerney, G.M. The Enigmatic Alphavirus Non-Structural Protein 3 (nsP3) Revealing Its Secrets at Last. Viruses 2018, 10, 105.

12. Mazzon, M.; Castro, C.; Thaa, B.; Liu, L.; Mutso, M.; Liu, X.; Mahalingam, S.; Griffin, J.L.; Marsh, M.; McInerney, G.M. Alphavirus-induced hyperactivation of PI3K/AKT directs pro-viral metabolic changes. PLoS Pathog. 2018, 14, e1006835.

13. Wu, H.; Windmiller, D.A.; Wang, L.; Backer, J.M. YXXM motifs in the PDGF-beta receptor serve dual roles as phosphoinositide 3-kinase binding motifs and tyrosine-based endocytic sorting signals. J. Biol. Chem. 2003, 278, 40425–40428.

14. Songyang, Z.; Shoelson, S.E.; Chaudhuri, M.; Gish, G.; Pawson, T.; Haser, W.G.; King, F.; Roberts, T.; Ratnofsky, S.; Lechleider, R.J.; et al. SH2 domains recognize specific phosphopeptide sequences. Cell 1993, 72, 767–778.

15. Thaa, B.; Biasiotto, R.; Eng, K.; Neuvonen, M.; Gotte, B.; Rheinemann, L.; Mutso, M.; Utt, A.; Varghese, F.; Balistreri, G.; et al. Differential Phosphatidylinositol-3-Kinase-Akt-mTOR Activation by Semliki Forest and Chikungunya Viruses Is Dependent on nsP3 and Connected to Replication Complex Internalization. J. Virol. 2015, 89, 11420–11437.

16. Shin, Y.K.; Liu, Q.; Tikoo, S.K.; Babiuk, L.A.; Zhou, Y. Influenza A virus NS1 protein activates the phosphatidylinositol 3-kinase (PI3K)/Akt pathway by direct interaction with the p85 subunit of PI3K. J. Gen. Virol. 2007, 88, 13–18.

17. Strunk, U.; Saffran, H.A.; Wu, F.W.; Smiley, J.R. Role of herpes simplex virus VP11/12 tyrosine-based motifs in binding and activation of the Src family kinase Lck and recruitment of p85, Grb2, Shc. J. Virol. 2013, 87, 11276–11286.

18. Sharma, A.; Bhomia, M.; Yeh, T.J.; Singh, J.; Maheshwari, R.K. Miltefosine inhibits Chikungunya virus replication in human primary dermal fibroblasts. F1000Res 2018, 7, 9.

19. Broeckel, R.; Sarkar, S.; May, N.A.; Totonchy, J.; Kreklywich, C.N.; Smith, P.; Graves, L.; DeFilippis, V.R.; Heise, M.T.; Morrison, T.E.; et al. Src Family Kinase Inhibitors Block Translation of Alphavirus Subgenomic mRNAs. Antimicrob. Agents Chemother. 2019, 63, e02325-18.

20. Das, I.; Basantray, I.; Mamidi, P.; Nayak, T.K.; Pratheek, B.M.; Chattopadhyay, S. Heat shock protein 90 positively regulates Chikungunya virus replication by stabilizing viral non-structural protein nsP2 during infection. PLoS ONE 2014, 9, e100531.

21. Sharma, A.; Balakathiresan, N.S.; Maheshwari, R.K. Chikungunya Virus Infection Alters Expression of MicroRNAs Involved in Cellular

Proliferation, Immune Response and Apoptosis. Intervirology 2015, 58, 332–341.

22. Wikan, N.; Khongwichit, S.; Phuklia, W.; Ubol, S.; Thonsakulprasert, T.; Thannagith, M.; Tanramluk, D.; Paemanee, A.; Kittisenachai, S.; Roytrakul, S.; et al. Comprehensive proteomic analysis of white blood cells from chikungunya fever patients of different severities. J. Transl. Med. 2014, 12, 96.
23. Jiang, T.; Qiu, Y. Interaction between Src and a C-terminal proline-rich motif of Akt is required for Akt activation. J. Biol. Chem. 2003, 278, 15789–15793.
24. Yori, J.L.; Lozada, K.L.; Seachrist, D.D.; Mosley, J.D.; Abdul-Karim, F.W.; Booth, C.N.; Flask, C.A.; Keri, R.A. Combined SFK/mTOR inhibition prevents rapamycin-induced feedback activation of AKT and elicits efficient tumor regression. Cancer Res. 2014, 74, 4762–4771.
25. Joubert, P.E.; Stapleford, K.; Guivel-Benhassine, F.; Vignuzzi, M.; Schwartz, O.; Albert, M.L. Inhibition of mTORC1 Enhances the Translation of Chikungunya Proteins via the Activation of the MnK/eIF4E Pathway. PLoS Pathog. 2015, 11, e1005091.
26. Joubert, P.E.; Werneke, S.W.; De la Calle, C.; Guivel-Benhassine, F.; Giodini, A.; Peduto, L.; Levine, B.; Schwartz, O.; Lenschow, D.J.; Albert, M.L. Chikungunya virus-induced autophagy delays caspase-dependent cell death. J. Exp. Med. 2012, 209, 1029–1047.
27. Varghese, F.S.; Thaa, B.; Amrun, S.N.; Simarmata, D.; Rausalu, K.; Nyman, T.A.; Merits, A.; McInerney, G.M.; Ng, L.F.; Ahola, T. The antiviral alkaloid berberine reduces chikungunya virus-induced mitogen-activated protein kinase (MAPK) signaling. J. Virol. 2015, 90, 9743–9757.
28. Oh, W.J.; Jacinto, E. mTOR complex 2 signaling and functions. Cell Cycl. 2011, 10, 2305–2316.
29. Scherbik, S.V.; Brinton, M.A. Virus-induced Ca2+ influx extends survival of west nile virus-infected cells. J. Virol. 2010, 84, 8721–8731.

30. Mohankumar, V.; Dhanushkodi, N.R.; Raju, R. Sindbis virus replication, is insensitive to rapamycin and torin1, suppresses Akt/mTOR pathway late during infection in HEK cells. Biochem. Biophys. Res. Commun. 2011, 406, 262–267.

31. Gharbi, S.I.; Zvelebil, M.J.; Shuttleworth, S.J.; Hancox, T.; Saghir, N.; Timms, J.F.; Waterfield, M.D. Exploring the specificity of the PI3K family inhibitor LY294002. Biochem. J. 2007, 404, 15–21.

32. Findlay, J.S.; Ulaeto, D. Semliki Forest virus and Sindbis virus, but not vaccinia virus, require glycolysis for optimal replication. J. Gen. Virol. 2015, 96, 2693–2696.

33. Silva da Costa, L.; Da Silva, A.P.P.; Da Poian, A.T.; El-Bacha, T. Mitochondrial bioenergetic alterations in mouse neuroblastoma cells infected with Sindbis virus: Implications to viral replication and neuronal death. PLoS ONE 2012, 7, e33871.

34. Dhanwani, R.; Khan, M.; Lomash, V.; Rao, P.V.; Ly, H.; Parida, M. Characterization of chikungunya virus induced host response in a mouse model of viral myositis. PLoS ONE 2014, 9, e92813.

35. Thio, C.L.; Yusof, R.; Abdul-Rahman, P.S.; Karsani, S.A. Differential proteome analysis of chikungunya virus infection on host cells. PLoS ONE 2013, 8, e61444.

36. El-Bacha, T.; Menezes, M.M.; Silva, M.C.A.e.; Sola-Penna, M.; Da Poian, A.T. Mayaro virus infection alters glucose metabolism in cultured cells through activation of the enzyme 6-phosphofructo 1-kinase. Mol. Cell Biochem. 2004, 266, 191–198.

37. Sanchez, E.L.; Lagunoff, M. Viral activation of cellular metabolism. Virology 2015, 479–480, 609–618.

38. Burke, J.D.; Platanias, L.C.; Fish, E.N. Beta interferon regulation of glucose metabolism is PI3K/Akt dependent and important for antiviral activity against coxsackievirus B3. J. Virol. 2014, 88, 3485–3495.

39. Chun, Y.; Kim, J. Autophagy: An Essential Degradation Program for Cellular Homeostasis and Life. Cells 2018, 7, E278.

40. Kim, J.; Kundu, M.; Viollet, B.; Guan, K.L. AMPK and mTOR regulate autophagy through direct phosphorylation of Ulk1. Nat. Cell Biol. 2011, 13, 132–141.

41. Echavarria-Consuegra, L.; Smit, J.M.; Reggiori, F. Role of autophagy during the replication and pathogenesis of common mosquito-borne flavi- and alphaviruses. Open Biol. 2019, 9, 190009.

42. Choi, Y.; Bowman, J.W.; Jung, J.U. Autophagy during viral infection—A double-edged sword. Nat. Rev. Microbiol. 2018, 16, 341–354.

43. Liang, Q.; Luo, Z.; Zeng, J.; Chen, W.; Foo, S.S.; Lee, S.A.; Ge, J.; Wang, S.; Goldman, S.A.; Zlokovic, B.V.; et al. Zika Virus NS4A and NS4B Proteins Deregulate Akt-mTOR Signaling in Human Fetal Neural Stem Cells to Inhibit Neurogenesis and Induce Autophagy. Cell Stem Cell 2016, 19, 663–671.

44. Cao, B.; Parnell, L.A.; Diamond, M.S.; Mysorekar, I.U. Inhibition of autophagy limits vertical transmission of Zika virus in pregnant mice. J. Exp. Med. 2017, 214, 2303–2313.

45. Zhou, Y.; Geng, P.; Liu, Y.; Wu, J.; Qiao, H.; Xie, Y.; Yin, N.; Chen, L.; Lin, X.; Liu, Y.; et al. Rotavirus-encoded virus-like small RNA triggers autophagy by targeting IGF1R via the PI3K/Akt/mTOR pathway. Biochim. Biophys. Acta Mol. Basis Dis. 2018, 1864, 60–68.

46. Yin, Y.; Dang, W.; Zhou, X.; Xu, L.; Wang, W.; Cao, W.; Chen, S.; Su, J.; Cai, X.; Xiao, S.; et al. PI3K-Akt-mTOR axis sustains rotavirus infection via the 4E-BP1 mediated autophagy pathway and represents an antiviral target. Virulence 2018, 9, 83–98.

47. Wang, R.; Zhu, Y.; Zhao, J.; Ren, C.; Li, P.; Chen, H.; Jin, M.; Zhou, H. Autophagy Promotes Replication of Influenza A Virus In Vitro. J. Virol. 2016, 2019, 93.

48. Krejbich-Trotot, P.; Gay, B.; Li-Pat-Yuen, G.; Hoarau, J.J.; Jaffar-Bandjee, M.C.; Briant, L.; Gasque, P.; Denizot, M. Chikungunya triggers an autophagic process which promotes viral replication. Virol. J. 2011, 8, 432.

49. Khongwichit, S.; Wikan, N.; Abere, B.; Thepparit, C.; Kuadkitkan, A.; Ubol, S.; Smith, D.R. Cell-type specific variation in the induction of ER stress and downstream events in chikungunya virus infection. Microb. Pathog. 2016, 101, 104–118.

50. Judith, D.; Mostowy, S.; Bourai, M.; Gangneux, N.; Lelek, M.; Lucas-Hourani, M.; Cayet, N.; Jacob, Y.; Prevost, M.C.; Pierre, P.; et al. Species-specific impact of the autophagy machinery on Chikungunya virus infection. EMBO Rep. 2013, 14, 534–544.

51. Abraham, R.; Mudaliar, P.; Padmanabhan, A.; Sreekumar, E. Induction of cytopathogenicity in human glioblastoma cells by chikungunya virus. PLoS ONE 2013, 8, e75854.

52. Shoji-Kawata, S.; Sumpter, R.; Leveno, M.; Campbell, G.R.; Zou, Z.; Kinch, L.; Wilkins, A.D.; Sun, Q.; Pallauf, K.; MacDuff, D.; et al. Identification of a candidate therapeutic autophagy-inducing peptide. Nature 2013, 494, 201–206.

53. Eng, K.E.; Panas, M.D.; Murphy, D.; Hedestam, G.B.K.; McInerney, G.M. Accumulation of autophagosomes in Semliki Forest virus infected cells is dependent on the expression of the viral glycoproteins. J. Virol. 2012, 86, 5674–5685.

54. Dahal, B.; Lin, S.C.; Carey, B.D.; Jacobs, J.L.; Dinman, J.D.; Van Hoek, M.L.; Adams, A.A.; Kehn-Hall, K. EGR1 upregulation following Venezuelan equine encephalitis viaarus infection is regulated by ERK and PERK pathways contributing to cell death. Virology 2020, 539, 121–128.

55. Cooray, S.; Jin, L.; Best, J.M. The involvement of survival signaling pathways in rubella-virus induced apoptosis. Virol. J. 2005, 2, 1.

56. Spuul, P.; Balistreri, G.; Kaariainen, L.; Ahola, T. Phosphatidylinositol 3-kinase-, actin-, microtubule-dependent transport of Semliki Forest Virus replication complexes from the plasma membrane to modified lysosomes. J. Virol. 2010, 84, 7543–7557.

57. Frolova, E.I.; Gorchakov, R.; Pereboeva, L.; Atasheva, S.; Frolov, I. Functional Sindbis virus replicative complexes are formed at the plasma membrane. J. Virol. 2010, 84, 11679–11695.

58. Kaur, S.; Sassano, A.; Joseph, A.M.; Majchrzak-Kita, B.; Eklund, E.A.; Verma, A.; Brachmann, S.M.; Fish, E.N.; Platanias, L.C. Dual regulatory roles of phosphatidylinositol 3-kinase in IFN signaling. J. Immunol. 2008, 181, 7316–7323.

09 THE RNA REPLICATION SITE OF TULA ORTHOHANTAVIRUS IS LOCATED WITHIN A MODIFIED GOLGI NETWORK

1. INTRODUCTION: AN OVERVIEW

The Orthohantavirus genus, belonging to the Hantaviridae family in the Bunyavirales order, consists of more than 36 species of enveloped segmented negative stranded RNA viruses [1]. These viruses are widely distributed worldwide and have the potential to cause severe human diseases [2]. Orthohantaviruses are classified into two clades, New World (NW) and Old World (OW), based on the country where they were isolated. The OW clade includes viruses that are widely distributed in Eurasia, while NW viruses are found exclusively in the Americas. Orthohantaviruses primarily infect a particular rodent host, leading to long-lasting infections [3]. On the other hand, numerous OW and NW orthohantaviruses are linked to different illness consequences in humans. Viruses belonging to the OW clade, such as Hantaan virus (HTNV) and Seoul virus (SEOV), are commonly linked to haemorrhagic fever with renal syndrome (HFRS). On the other hand, NW viruses like Andes virus (ANDV) and Sin Nombre virus (SNV) are responsible for causing hantavirus cardiopulmonary syndrome (HCPS). Both disorders are distinguished by significant arterial leakage, with human fatality rates varying from 0.1–10% for HFRS to over 40% for HCPS [4]. In rodent hosts, there are few observable changes in cell structure and function, and the animals have asymptomatic or subclinical long-lasting infections [5]. The ability of hantaviruses to avoid pathogen surveillance and innate immune defenses is reflected in cell culture systems that become persistently infected [6].

The orthohantavirus genome consists of three virion-associated negative sense RNA segments (vRNAs) known as small (S), medium (M), and large (L) based on their respective sizes [7]. The S segment contains the genetic information for the nucleocapsid protein (NP), which covers the viral RNAs and forms helical ribonucleoprotein (RNP) complexes. These complexes serve as templates for the production of RNA. The M segment contains a glycoprotein precursor (GPC) which is cleaved by cellular proteases to produce Gn and Gc proteins. These proteins form envelope spikes that play a crucial role in the attachment and entry of the virus. The L segment contains the genetic information for the large (L) protein, which acts as the catalytic part of the RNA-dependent RNA polymerase (RdRp). This protein is responsible for two different activities related to RNA synthesis. First, it transcribes mRNA from each vRNA template, resulting in a single 5' capped mRNA. Second, it replicates the three vRNA segments by creating complementary (cRNA) intermediates. The process of Hantavirus transcription requires the RdRp-associated endonuclease acquiring short 5′ capped oligoribonucleotides from host cell mRNAs. These oligoribonucleotides are subsequently utilized by the RdRp to initiate mRNA synthesis through a 'prime-and-realign' method [8]. It is believed that the primers for hantaviruses come from processing bodies (PBs). PBs are non-enveloped cytoplasmic RNP granules where non-sense mRNAs are held until they are broken down. It has been demonstrated that the SNV NP strongly associates with PBs [9]. It is believed that NP binds to the mRNA cap and shields it from being broken down by the de-capping enzymes DCP1a and DCP2, which are present in the PB. These enzymes are necessary for the survival of other viruses in the Bunyavirales group [10].

Hantaviruses, like all bunyaviruses, are believed to replicate inside cytoplasmic viral factories [11]. The reproduction factory of the prototype bunyavirus, Bunyamwera virus (BUNV), has been observed and extensively documented. BUNV in mammalian cells forms extensive virus factories near the Golgi stack, enlisting organelles, cytoskeletal proteins, and other cellular components to create elongated membranous structures called viral tubes [12]. The tubes contain several polymerase molecules that are enclosed by membranes. These polymerase molecules are responsible for synthesizing viral RNA, which then leads to the formation of RNPs for the assembly of new viral particles. It is believed that RNPs move to nearby assembly sites in the Golgi using actomyosin motors [12]. There, they engage with viral glycoproteins and form buds inside the Golgi lumen. Virions undergo maturation as they are transported within vesicles produced from the Golgi apparatus to the plasma membrane for release. The replication and assembly of BUNV in insect cells, where the virus creates a long-lasting infection, also takes place in the Golgi [13]. Nevertheless, it has been discovered that in mosquito cells, viruses do not gather within the cell and instead leave promptly after being formed, while NP and L proteins concentrate in cytoplasmic aggregates that are not bound by membranes [13].

Unlike BUNV, the specific location within the cell where hantavirus replicates has not been determined. Since NP is the primary protein component of RNPs, it is anticipated that its position would align with areas where RNA synthesis occurs. Previous studies have demonstrated that NP from the NW hantaviruses, ANDV and Black Creek Canal virus (BCCV), gathers within membrane structures that surround the Golgi [14,15]. According to this, the nucleoprotein (NP) from the OW hantavirus Tula virus (TULV) is found in perinuclear regions. Additionally, the NP of HTNV is located in both perinuclear regions and the endoplasmic reticulum (ER)–Golgi intermediate compartment. Additional evidence regarding the whereabouts of hantaviral RNA synthesis sites was discovered through the observation that the GFP-tagged L protein of TULV was found in the same location as the Golgi [19]. Additionally, it was observed that a recombinant L protein from ANDV gathered in perinuclear regions, co-localizing with both its corresponding NP and the PB marker DCP1a [20]. In general, hantavirus assembly is believed to happen through a process called budding into the Golgi, with a few exceptions related to NW hantavirus. This suggests that the information provided above on BUNV replication and assembly sites may also be applicable to hantaviruses in general. This study employed confocal immunofluorescence microscopy, fluorescent single molecule in-situ hybridization (FISH), and electron microscopy (EM) to precisely identify and analyze the specific cellular locations where NP and S segment RNAs were found during the entire duration of an infection caused by the OW hantavirus model, TULV. Our findings indicate that during the most severe and long-lasting stages of infection, NP exists in the form of large filamentous and tubular structures. These structures are found in close proximity to TIA-1, a marker for stress granules, as well as the Golgi. Additionally, the FISH analysis strongly suggests that these structures, which contain TIA-1 and originate from the Golgi, are the locations where TULV RNA synthesis occurs.

2. METHODS

2.1. Virus and Cells

Tula virus (TULV) strain Moravia/5302v/95 was obtained from The National Collection of Pathogenic Viruses, PHE, UK. Stocks were shown to be mycoplasma free using MycoAlert (Lonza group AG, Basel, Switzerland). Vero E6 cells are a clone of Vero 76 cells isolated from Cercopithecus aethiops (African green monkey) kidney cells, and were obtained from the European Collection of Authenticated Cell Cultures (ECACC 85020206). These cells were selected as previous findings show that TULV-infection allows the establishment of a persistently infected state [23]. Virus titres were calculated by indirect staining using the anti-NP antibody in conjunction with an Alexa Fluor 488 secondary antibody (Thermo Fisher Scientific, Waltham, MA USA) as previously described [23].

2.2. Antibody Production and Validation

Antisera specific for TULV NP were generated in house as previously described [23], using bacterially expressed SEOV NP core domain as an antigen, purified as previously described, and used to immunize a sheep (AltaBiosciences, Redditch, UK). Antibodies were validated by Western blot analysis.

2.3. TULV Infections of Vero E6 Cells

Briefly, Vero E6 cells were plated out at 1 × 106 cells per 25 cm2 flask or at 1 × 105 cells/well of a 12-well plate onto glass coverslips, sterilised by submerging into 100% ethanol and washing with sterile PBS. Cells were maintained in a humidified incubator at 37 °C with 5% CO2. TULV infections were carried out at a multiplicity of infection (MOI) of 0.1–0.5 in serum-free DMEM. Virus was adsorbed for 90 min before the addition of DMEM supplemented with 2% foetal bovine serum, 100 U/mL penicillin and 100 μg/mL streptomycin. For long-term infections, TULV infections were carried out in 25 cm2 flask and seeded onto coverslips 24 h prior to fixation. Fixation was carried out using 4% paraformaldehyde at 36 h post infection, 7 days post infection and 30 days post infection to represent early, peak and persistent infections.

2.4. Indirect Immunofluorescence

Mock-infected (media alone) or TULV-infected monolayers on coverslips were permeabilized in ice-cold methanol before washing in 1X PBS and blocked in 5% BSA in 1×PBS for 30 min. Primary staining was carried out using the following antibodies and reagents: in house anti-TULV NP [23], Golgi (Santa Cruz Biotechnology Inc, Dallas, TX AE6, USA), ER (Thermo Fisher Scientific Concanavalin A), β-actin (New England Biolabs (NEB), Ipswich, MA, USA D6A8), β-tubulin (NEB 9F3), vimentin (NEB D21H3), Rabs 5, 7 and 11 (NEB 2143, D95F2, D4F5), LAMP1 (NEB D2D11), clathrin heavy chain (NEB 4796), TIA-1 (Santa Cruz G-3) and DCP1a (Santa Cruz 56-Y). Incubations were carried out for 2 h at room temperature, followed by thorough washing in 1X PBS. Secondary antibody staining was carried out using donkey anti-goat Alexa Fluor 488 (Thermo Fisher Scientific A-11055), donkey anti-rabbit Alexa Fluor 555 (Thermo Fisher Scientific A-31572), donkey anti-mouse Alexa Fluor 555 (Thermo Fisher Scientific A-31570), donkey anti-rabbit Alexa Fluor 647 (Thermo Fisher Scientific A-31573), donkey anti-mouse Alexa Fluor 647 (Thermo Fisher Scientific A-31571) at a 1:500 dilution for 1 h. Nuclear staining was carried out by incubating with 300 nM DAPI for 3–5 min at room temperature. Cells to be visualised by confocal microscopy were mounted onto slides using Vectashield (Vector laboratories, Burlingame, CA, USA) hard set anti-fade mounting media and cured for 15 min before storing at 4 °C overnight. Cells to be further analysed by RNA-FISH were fixed in 4% formaldehyde for 20 min and stored in 1×PBS supplemented with 0.1 % diethylprocarbonate (DEPC).

2.5. TEM Preparation of TULV-Infected Vero E6 Cells

Vero E6 cells were fixed by adding 5% glutaraldehyde (v/v in 0.1 M phosphate buffer (PB) Agar Scientific, Stansted, UK) to an equal amount of cell media in the tissue culture flask, resulting in a final glutaraldehyde concentration of 2.5% (v/v). Cells were left for 2.5 h to fix at room temperature. After fixation, the cells were removed from the tissue culture flask by scraping and washed in 0.1 M PB for 30 min, twice. The cells were then post-fixed in 0.1 M PB with 1% osmium tetroxide (w/v) (Agar Scientific) for 1 h followed by four washes in 0.1 M PB. Subsequently, the post-fixed cells were dehydrated using an ascending ethanol series (40%, 60%, 80% and 2×100%) for 20 min, and then twice in propylene oxide (Agar Scientific) for 20 min. Dehydrated cells were incubated with a 1:1 mixture of propylene oxide and araldite epoxy resin (araldite CY212, dodecenyl succinic anhydride, and tri-dimethylaminomethyl phenol (Agar Scientific)) overnight, after which the mixture was replaced by pure araldite and incubated for 8 h. To ensure adequate embedding, new resin was added four times, after which the cells were polymerized at 60 °C for 24 to 48 h. Embedded cells were then sectioned in the ultra-thin range (80 to 100 nm) using a EM UC7 Ultramicrotome (Leica, Wetzlar, Germany). Ultra-thin sections were then placed upon 200 mesh carbon-coated copper-formvar grids and stained with lead citrate and uranyl acetate (UA). Sectioned cells were stained in 8% saturated (w/v) UA (Agar Scientific) for 1 h. UA-stained grids were then washed in dH2O for 5 min (×5). Afterwards, the sections were stained with lead citrate (Agar Scientific) for 10 min and washed in 0.02 M NaOH for 1 min (×3), and in dH2O for 1 min (×5). Sections were then imaged at 120 kV using a FEI Tecnai G2-Spirit electron microscope (Thermo Fisher Scientific) with a Gatan US4000/SP 4k × 4k CCD camera (Gatan, Pleasanton, CA, USA).

2.6. Fluorescent In-Situ Hybridisation (FISH) Probes

Fluorescent in-situ hybridisation DNA probes were designed against the sense and anti-sense TULV S segment ORF sequence using the Stellaris Probe Designer version 4.2 (LGC Biosearch Technologies, Novato, CA, USA). The Stellaris Probe Designer provided 48 oligonucleotide probe sequences that were 20 nucleotides in length with a minimum spacing of 2 nucleotides. Target TULV S segment ORF sequence was obtained from sequencing viral RNA isolated from TULV-infected cell culture medium (Genewiz Inc, South Plainfield, NJ, USA). Probes were assessed for cross-hybridization using NCBI Nucleotide Blast against C. aethiops (taxid:9534) mRNAs and against the TULV M and L segment sense and anti-sense sequences (strain: Tula/Moravia/5302v/95). Any probes with more than 15 consecutive complimentary nucleotides were discarded. TULV S segment sense and anti-sense probes were labelled with Quasar670. Strand sets were kept separate throughout this study.

2.7. Fluorescent In-Situ Hybridisation (FISH)

Probe hybridisation was carried out according to manufacturers' instructions. Briefly, each set of strand-specific FISH probes were reconstituted to 12.5 mM in RNase free 1× TE buffer (10 mM Tris-HCl, 1 mM EDTA pH 8.0 (G Biosciences) diluted in UltraPure DNase/RNase-Free Distilled water (Thermo Fisher Scientific) and stored separately at -20 °C. For each coverslip, 1 µL probe stock was used per 100 µL hybridisation buffer (Stellaris RNA FISH hybridisation buffer diluted with deionised formamide (Severn Biotech Ltd, Kidderminster, UK) in a 9:1 ratio).

Cells were incubated in Stellaris RNA FISH wash buffer A diluted with deionized formamide and UltraPure DNase/RNase-Free Distilled water in a 2:7:1 ratio for 5 min at room temperature. Additionally, 100 µL hybridisation buffer with probes was dispensed into a droplet onto parafilm in a humidified chamber. Coverslips were incubated cell-side down onto probe mix for 4 h at 37 °C, protected from light. Cells were incubated in 1 mL Stellaris RNA FISH wash buffer A for 30 min at 37 °C, protected from light and then incubated in 1 mL Stellaris RNA FISH wash buffer A with 300 nM DAPI for 5 min at room temperature. The buffer was removed and the cells were incubated in 1 mL Stellaris RNA FISH wash buffer B for 5 min before mounting onto glass microscope slides using Vectashield hard-set antifade mounting media (Vector laboratories). Slides were cured for 45 min at room temperature, protected from light, and imaged within 24 h.

2.8. Quantitative RT-PCR

TULV RNA was isolated from infected cells or supernatant using the QIAamp viral RNA mini kit (Qiagen, Hilden, Germany), and the quantitative reverse transcriptase real-time PCR was carried out using the 1-step RT-qPCR kit (Promega, Madison, WI, USA), both as previously described [23]. For each qRT-PCR experiment, in vitro transcribed TULV S segment RNAs were diluted to generate 108, 106, 104, 102 genome copies/5 μL and used to build a standard curve from which the sample relative genome copies per mL could be calculated, as previously described [23].

2.9. Inhibition of Intermediate Filaments

Vero E6 were infected with TULV at a MOI of 0.5 in a 6-well plate. Infected cells were harvested for immunofluorescence and q-RT PCR analysis at 30 days post infection. Prior to harvesting, the cells were treated with the following cytoskeletal inhibitors; 17 μM nocodazole (Sigma-Aldrich, St. Louis, MO, USA) for 60 min and 400 nM okadaic acid (OA) (Sigma-Aldrich) for 30, 60 or 90 min. Infected Vero E6 cells treated with OA were also recovered by removing the inhibitor and incubating overnight with fresh media before harvesting. Mock treatments were carried out using an equal volume of solvent (dimethyl sulfoxide (DMSO)/ethanol).

2.10. Laser Scanning Confocal Microscopy

Laser scanning confocal imaging was carried out using the Zeiss LSM700 confocal microscope using the 40×/1.3 Oil DIC Plan Apochromat, M27 objective lens with the diode 405 nm, 488 nm, 55 nm and 639 nm lasers or the Zeiss LSM880 + Airyscan upright confocal microscope using the Plan-Apochromat 40×/1.4 Oil DIC objective lens with the diode 405 nm, argon 488 nm, DPSS 561 nm and HeNe 633 nm lasers. Image analysis was carried out in Fiji Is Just ImageJ (FIJI).

3. RESULTS

3.1. TULV NP Forms Both Punctate and Tubular Structures during Infection

We first analysed the intracellular distribution of TULV NP during an extended time course of up to 30 days, with time points of 36 h post infection (hpi), 7 days post infection (dpi) and 30 dpi chosen to represent the early, peak and persistent stages of TULV replication, respectively, as previously reported [23]. At the 36 hpi time point, NP mostly localised within discrete perinuclear puncta, with a minor component forming larger filamentous and tube-like structures (Figure 1A, top row). At this early time point, NP compartments had an average area of 0.54 μm2 (14 cells and 1248 compartments analysed) with the largest structure within a single focal plane having an area of 11.8 μm2. After 7 dpi, the NP compartments had an average area of 0.96 μm2 (31 cells and 2959 compartments analysed) with the largest structure having an area of 154 μm2 (Figure 1A, middle row). At 30 dpi, when fewer cells were present within the culture, but when surviving cells were still producing infectious TULV [23], most of the NP was present as tubes/filaments, reaching a maximum area of 513 μm2 (15 cells and 8615 compartments analysed), with only low levels of punctate NP visible (Figure 1A, bottom row). Across the time course, the length of the filamentous/tubular structures increased, and their diameters remained relatively constant (Figure 1A).

To reveal further information of the three-dimensional morphology of the large NP-stained structures, successive Z axis sections of persistently infected cells (0.25 μm per stack) were reconstructed to a generate a three-dimensional image. These revealed that the long tubular and filamentous structures were connected and appeared to wrap around the nucleus (Figure 1B).

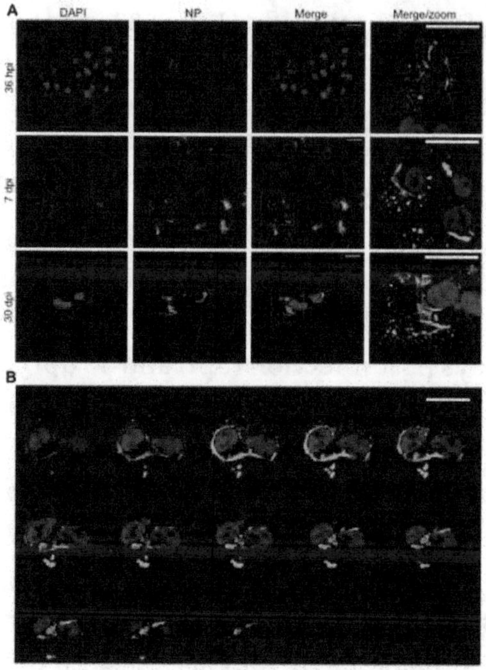

Figure 1. Confocal images showing the characteristic morphology of Tula virus (TULV) nucleocapsid protein (NP) -stained structures at early, peak and persistent stages of TULV infection in Vero E6 cells. (A) Vero E6 cells were infected with TULV at a multiplicity of infection of 0.1, and the images were taken at 36 h post infection (hpi), 7 days post infection (dpi) and 30 dpi using 40× magnification. Nuclei were stained with DAPI (blue) and TULV NP, detected using NP antisera (green), revealing the characteristic NP punctate and filamentous/tubular structures in the perinuclear region. (B) Successive Z sections of TULV-infected cells at 30 dpi show NP filamentous/tubular structures extending around the nucleus. The scale bars represent 30 μm.

3.2. Imaging of TULV-Induced Filamentous Cytoplasmic Structures Using Transmission Electron Microscopy

To further characterise these TULV NP-associated filamentous tubular structures, we then used transmission electron microscopy (TEM) to visualize the sections of TULV-infected Vero E6 cells at 36 hpi, 7 dpi and 30 dpi time points (Figure 2). During the early 36 hpi time point, no clear differences were detected between the TULV- and mock-infected cells (data not shown), most likely due to the relatively small size of TULV NP-associated structures, previously viewed (Figure 1).

However, after 7 days, long filamentous and tubular structures were observed in 6% of the infected cells (eight out of 134 cells; Figure 2B), that were absent in non-infected cells, with no filaments detected in any of 61 non-infected cells that were imaged (Figure 2A). The relatively small percentage of infected cells containing filaments is explained because the EM images represent thin sections (~100 nm thick), suggesting that the filaments are concentrated in specific locations within the infected cell. Such filaments comprised clusters of small bundles, with each comprising multiple slender filamentous structures. Filament bundles were around ~100 to 1000 nm in length and spread over a large area within the cell, forming clusters that reached up to 4 μm in length. While filaments within the bundles were aligned in similar directions, the separate clusters often protruded in different orientations.

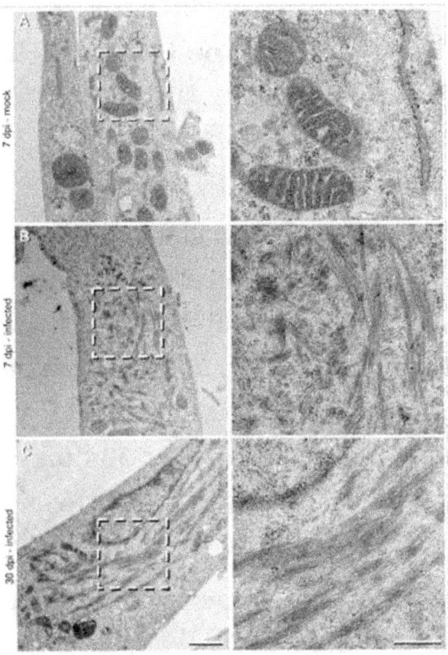

Figure 2. Transmission electron microscopy (TEM) images showing the filamentous structures within the TULV-infected Vero E6 cells. Vero E6 cells, (A) mock-infected, or infected with TULV at a MOI of 0.1 at either (B) 7 days post infection (dpi) or (C) 30 dpi. Higher magnification images from the marked inset regions are shown in the right column. Scale bars represent 500 nm and 1 μm on low and high magnification micrographs, respectively.

At 30 dpi, the filamentous clusters were present, but had approximately doubled in their frequency (to 10%; out of ~60 imaged cells) and had dramatically increased in size to ~5 μm in total length (Figure 2C). Individual bundles of tubular filaments were still evident, but these were less frequently observed. Instead, the bundles were longer (500–3000 nm) and shared a similar orientation. The dimensions and overall morphology of these TULV-induced structures at both the 7 dpi and 30 dpi time points coincided with those filamentous tubular structures stained for TULV NP and visualized by light microscopy, shown above (Figure 1).

3.3. Co-Localization of TULV NP with Cytoskeletal Components

Previous studies have shown that HTNV NP associates with actin [15], tubulin and vimentin [18], the latter forming cages surrounding HTNV NP in the vicinity of the ER-Golgi intermediate compartment (ERGIC). Consequently, we then determined the distribution of the punctate and filamentous forms of TULV NP in relation to these cytoskeletal components.

At 36 hpi, no association of either punctate or filamentous NP with actin, tubulin, or vimentin were observed (Figure 3A). However, at 7 dpi, vimentin was redistributed and co-localized with NP in the larger filamentous/tubular structures (Figure 3B, line scan 1) although vimentin/NP co-localization was less apparent in punctate compartments (Figure 3B, line scan 2). At 30 dpi, vimentin appeared to surround the NP structures (Figure 3C). Neither actin nor tubulin co-localized with NP at any time point.

To further characterise the association between vimentin and NP, the infected cells were treated with okadaic acid (OA), an inhibitor of protein phosphatase 1 and 2A [24] that promotes vimentin disassembly [25]. OA treatment led to the disappearance of the typical NP structures, which rapidly reformed upon OA removal. In addition, OA treatment of TULV-infected cultures at 30 dpi led to reduced TULV genome copy numbers, suggesting that the disruption of the vimentin network was detrimental to TULV RNA replication. In contrast, nocodazole (NOC) caused no changes to the tubulin networks and did not influence TULV replication.

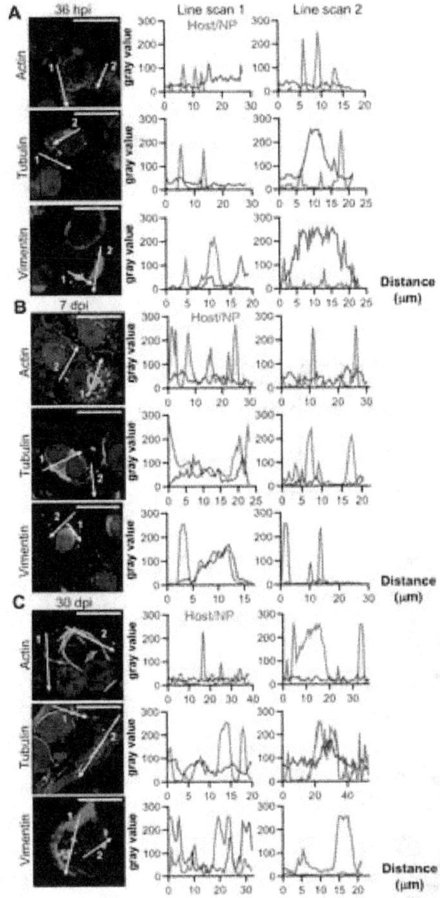

Figure 3. Co-localisation between the TULV NP and cytoskeletal filaments at the early, peak and persistent stages of TULV infection in Vero E6 cells. The spatial distribution of the cytoskeletal markers actin, tubulin and vimentin (magenta) was observed alongside TULV NP (green) in Vero E6 cells infected with TULV at a MOI of 0.1 by confocal microscopy at 40× magnification at (A) 36 h post infection (hpi), (B) 7 days post infection (dpi), and (C) 30 dpi. The scale bar represents 30 µm. Fluorescent line scans were taken using Fiji software with the designated number labels corresponding to adjacent line scan plots. Nuclei were stained with DAPI (blue), the TULV NP was detected using NP antisera and the cytoskeleton was detected using primary antibodies against β-actin, β-tubulin and vimentin. Images shown in panels A, B and C show the representative analysis of 87, 109 and 58 cells, respectively.

3.4. Stress Granules (SGs) Are Recruited to Filamentous Structures during TULV Infection

PBs and stress granules (SGs) are non-membrane bound, liquid phase compartments that have been shown to stimulate the RNA synthesis of some negative sense RNA viruses and closely associate with their corresponding replication factories [26,27,28,29]. To examine the possible association of TULV NP and these compartments, the infected cells were stained with the PB marker DCP1a, a PB-resident mRNA de-capping enzyme, or the SG marker TIA-1, an RNA binding protein that scaffolds SG formation.

At the early, peak and persistent stages of the time course visualized, DCP1a-stained puncta were abundant (Figure 4A), but were spatially-distinct from the large filamentous/tubular NP-stained structures (Figure 4B,C) and with fewer than 4% co-localizing with NP (Figure 4D).

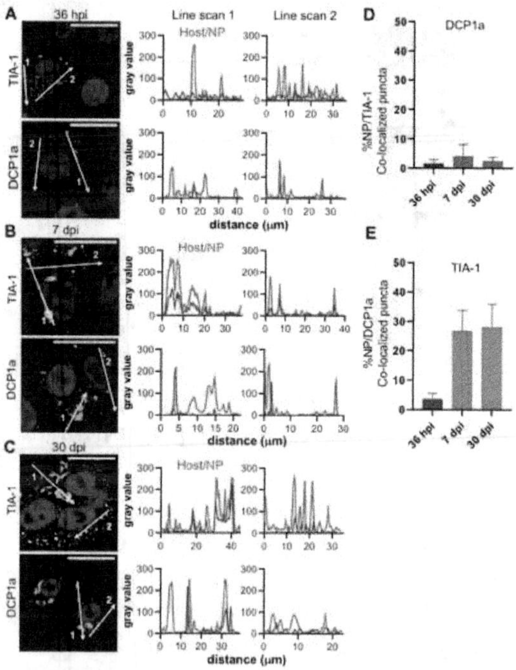

Figure 4. Co-localisation between TULV NP and either the stress granules (SGs) or the processing bodies (PBs) at the early, peak and persistent stages of TULV infection in Vero E6 cells. The spatial distribution of the SG marker TIA-1 and the PB marker DCP1a (magenta) was observed alongside TULV NP (green) in Vero E6 cells infected with TULV at a MOI

of 0.1 by confocal microscopy at 40× magnification at the following time points; (A) 36 h post infection (hpi), (B) 7 days post infection (dpi) and (C) 30 dpi. The scale bars represent 30 μm. Fluorescent line scans were taken using Fiji software with the designated number labels corresponding to the adjacent line scan plots. Nuclei were stained with DAPI (blue), TULV NP was detected using NP antisera, and the SGs and PBs were detected using specific antisera against TIA-1 and DCP1a, respectively. Percentage co-localization of D) TIA-1 and E) DCP1a puncta co-localizing with NP at all three time points were quantified and represented as separate histograms. Images shown in panels A, B and C show a representative analysis of 7, 25 and 23 cells, respectively.

In contrast, the staining pattern of TIA-1 was different; at 36 hpi, only low numbers of TIA-1 puncta were detectable and their co-localization with NP was relatively low (Figure 4A). However, at 7 dpi and 30 dpi, the abundance of TIA-1 increased in dense cytoplasmic structures that co-localized with both filamentous/tubular NP compartments, as well as NP puncta (Figure 4B–C). The line scans associated with these later time points clearly revealed an overlap of the TIA-1 signal with both the filamentous/tubular and punctate NP signals, with their peak intensities coinciding (Figure 4B,C), corroborated by almost 30% of the TIA-1-stained compartments co-localizing with NP puncta (Figure 4E). Taken together, these results suggest that SG formation is simulated during the later stages of TULV infection, and that SGs are recruited to NP positive filamentous/tubular compartments.

3.5. TULV NP Associates with the Golgi during Infection

For many hantaviruses of both OW and NW clades, their corresponding NPs accumulate at sites within and adjacent to the Golgi, suggesting that this compartment may play an important role in hantavirus RNA synthesis. To better understand the intracellular distribution of TULV NP, the proximity of known ER and Golgi markers were assessed relative to NP in TULV-infected cells.

At 36 hpi, the Golgi appeared with the classical morphology of elongated and stacked vesicular compartments that were distinct from NP (Figure 5A). In contrast, at both 7 dpi and 30 dpi, there was extensive overlap with NP and the Golgi, in both large filamentous/tubular NP structures, and also punctate NP, with closely coinciding peaks on the associated line scans (Figure 5B,C). No such co-localisation between NP and the ER marker concanavalin A was observed at any of the specified time points (Figure 5A–C). In addition, NP showed minimal co-staining with clathrin, Rab5, Rab7, Rab11 or LAMP 1 markers, implying that the punctate NP structures were not endosomal in nature.

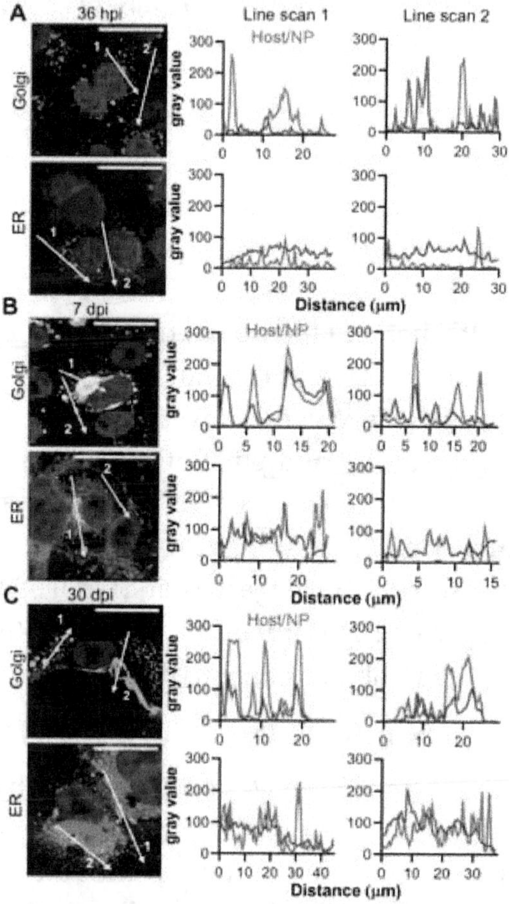

Figure 5. Co-localisation between the TULV NP and either the endoplasmic reticulum (ER) or the Golgi compartments at the early, peak and persistent stages of TULV infection in Vero E6 cells. The spatial distribution of the endoplasmic reticulum and the Golgi network (both magenta) was observed alongside TULV NP (green) in Vero E6 cells infected with TULV at an MOI of 0.1 by confocal microscopy at 40 X magnification at (A) 36 h post infection (hpi), (B) 7 days post infection (dpi) and (C) 30 dpi. The scale bars represent 30 µm. Fluorescent line scans were taken using Fiji software

with the designated number labels corresponding to the adjacent line scan plots. Nuclei were stained with DAPI (blue), TULV NP was detected using NP antisera, with the ER detected using concanavalin-A and the Golgi was detected by specific antisera. Images shown in panels A, B and C show a representative analysis of the 19, 27 and 13 cells, respectively.

3.6. Products of TULV RNA Synthesis co-Localize with NP

The results presented thus far show that NP exhibits a close co-localization with the Golgi and SGs, within large filamentous and tubular structures. As NP is the major protein component of TULV RNPs, which act as templates for TULV mRNA transcription and RNA replication, we next investigated whether these NP-associated compartments were also stained for the products of TULV RNA synthesis.

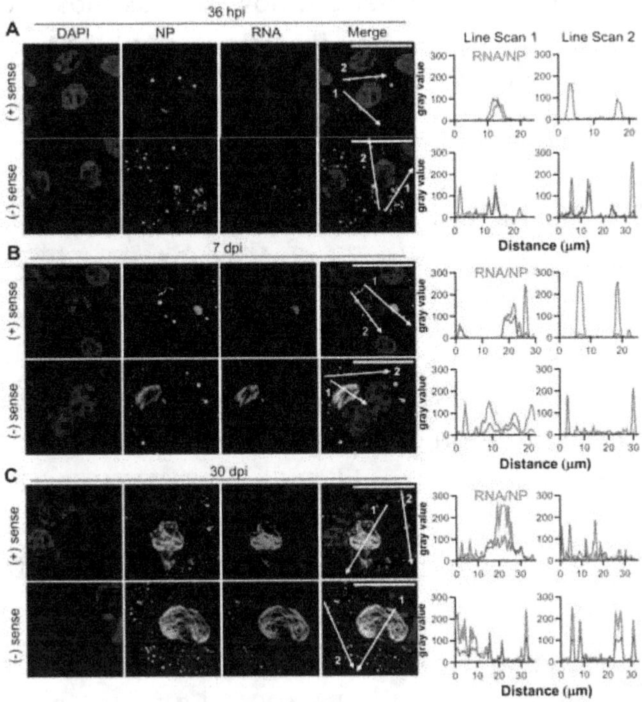

Figure 6. Co-localisation between the TULV NP and the TULV small (S) segment RNAs at the early, peak and persistent stages of TULV infection in Vero E6 cells. The spatial distribution of the TULV NP (green) with positive and negative sense S segment RNA (red) detected with strand-specific probe sets was observed using confocal microscopy at (A) 36 h post infection (hpi), (B) 7 days post infection (dpi) and (C) 30 dpi. The size bars represent 30 μm. Fluorescent line scans were taken using Fiji software

with the designated number labels corresponding to the adjacent line scan plots. Nuclei were stained with DAPI (blue), TULV NP was detected using NP antisera and S segment RNA was detected using sense or anti-sense RNA FISH probes. Images presented in panels A, B and C show a representative analysis of 14, 25 and 8 cells, respectively.

Two independent FISH probe sets were designed comprising multiple non-overlapping fluorescently labelled short oligonucleotides with near complete coverage of the TULV S segment RNA target. One set was designed to detect negative sense TULV S segment RNAs, comprising the vRNA genome alone, whereas a second set was designed to detect positive sense S segment RNAs. The positive sense probe set would unavoidably hybridize to both S segment mRNAs and cRNAs due to their nested genetic organisation, however, as most positive sense RNAs generated during bunyavirus infection are mRNA transcripts [30], the majority of the detected positive sense signal could be attributed to mRNA transcripts, with only a minor contribution from cRNA replication products.

At the early, peak and persistent time points, the cells were fixed and stained for TULV NP along with the individual probe sets. Only the cells staining positive for NP were also stained with the FISH-probe sets, showing that the probes were specific for viral RNA targets. Interestingly, the ability of the probe sets to detect negative sense RNAs showed that the association of NP with RNA within assembled RNPs did not prevent probe binding. In addition, the TULV RNA signal was not detected outside regions positive for NP, strongly suggesting that, as expected, NP is required for all TULV RNA synthesis activities.

At 36 hpi, 7 dpi and 30 dpi time points, the signal corresponding to both the positive and negative sense RNAs co-localized with NP, with the extremely close co-localization of NP and RNA within the filamentous/tubular objects (Figure 6A–C). RNA and NP also co-localized within some NP puncta of various sizes, although with a less consistent level of abundance. Line scan analysis of the filamentous/tubular and punctate NP structures supported these observations, with consistently high levels of the NP and RNA signal within the filamentous/tubular structures, but often lower levels of RNA in NP puncta.

At all of the time points assessed, there were no consistent differences between the distribution of positive and negative sense RNAs in relation to the NP structures, suggesting that the synthesis of mRNA/cRNAs and vRNAs is not segregated.

3.7. Concurrent Detection of TULV and Cellular Components in Persistently Infected Cells

To confirm the above findings, and unequivocally show that the staining of viral and host components were located within the same compartments, 30 dpi persistently infected cells were co-stained for either NP/RNA/TIA-1 or NP/RNA/Golgi, concurrently (Figure 7).

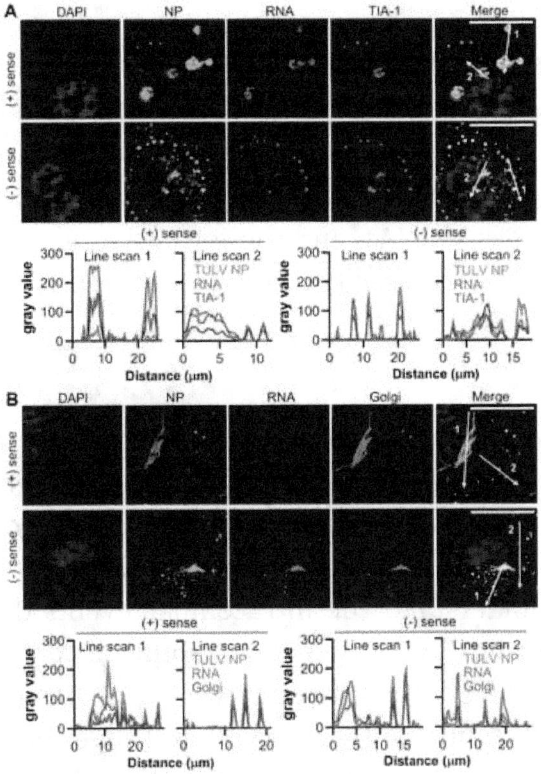

Figure 7. Co-localisation between the TULV NP, the TULV S segment positive and the negative sense RNAs and host cell components TIA-1 and the Golgi marker in Vero E6 cells at 30 days post infection. The spatial distribution of the TULV NP (green), the S segment RNA (magenta) and the host cell markers (A) TIA-1, or (B) Golgi (both red) were observed during the persistent infections using confocal microscopy at 40× magnification.

Fluorescent line scans were taken using Fiji software with designated number labels corresponding to the adjacent line scan plots. The nuclei were stained with DAPI (blue), TULV NP was detected using NP antisera, with TIA-1 and Golgi detected using specific antisera. Scale bar represents 30 µm. TULV NP was detected using NP antisera, and S segment RNAs were detected using separate sets of positive sense or negative-sense RNA FISH probes. Images shown in panels A and B show a representative analysis of 8 and 20 cells, respectively.

Merging the signals from all three channels representing NP/RNA/TIA-1 confirmed that the NP, TIA-1 and both polarities of TULV RNA were simultaneously present within both filamentous and punctate structures. As shown above (Figure 6), all RNA staining coincided with that of NP, however, some of these regions were devoid of TIA-1. Similarly, merging the signals for NP/RNA/Golgi revealed high levels of co-localization for all three markers within the filamentous/tubular compartments (Figure 7B). Along with the results shown in Figure 5, these findings strongly suggest that TULV infection leads to the redistribution of the Golgi to coincide with TULV RNAs, and the formation of structures that we suggest most likely represent the viral replication factories, the major sites of TULV RNA synthesis.

4. DISCUSSION

Many hantaviruses establish persistent infections within their rodent hosts by manipulating the host cellular environment to allow the abundant synthesis of viral components for an extended time period, while also evading or minimizing the cellular innate immune response [6,7]. To better understand the intracellular events during long-term hantavirus infections, we used immunofluorescent microscopy to investigate the intracellular distribution of TULV NP and RNA, alongside multiple host cell components during an extended time course of up to 30 days, during which time the persistent infection by TULV has been previously demonstrated [23]. This was combined with EM imaging to understand the morphological changes that TULV induces in infected cells. Our findings revealed that as the infection progressed, the NP formed increasingly large ultrastructural assemblies, mostly with an overall filamentous and tubular appearance, but also as discrete puncta

(Figure 1). At all the time points examined, these NP assemblies closely co-localized with TULV RNA, regardless of its negative or positive sense polarity (Figure 6). This close association between all hantavirus RNA and NP is consistent with the known structure of the hantavirus RNP segments, in which hantaviral vRNAs and cRNAs are always found in close association with NP, forming extended helical assemblies [31]. Interestingly, our observations also suggest that TULV mRNAs, which will likely comprise the majority of positive sense TULV-specific RNA in the infected cell [30], also remain in close proximity with NP. This may reflect the organisation of the replication factories, with newly transcribed S, M and L segment mRNAs remaining in the vicinity of their respective NP enwrapped vRNA templates. Alternatively, it may reflect the previous findings that the hantavirus NP binds the 5′ mRNA cap at a separate RNA binding site [32], and this interaction is proposed to enable mRNA circularization, so as to promote eIF4F independent translation [33]. The lack of distinction between the sites of positive sense and negative sense RNA staining at the resolution used here also suggests that TULV RNA replication and mRNA transcription processes are not physically separated into distinct compartments.

Co-staining investigations utilizing markers for intracellular compartments demonstrated that NP and TULV RNA were present in the same location within structures that were in close proximity to the Golgi (Figure 5). During the initial phase of infection (36 hours post-infection), NP, RNA, and Golgi components were found in both punctate and irregular filamentous structures inside and around compartments that had a typical Golgi architecture. At the latter time periods of 7 dpi and 30 dpi, there was an increased abundance of NP and RNA within a single filamentous/tubular structure of considerable dimensions. This structure was tightly associated with the Golgi, as shown in Figure 5B–C. The structure stained with filamentous/tubular NP/RNA/Golgi exhibited an unusual architecture that deviated from the typical compact ribbon structure of Golgi cisternae. This indicates that the Golgi compartment underwent remodeling, possibly due to TULV-related activities. Given that both TULV NP and RNA components are present in these compartments (Figure 7), it is reasonable to conclude that the remodelled Golgi serves as the location for TULV RNA synthesis and accumulation. Furthermore, it is highly probable that these compartments function as TULV replication factories. The observation of elongated filamentous formations at both 7 dpi and 30 dpi time points in TULV-infected cells, as seen using TEM, corresponded with our results obtained through light microscopy (Figure 2). The cell sections infected with TULV exhibited a higher level of detail in the stained structures of filamentous/tubular NP/RNA/Golgi, thereby confirming their abnormal shape. Their apparent lack of presence in certain sections of 36 hours post-infection (hpi) cells is most likely due to their low quantity and the resulting high probability that they would not be visible in the chosen sections. Only a limited number of research have used electron microscopy (EM) to examine cells infected with hantavirus. These investigations mostly focused on analyzing the structure and appearance of hantavirus particles associated with the cells [34,35,36]. Nevertheless, the filamentous structures we saw in this study have similarities to the filamentous and granular structures that were caused by the Sin Nombre virus [21]. These structures were likewise found in close proximity to the Golgi. Nevertheless, our investigation demonstrated that these elongated structures also contain a substantial amount of viral RNA. If the filamentous/tubular NP-stained structures serve as the locations for RNA synthesis and accumulation, then the purpose of the punctate NP structures observed throughout the infection time course remains uncertain. Since they exhibit minimal affinity for RNA staining, we suggest that these sites are improbable locations for active and prolific RNA synthesis. One potential explanation is that these structures could be Golgi-derived particles, released as a result of Golgi fragmentation. However, another possibility that cannot be dismissed is that the smallest of these structures might be exocytic vesicles carrying fully assembled viruses, which are in the process of being transported to the plasma membrane for release. In contrast to the SG marker TIA-1, the degree of co-localization between the NP and the PB marker DCP1a was notably reduced (Figure

4). This discovery was surprising because earlier research has indicated a significant correlation between DCP1a and the NP of the NW SNV. It is hypothesized that the NP binds and protects capped RNAs, which are then delivered to the transcribing RdRp. These findings suggest notable differences in the interactions between the host and pathogen for hantavirus species originating from the New World (NW) and Old World (OW). OW hantaviruses, such as TULV, may not acquire or process cap moieties from host cells in a similar fashion. Given the robust link we have demonstrated between TULV NP and TIA-1, it is possible that SGs might be utilized as an alternate reservoir of caps. The observation that the SG marker TIA-1 exhibited a punctate distribution in the early stages of virus infection, but later showed a filamentous/tubular distribution, suggests that it was redistributed into the Golgi-derived filamentous/tubular compartments (Figure 4A–D). According to our current knowledge, there have been no earlier reports of significant co-localization between SGs and the Golgi for a natural reason. Hence, we propose that this shift is a response to TULV infection. The existing knowledge on whether this response is supportive of viral replication or inhibitory to viral replication remains ambiguous. There is substantial data suggesting that SGs function as inherent platforms for immunological signaling [37]. One possible reason is that when viruses are confined within TULV replication factories, it may impede or delay the host cell's immune response to the infection. This immune evasion strategy may contribute to the development of hantavirus persistence, a common disease.

CONCISE SUMMARY

The Hantaviridae family, which belongs to the Bunyavirales order, consists of RNA viruses with three segments that have a negative sense. Many of these viruses are transmitted by rodents and are known to cause severe and often deadly diseases in humans. On the other hand, when rodents that carry hantavirus are infected, they experience either no symptoms or the infection remains dormant. This can be replicated in laboratory cells, which then become chronically infected. This study employed Tula virus (TULV) to examine the site of hantavirus replication at different stages of infection, including the early, peak, and chronic phases, over a period of 30 days. Through the utilization of immunofluorescent (IF) microscopy, we have demonstrated that the TULV nucleocapsid protein (NP) is present in both punctate and filamentous formations. Furthermore, the size of the filamentous structures increases as the infection advances. Transmission electron imaging of cell sections infected with TULV showed that the filamentous structures consisted of clusters of aligned filament bundles. Throughout the infection time period, the filamentous structures associated with NP showed a growing tendency to be located in close proximity to the Golgi and the stress granule marker TIA-1. This indicates a redistribution of these cellular organelles. Fluorescent in-situ hybridization was used to analyze the intracellular distribution of TULV RNAs. The results showed that both genomic and mRNAs were found in the same location as Golgi-associated filamentous compartments that tested positive for TIA. The results demonstrate that TULV causes a significant restructuring of the internal cellular environment, leading to the creation of TULV RNA synthesis factories within modified Golgi compartments.

REFERENCES

1. Abudurexiti, A.; Adkins, S.; Alioto, D.; Alkhovsky, S.V.; Avsic-Zupanc, T.; Ballinger, M.J.; Bente, D.A.; Beer, M.; Bergeron, E.; Blair, C.D.; et al. Taxonomy of the order Bunyavirales: Update 2019. Arch. Virol. 2019, 164, 1949–1965.

2. Spiropoulou, C.F.; Bente, D.A. Orthohantavirus, Orthonairovirus, Orthobunyavirus and Phlebovirus. In Fields Virology; Emerging Viruses, 7th ed.; Knipe, D., Howley, P., Eds.; Lippincott, Williams & Wilkins: Philadelphia, PA, USA, 2020; Volume 1, p. 750.

3. Schountz, T.; Prescott, J. Hantavirus Immunology of Rodent Reservoirs: Current Status and Future Directions. Viruses 2014, 6, 1317–1335.

4. Jonsson, C.B.; Figueiredo, L.T.M.; Vapalahti, O. A Global Perspective on Hantavirus Ecology, Epidemiology, and Disease. Clin. Microbiol. Rev. 2010, 23, 412–441.
5. Meyer, B.J.; Schmaljohn, C.S. Persistent hantavirus infections: Characteristics and mechanisms. Trends Microbiol. 2000, 8, 61–67.
6. Meyer, B.J.; Schmaljohn, C.S. Accumulation of Terminally Deleted RNAs May Play a Role in Seoul Virus Persistence. J. Virol. 2000, 74, 1321–1331.
7. Barr, J.N.; Weber, F.; Schmaljohn, C.S. Bunyavirlales; The viruses and thier replication. In Fields Virology; Emerging Viruses, 7th ed.; Knipe, D., Howley, P., Eds.; Lippincott, Williams & Wilkins: Philadelphia, PA, USA, 2020; Volume 1, p. 706.
8. Garcin, D.; Lezzi, M.; Dobbs, M.; Elliott, R.M.; Schmaljohn, C.; Kang, C.Y.; Kolakofsky, D. The 5′ ends of Hantaan virus (Bunyaviridae) RNAs suggest a prime-and-realign mechanism for the initiation of RNA synthesis. J. Virol. 1995, 69, 5754–5762.
9. Mir, M.A.; Duran, W.A.; Hjelle, B.L.; Ye, C.; Panganiban, A.T. Storage of cellular 5′ mRNA caps in P bodies for viral cap-snatching. Proc. Natl. Acad. Sci. USA 2008, 105, 19294–19299.
10. Hopkins, K.C.; McLane, L.M.; Maqbool, T.; Panda, D.; Gordesky-Gold, B.; Cherry, S. A genome-wide RNAi screen reveals that mRNA decapping restricts bunyaviral replication by limiting the pools of Dcp2-accessible targets for cap-snatching. Genes Dev. 2013, 27, 1511–1525.
11. Novoa, R.R.; Calderita, G.; Arranz, R.; Fontana, J.; Granzow, H.; Ortiz, C.R. Virus factories: Associations of cell organelles for viral replication and morphogenesis. Boil. Cell 2012, 97, 147–172.
12. Fontana, J.; Lopez-Montero, N.; Elliott, R.M.; Fernández, J.J.; Ortiz, C.R. The unique architecture of Bunyamwera virus factories around the Golgi complex. Cell. Microbiol. 2008, 10, 2012–2028.
13. Lopez-Montero, N.; Ortiz, C.R. Self-protection and survival of arbovirus-infected mosquito cells. Cell. Microbiol. 2010, 13, 300–315.

14. Rowe, R.K.; Suszko, J.W.; Pekosz, A. Roles for the recycling endosome, Rab8, and Rab11 in hantavirus release from epithelial cells. Virology 2008, 382, 239–249.

15. Ravkov, E.V.; Compans, R.W. Hantavirus Nucleocapsid Protein Is Expressed as a Membrane-Associated Protein in the Perinuclear Region. J. Virol. 2001, 75, 1808–1815.

16. Kaukinen, P.; Vaheri, A.; Plyusnin, A. Non-covalent interaction between nucleocapsid protein of Tula hantavirus and small ubiquitin-related modifier-1, SUMO-1. Virus Res. 2003, 92, 37–45.

17. Maeda, A.; Lee, B.H.; Yoshimatsu, K.; Saijo, M.; Kurane, I.; Arikawa, J.; Morikawa, S. The Intracellular Association of the Nucleocapsid Protein (NP) of Hantaan Virus (HTNV) with Small Ubiquitin-like Modifier-1 (SUMO-1) Conjugating Enzyme 9 (Ubc9). Virology 2003, 305, 288–297.

18. Ramanathan, H.N.; Chung, D.-H.; Plane, S.J.; Sztul, E.; Chu, Y.-K.; Guttieri, M.C.; McDowell, M.; Ali, G.; Jonsson, C.B. Dynein-Dependent Transport of the Hantaan Virus Nucleocapsid Protein to the Endoplasmic Reticulum-Golgi Intermediate Compartment. J. Virol. 2007, 81, 8634–8647.

19. Kukkonen, S.K.J.; Vaheri, A.; Plyusnin, A. Tula hantavirus L protein is a 250 kDa perinuclear membrane-associated protein. J. Gen. Virol. 2004, 85, 1181–1189.

20. Heinemann, P.; Schmidt-Chanasit, J.; Günther, S. The N Terminus of Andes Virus L Protein Suppresses mRNA and Protein Expression in Mammalian Cells. J. Virol. 2013, 87, 6975–6985.

21. Goldsmith, C.S.; Elliott, L.H.; Peters, C.J.; Zaki, S.R. Ultrastructural characteristics of Sin Nombre virus, causative agent of hantavirus pulmonary syndrome. Arch. Virol. 1995, 140, 2107–2122.

22. Ravkov, E.V.; Nichol, S.T.; Compans, R.W. Polarized entry and release in epithelial cells of Black Creek Canal virus, a New World hantavirus. J. Virol. 1997, 71, 1147–1154.

23. Davies, K.; Afrough, B.; Mankouri, J.; Hewson, R.; Edwards, T.A.; Barr, J.N. Tula orthohantavirus nucleocapsid protein is cleaved in infected cells and may sequester activated caspase-3 during persistent infection to suppress apoptosis. J. Gen. Virol. 2019, 100, 1208–1221.

24. Yatsunami, J.; Fujiki, H.; Suganuma, M.; Yoshizawa, S.; Eriksson, J.E.; Olson, M.O.; Goldman, R.D. Vimentin is hyperphosphorylated in primary human fibroblasts treated with okadaic acid. Biochem. Biophys. Res. Commun. 1991, 177, 1165–1170.

25. Lee, W.-C.; Yu, J.-S.; Yang, S.-D.; Lai, Y.-K. Reversible hyperphosphorylation and reorganization of vimentin intermediate filaments by okadaic acid in 9L rat brain tumor cells. J. Cell. Biochem. 1992, 49, 378–393.

26. Dinh, P.X.; Beura, L.K.; Das, P.B.; Panda, D.; Das, A.; Pattnaik, A.K. Induction of Stress Granule-Like Structures in Vesicular Stomatitis Virus-Infected Cells. J. Virol. 2012, 87, 372–383.

27. Lindquist, M.E.; Lifland, A.W.; Utley, T.J.; Santangelo, P.J.; Crowe, J.E. Respiratory Syncytial Virus Induces Host RNA Stress Granules To Facilitate Viral Replication. J. Virol. 2010, 84, 12274–12284.

28. Nelson, E.V.; Schmidt, K.M.; Deflubé, L.R.; Doğanay, S.; Banadyga, L.; Olejnik, J.; Hume, A.; Ryabchikova, E.; Ebihara, H.; Kedersha, N.; et al. Ebola Virus Does Not Induce Stress Granule Formation during Infection and Sequesters Stress Granule Proteins within Viral Inclusions. J. Virol. 2016, 90, 7268–7284.

29. Nikolic, J.; Civas, A.; Lama, Z.; Lagaudriere-Gesbert, C.; Blondel, D. Rabies Virus Infection Induces the Formation of Stress Granules Closely Connected to the Viral Factories. PLOS Pathog. 2016, 12, e1005942.

30. Barr, J.N.; Elliott, R.M.; Dunn, E.F.; Wertz, G.W. Segment-specific terminal sequences of Bunyamwera bunyavirus regulate genome replication. Virology 2003, 311, 326–338.

31. Arragain, B.; Reguera, J.; Desfosses, A.; Gutsche, I.; Schoehn, G.; Malet, H. High resolution cryo-EM structure of the helical RNA-bound Hantaan virus nucleocapsid reveals its assembly mechanisms. eLife 2019, 8.

32. Mir, M.A.; Sheema, S.; Haseeb, A.; Haque, A. Hantavirus Nucleocapsid Protein Has Distinct m7G Cap- and RNA-binding Sites. J. Boil. Chem. 2010, 285, 11357–11368.

33. A Mir, M.; Panganiban, A.T. A protein that replaces the entire cellular eIF4F complex. EMBO J. 2008, 27, 3129–3139.

34. Gu, S.H.; Kumar, M.; Sikorska, B.; Hejduk, J.; Markowski, J.; Markowski, M.; Liberski, P.P.; Yanagihara, R. Isolation and partial characterization of a highly divergent lineage of hantavirus from the European mole (Talpa europaea). Sci. Rep. 2016, 6, 21119.

35. Elliott, L.H.; Goldsmith, C.S.; Childs, J.; Humphrey, C.D.; Nichol, S.T.; Peters, C.J.; Zaki, S.R.; Ksiazek, T.G.; Rollin, P.E.; Krebs, J.W.; et al. Isolation of the Causative Agent of Hantavirus Pulmonary Syndrome. Am. J. Trop. Med. Hyg. 1994, 51, 102–108.

36. Xu, F.; Yang, Z.; Wang, L.; Lee, Y.-L.; Yang, C.-C.; Xiao, S.-Y.; Xiao, H.; Wen, L. Morphological Characterization of Hantavirus HV114 by Electron Microscopy. Intervirology 2007, 50, 166–172.

37. McCormick, C.; Khaperskyy, D.A. Translation inhibition and stress granules in the antiviral immune response. Nat. Rev. Immunol. 2017, 17, 647–660.

10 THE COADMINISTRATION OF IFN AND TNF TRIGGERS AN ANTIVIRAL AND IMMUNOREGULATORY RESPONSE THROUGH NON-CANONICAL PATHWAYS THAT INVOLVE STATE AND IRFß

1. INTRODUCTION: AN OVERVIEW

Interferon (IFN) β is crucial in the initial defense against viruses, as it has the capacity to trigger a comprehensive antiviral transcriptional response in almost all types of cells [1]. IFNβ also has crucial immunoregulatory roles that influence the outcome of the adaptive immune response to infections [1,2]. Over the years, researchers have mostly conducted in vitro and in vivo experiments to understand the mechanisms and functional effects of IFNβ signaling, focusing mainly on the activation of a single cytokine. This is unlikely to accurately represent physiological conditions, as a wide range of cytokines are released in a certain scenario. Therefore, a cell responds concurrently to a mixture of cytokines in order to promote the right transcriptional pathway. The response to IFNβ is highly reliant on the specific circumstances, especially in relation to its interaction with other cytokines. During the immune response to viral infections, there is an increase in the levels of IFNβ and Tumor Necrosis Factor (TNF). Elevated levels of these cytokines are also linked to many autoinflammatory and autoimmune illnesses, including Systemic Lupus Erythematosus (SLE), psoriasis, and Sjögren's syndrome [3]. Although the reciprocal regulation of IFNβ and TNF has been extensively studied [4,5,6], the specific interaction between these two cytokines is still not well understood.

IFNβ primarily functions by binding to the IFNAR receptor (IFNAR1 and IFNAR2), which triggers the Janus kinase (JAK)/signal transducer and activator of transcription (STAT) pathway. This pathway involves the phosphorylation of STAT1 and STAT2 by JAK1 and Tyk2, respectively. Additionally, other STAT members may also be phosphorylated to a lesser degree, depending on the specific type of cell [7,8]. The phosphorylated forms of STAT1 and STAT2, along with IFN Regulatory Factor (IRF) 9, combine to create the IFN-stimulated gene factor 3 (ISGF3) complex. This complex then attaches to the consensus IFN-stimulated response element (ISRE) sequences found in the promoter region of numerous IFN stimulated genes (ISGs) [9]. The formation of the ISGF3 complex is seen as a clear indication of the activation of the type I IFN response. Consequently, the presence of STAT1 in a particular context has become a reliable indicator of the activation of type I IFN signaling. Nevertheless, in recent years, there has been a growing body of research that challenges the prevailing belief by establishing the presence of non-canonical JAK-STAT signaling pathways that regulate type I IFN responses [8,11]. The combined effect of IFNβ and TNF was demonstrated to strengthen the antiviral response against Vesicular Stomatitis Virus (VSV), Myxoma virus, and paramyxovirus infections [12,13,14]. Gene expression analyses demonstrated that the combination of IFNβ and TNF has a synergistic effect on the regulation of numerous genes that are induced by each cytokine individually. Additionally, this combination also triggers a distinct transcriptional program that consists of genes that do not respond to IFNβ or TNF alone, or only respond to one of the cytokines. The specific signaling mechanisms activated after the costimulation with IFNβ and TNF are still unknown, however it is generally believed that the gene expression response depends on the considerable interaction between the signaling pathways generated by both IFNβ and TNF. The analysis of the enrichment of specific transcription factors binding sites in the promoters of a group of genes that are triggered by both IFNβ and TNF did not provide any insight into the specific regulation of these genes [13]. Our earlier research demonstrated that the DUOX2 gene is classified as a delayed gene, meaning it is significantly activated at high levels when exposed to both IFNβ and

TNF in lung epithelial cells [8]. Our findings indicate that the production of DUOX2 relies on the presence of STAT2 and IRF9, but not STAT1. This suggests that STAT2 and IRF9 may function in a separate pathway that is independent of STAT1, and this pathway might potentially play a role in regulating genes downstream of IFNβ and TNF [14]. The objective of this study was to thoroughly analyze the gene expression pattern of the delayed response to IFNβ and TNF, which occurs without the involvement of STAT1. Additionally, we intended to assess the contribution of STAT2 and IRF9 in regulating this response. In our study, we discovered that the combined action of IFNβ and TNF leads to the activation of a wide range of genes involved in antiviral defense and immune regulation, even in the absence of STAT1. We additionally document the differential control of certain groups of genes activated by IFNβ and TNF, with STAT2 and IRF9 playing separate regulatory roles. Although IFNβ and TNF exert their effects through the combined action of STAT2 and IRF9, distinct groups of genes were only controlled by either STAT2 or IRF9. In summary, our research has shown alternative STAT2 and/or IRF9-dependent pathways that work together to control specific sets of antiviral and immunoregulatory genes in the presence of IFNβ and TNF communication.

2. MATERIALS AND METHODS

2.1. Cell Culture and Stimulation

A549 cells (American Type Culture Collection, ATCC) were grown in F-12 nutrient mixture (Ham) medium supplemented with 10% heat-inactivated fetal bovine serum (HI-FBS) and 1% L-glutamine. The 2ftGH fibrosarcoma cell line and the derived STAT1-deficient U3A cell line, a generous gift from Dr. G. Stark, Cleveland, USA [15], were grown in DMEM medium supplemented with 10% HI-FBS or HI-Fetal Clone III (HI-FCl) and 1% L-glutamine. U3A cells stably expressing STAT1 were generated by transfection of the STAT1 alpha flag pRc/CMV plasmid (Addgene plasmid #8691; a generous gift from Dr. J. Darnell, Rockfeller University, USA [16,17]) and selection with 800 μg/ml Geneticin (G418). Monoclonal populations of U3A stably expressing STAT1 cells were isolated. A pool of two clones, referred to as U3A-STAT1, was used in the experiments to mitigate the clonal effects. U3A-STAT1 cells were maintained in culture in DMEM supplemented with 10% HI-FCl, 1% Glu, and 200 μg/mL G418. All cell lines were cultured without antibiotics except for the selection of stable cells. All media and supplements were from Gibco (Life Technologies, Grand Island, NY, USA), with the exception of HI-FCl, which was from HyClone (Logan, UT, USA). Mycoplasma contamination was excluded by regular analysis using the MycoAlert Mycoplasma Detection Kit (Lonza, Basel, Switzerland). Cells were stimulated with IFNβ (1000 U/mL, PBL Assay Science, Piscataway, NJ, USA), TNF (10 ng/mL, R&D Systems, Minneapolis, MN, USA) or IFNβ (1000 U/mL) +TNF (10 ng/mL) for the indicated times.

2.2. siRNA Transfection

The sequences of non-targeting control (Ctrl) and STAT2- and IRF9-directed RNAi oligonucleotides (Dharmacon, Lafayette, CO, USA) have previously been described in [14]. U3A cells at 30% confluency were transfected using the Oligofectamine transfection reagent (Life Technologies-Thermofisher, Carlsbad, CA, USA). RNAi transfection was pursued for 48 h before stimulation.

2.3. Immunoblot Analysis

Cells were lysed on ice using Nonidet P-40 lysis buffer as fully detailed in [18]. Whole-cell extracts (WCE) were quantified using the Bradford protein assay (Bio-Rad, Hercules, CA, USA), resolved by SDS-PAGE and transferred to nitrocellulose membrane before analysis by immunoblot. Membranes were incubated with the following primary antibodies: anti-actin Cat #MAB1501 from Millipore (Burlington, MA, USA), anti-IRF9 Cat #610285 from BD Transduction Laboratories (San Jose, CA, USA), and anti-STAT1-P-Tyr701 Cat #9171, anti-STAT2-P-Tyr690 Cat #4441, anti-STAT1 Cat #9172, anti-STAT2 Cat #4594, all from Cell Signaling (Danvers, MA, USA), before further incubation with horseradish peroxidase (HRP)-conjugated secondary antibodies (KPL, Gaithersburg, MD, USA or Jackson Immunoresearch Laboratories, West Grove, PA, USA). Antibodies were diluted in PBS containing 0.5% Tween and either

5% nonfat dry milk or BSA. Immunoreactive bands were visualized by enhanced chemiluminescence (Western Lightning Chemiluminescence Reagent Plus, Perkin-Elmer Life Sciences Waltham, MA, USA) using a LAS4000mini CCD camera apparatus (GE Healthcare, Mississauga, ON, Canada).

2.4. RNA Isolation and qRT-PCR Analyses

Total RNA was prepared using the RNAqueous-96 Isolation Kit (Invitrogen-Thermo Fisher, Carlsbad, CA, USA) following the manufacturer's instructions. Total RNA (1 μg) was subjected to reverse transcription using the QuantiTect Reverse Transcription Kit (Qiagen, Toronto, ON, Canada). Quantitative PCR were performed using either Fast start SYBR Green Kit (Roche, Indianapolis, IN, USA) for MX1, IDO, APOBEC3G, CXCL10, NOD2, PKR, IRF1, IFIT1 and IL8 or TaqMan Gene Expression Assays (Life Technologies-Thermo Fisher) for DUOX2, IFI27, SERPINB2, IL33, CCL20, ISG20. Sequences of oligonucleotides and probes used in PCR reactions. Data collection was performed on a Rotor-Gene 3000 Real Time Thermal Cycler (Corbett Research, Mortlake, Australia). Gene inductions were normalized over S9 levels, measured using Fast start SYBR Green Kit or TaqMan probe as necessary. Fold induction of genes was determined using the ΔΔCt method [19]. All qRT-PCR data are presented as the mean ± standard error of the mean (SEM).

2.5. RNA-Sequencing (RNASeq)

Total RNA prepared as described above was quantified using a NanoDrop Spectrophotometer ND-1000 (NanoDrop Technologies, Inc., Wilmington, DE, USA) and its integrity was assessed using a 2100 Bioanalyzer (Agilent Technologies, Santa Clara, CA, USA). Libraries were generated from 250 ng of total RNA using the NEBNext poly(A) magnetic isolation module and the KAPA stranded RNA-Seq library preparation kit (Kapa Biosystems, Wilmington, MA, USA), as per the manufacturer's recommendations. TruSeq adapters and PCR primers were purchased from IDT. Libraries were quantified using the Quant-iT™ PicoGreen® dsDNA Assay Kit (Molecular Probes, Eugene, OR, USA) and the Kapa Illumina GA with Revised Primers-SYBR Fast Universal kit (Kapa

Biosystems). Average size fragment was determined using a LabChip GX (Perkin-Elmer Life Sciences, Waltham, MA, USA) instrument. Massively parallel sequencing was carried out on an Illumina HiSeq 2500 sequencer (Illumina Inc., San Diego, CA, USA). Read counts were obtained using HTSeq. Reads were trimmed from the 3' end to have a Phred score of at least 30. Illumina sequencing adapters were removed from the reads and all reads were required to have a length of at least 32bp. Trimming and clipping was performed using Trimmomatic [20]. The filtered reads were aligned to the Homo-sapiens assembly GRCh37 reference genome. Each read set was aligned using STAR [21] and merged using Picard (http://broadinstitute.github.io/picard/). For all samples, the sequencing resulted in more than 29 million clean reads (ranging from 29 to 44 million reads) after removing low quality reads and adaptors. The reads were mapped to the total of 63,679 gene biotypes including 22,810 protein-coding genes. The non-specific filter for 1 count-per million reads (CPM) in at least three samples was applied to the reads and 14,254 genes passed this criterion.

2.6. Bioinformatics Analysis

Differential transcripts analysis. A reference-based transcript assembly was performed, which allows the detection of known and novel transcripts isoforms, using Cufflinks [22], merged using Cuffmerge (cufflinks/AllSamples/merged.gtf) and used as a reference to estimate transcript abundance and perform differential analysis using Cuffdiff and Cuffnorm tool to generate a normalized data set that includes all the samples. The fragments per kilobase million (FPKM) values calculated by Cufflinks were used as input. The transcript quantification engine of Cufflinks, Cuffdiff, was used to calculate transcript expression levels in more than one condition and test them for significant differences. To identify a transcript as being differentially expressed, Cuffdiff tests the observed log-fold-change in its expression against the null hypothesis of no change (i.e., the true log-fold-change is zero). Because of measurement errors, technical variability, and cross-replicate biological variability might result in an observed log-fold-change that is non-zero, Cuffdiff assesses significance using a model of variability in the log-fold-change under the null hypothesis. This model is described in detail in [23]. The differential gene expression analysis was performed using DESeq [24] and edgeR [25] within the R Bioconductor packages. Genes were considered differentially expressed between two group if they met the following requirement: fold change (FC) > ±1.5, $p < 0.05$, false discovery rate (FDR) < 0.05.

Enrichment of gene ontology (GO). GO enrichment analysis amongst differentially expressed genes (DEGs) was performed using Goseq [26] against the background of full human genome (hg19). GO-terms with adjusted p value < 0.05 were considered significantly enriched.

Clustering of DEGs. We categorized the DEGs according to their response upon silencing of siSTAT2 and siIRF9; categories are listed as A to I (Figure 2E). Then to determine relationship between these categories, we calculated the distance of centers of different categories. For each gene, we transformed siSTAT2 and siIRF9 FC to deviation from the mean FC of the category the respective gene belongs to using the equation: $FCnew = FCold - \varepsilon (FCcategory)$. The parameter ε was estimated to give the perfect match between predefined categories (A to I) and clustering based on Euclidean distance. Results were plotted as a heatmap.

Modular transcription analysis. The tmod package in R [27] was used for modular transcription analysis. In brief, each transcriptional module is a set of genes that shows coherent expression across many biological samples [28,29]. Modular transcription analysis then calculates significant enrichment of a set of foreground genes, here DEGs, in pre-defined transcriptional module compared to a reference set. For transcriptional modules, we used a combined list of 606 distinct functional modules encompassing 12,712 genes, defined by Chaussabel et al. [30] and Li et al. [31], as the reference set in tmod package. The hypergeometric test devised in tmodHGtest was used to calculate enrichments and p-values employing Benjamini-Hochberg correction [32] for multiple sampling. All the statistical analyses and graphical presentations were performed in R [33].

2.7. Virus Titration by Plaque Assay

Quantification of VSV infectious virions was achieved through methylcellulose plaque forming unit assays. U3A and U3A-STAT1 cells were either left untreated or stimulated with IFNβ or IFNβ + TNF for 30 h. Cells were then infected with Vesicular Stomatitis Virus (VSV)-GFP (kindly provided by Dr. J. Bell, University of Ottawa, Canada) at a multiplicity of infection (MOI) of 5 for 1 h in serum-free medium (SFM). Cells were then washed twice with SFM and further cultured in DMEM medium containing 2% HI-FCl. The supernatants were harvested at 12 h post-infection and serial dilutions were used to infect confluent Vero cells (ATCC) for 1 h in SFM. The medium was then replaced with 1% methylcellulose in DMEM containing 10% HI-FCl. Two days post-infection, GFP-positive plaques were detected using a Typhoon Trio apparatus and quantified using the ImagequantTL software (GE Healthcare, Mississauga, ON, Canada).

2.8. Luciferase Gene Reporter Assay

U3A or U3A-STAT1 cells at 90% confluency were cotransfected with 100 ng of one of the following CXCL10 promoter containing firefly luciferase reporter plasmids (generously donated by Dr. David Proud, Calgary, [34]), CXCL10prom-972pb-pGL4 (full length -875/+97 promoter), CXCL10prom-376pb-pGL4 (truncated -279/+97 promoter), CXCL10prom972pb-ΔISRE(3)-pGL4 (full length promoter with ISRE(3) site mutated), together with 50 ng of pRL-null renilla-luciferase expressing plasmid (internal control). Transfection was performed using Lipofectamine 2000 (Life Technologies-Thermo Fisher) using a 1:2 DNA to lipofectamine ratio. At 8 h post-transfection, cells were stimulated for 16 h with either IFNβ or IFNβ + TNF. Firefly and renilla luciferase activities were quantified using the Dual-luciferase reporter assay system (Promega Corporation, Madison, WI, USA). Luciferase activities were calculated as the luciferase/renilla ratio and were expressed as fold over the non-stimulated condition.

2.9. Statistical Analyses

Statistical analyses of qRT-PCR and luciferase assay results were performed using the Prism v7 or v8 software (GraphPad Software, San Diego, CA, USA) using the tests indicated in the figure legends. Statistical significance was evaluated using the following p values: $p < 0.05$ (*), $p < 0.01$ (**), $p < 0.001$ (***) or $p < 0.0001$ (****). Differences with a p-value < 0.05 were considered significant. Statistical analysis of the RNASeq data is described in the Bioinformatics analysis section above.

2.10. Data and Software Availability

The entire set of RNAseq data has been submitted to the Gene Expression Omnibus (GEO) database (http://www.ncbi.nlm.nih.gov/geo) under accession number GEO: GSE111195.

3. RESULTS

3.1. Distinct Induction Profiles of Antiviral and Immunoregulatory Genes in Response to IFNβ, TNF and IFNβ + TNF

First, to validate previous observations, we sought to determine the induction profile by the combination of IFNβ and TNF of a selected panel of immunoregulatory and antiviral genes that had previously been shown to be regulated by IFNβ or TNF alone. A549 cells, in which we previously documented the synergistic action of IFNβ and TNF on the DUOX2 gene, were stimulated either with IFNβ, TNF, or IFNβ + TNF for various times between 3–24 h and the relative mRNA expression levels were quantified by qRT-PCR. Analysis of the expression of the selected genes revealed distinct profiles of response to IFNβ, TNF, or IFNβ + TNF (Figure 1). IDO, DUOX2, CXCL10, APOBEC3G, ISG20, and IL33 exhibited synergistic induction in response to IFNβ + TNF compared to IFNβ or TNF alone. Expression in response to IFNβ + TNF increased over time, with maximum expression levels observed between 16 and 24 h. While NOD2 and IRF1 induction following stimulation with IFNβ + TNF was also significantly higher than upon IFNβ or TNF single cytokine stimulation, they exhibited a steady-state induction profile starting as early as 3 h. MX1 and PKR, two typical IFNβ-inducible ISGs, were found induced by IFNβ + TNF similarly to IFNβ alone. CCL20 responded to IFNβ + TNF with a kinetic and amplitude similar to TNF, but was not responsive to IFNβ alone. IL8 expression was not induced by IFNβ, but was increased by TNF starting at 3 h and remained steady until 24 h. In contrast to other genes, IL8 induction in response to IFNβ + TNF was significantly decreased compared to TNF alone. Overall, these results confirm previous reports that a subset of antiviral and immunoregulatory genes is greatly increased in response to IFNβ + TNF compared to either IFNβ or TNF alone.

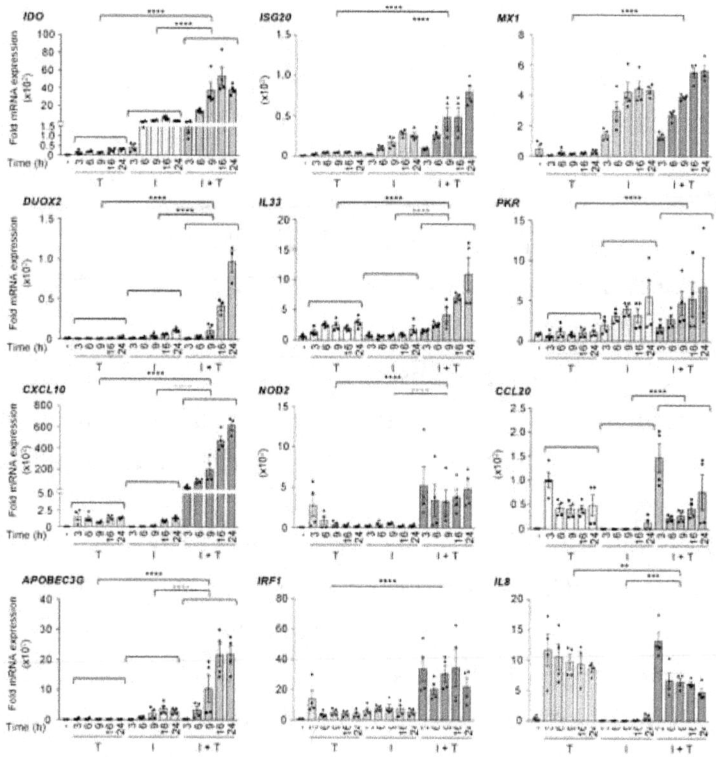

Figure 1. Expression of a panel of antiviral and immunoregulatory genes in response to IFNβ, TNF, and IFNβ + TNF. A549 cells were stimulated with either TNF (T), IFNβ (I), or costimulated with IFNβ + TNF (I + T) for the indicated times. Quantification of mRNA was performed by qRT-PCR and expressed as fold expression after normalization to the S9 mRNA levels using the ΔΔCt method. Mean +/− SEM, n ≥ 3. Statistical comparison of TNF vs. IFNβ + TNF and IFNβ vs. IFNβ + TNF was conducted using two-way ANOVA with Tukey's post-test. $p < 0.01$ (**), $p < 0.001$ (***) or $p < 0.0001$ (****).

3.2. Workflow for Genome-Wide Characterization of the Delayed Transcriptional Program Induced by IFNβ+TNF in the Absence of STAT1

In a previous study, we provided evidence supporting the existence of a STAT1-independent, but STAT2- and IRF9-dependent, pathway engaged downstream of IFNβ + TNF [14]. Here, using the STAT1-deficient human U3A cell line [15], we aimed to fully characterize the STAT1-independent transcriptional program induced by IFNβ + TNF. The human U3A cell line was derived by mutagenesis from the 2ftGH cells [15], in which a synergistic response to IFNβ + TNF can be observed on a subset of genes (Figure 2A). Two hallmarks of STAT2 and IRF9 activation, i.e., STAT2 Tyr690 phosphorylation and induction of IRF9, were observed in the U3A cells following stimulation with IFNβ + TNF, although to reduced levels compared to the parental 2ftGH cells expressing endogenous STAT1 (Figure 2B). This observation implies that the activation of STAT2 and IRF9 depends to a large extent on the STAT1-dependent canonical pathway, but that a significant response occurs in the absence of STAT1. Therefore, the human U3A cell model is suitable for specifically studying STAT1-independent, but STAT2- and IRF9-dependent, gene expression in response to IFNβ + TNF.

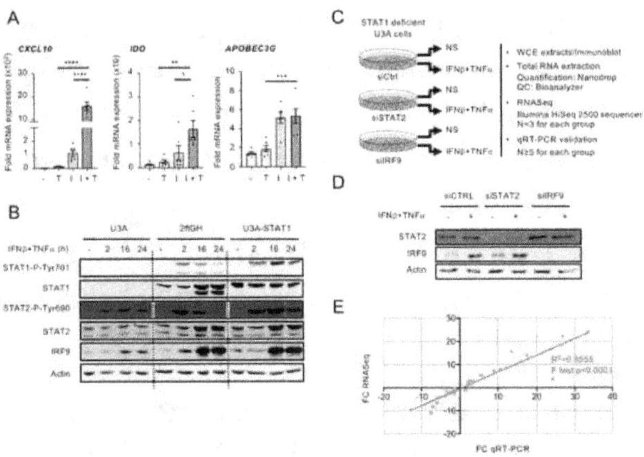

Figure 2. Experimental design used to study the STAT1-independent delayed transcriptional program induced by the combination of IFNβ and TNF. (A) 2ftGH cells were stimulated with either TNF (T), IFNβ (I), or costimulated with IFNβ + TNF (I + T) for 24 h. Quantification of mRNA was performed by qRT-PCR and expressed as fold expression after

normalization to the S9 mRNA levels using the ΔΔCt method. Mean +/- SEM, n ≥ 5. Statistical comparison was conducted using one-way ANOVA with Tukey's post-test. $p < 0.05$ (*), $p < 0.01$ (**), $p < 0.001$ (***), or $p < 0.0001$ (****). (B) U3A (STAT1-deficient), 2ftGH (parental STAT1-positive) cells and U3A-STAT1 cells (U3A cells stably reconstituted with STAT1) were left untreated or stimulated with IFNβ + TNF for the indicated times. WCE (whole cell extracts) were analyzed by SDS-PAGE followed by immunoblot using anti STAT1-P-Tyr701, total STAT1, STAT2-P-Tyr690, total STAT2, IRF9, or actin antibodies. (C–E) U3A cells were transfected with siCTRL, siSTAT2, or siIRF9 before being left untreated (NS) or stimulated with IFNβ + TNF for 24 h. (C) The schematic describes the workflow of sample preparation and analysis. (D) WCE were analyzed by SDS-PAGE followed by immunoblot using anti STAT2, IRF9, and actin antibodies. (E) Graph showing the correlation between fold-changes (FC) measured by RNASeq and qRT-PCR for 13 randomly selected genes. Data from siCTRL NS vs. siCTRL IFNβ +TNF, siSTAT2 IFNβ +TNF vs. siCTRL IFNβ + TNF, siIRF9 IFNβ + TNF vs. siCTRL IFNβ + TNF conditions were used.

To profile the genome wide transcriptional program induced by the combination of IFNβ and TNF in the absence of STAT1 and define the role of STAT2 and IRF9, the U3A cells were transfected with Control (Ctrl)-, STAT2- or IRF9-RNAi and further left untreated or stimulated with IFNβ (1000 U/mL) + TNF (10 ng/mL) for 24 h (Figure 2C). Efficient silencing was confirmed by immunoblot (Figure 2D). Total RNA was isolated and analyzed by RNA sequencing (n = 3 for each group detailed in Figure 2B) on an Illumina HiSeq2500 platform. To validate the expression profile of genes in RNASeq results, the fold changes (FC) of 13 genes randomly selected were analyzed by qRT-PCR in each experimental groups, i.e., siCTRL non-stimulated (NS) vs. siCTRL IFNβ + TNF, siCTRL IFNβ + TNF vs. siSTAT2 IFNβ + TNF and siCTRL IFNβ + TNF vs. siIRF9 IFNβ + TNF. A positive linear relationship between RNASeq and qRT-PCR FC was observed (Figure 2E).

3.3. A Broad Antiviral and Immunoregulatory Transcriptional Signature is Induced by IFNβ + TNF in the Absence of STAT

To identify differentially expressed genes (DEGs) upon IFNβ + TNF stimulation in the absence of STAT1, comparison between non-stimulated (NS) and IFNβ + TNF-stimulated control cells was performed (Figure 3A). In total, 612 transcripts, including protein-coding transcripts, pseudogenes and long non-coding RNA (lncRNA), were significantly different (FC > 1.5, p < 0.05, FDR < 0.05) in IFNβ + TNF vs. NS. Among these, 482 DEGs were upregulated and 130 were downregulated (Figure 3B.

To identify the Biological Processes (BP) and Molecular Functions (MF) induced by IFNβ + TNF independently of STAT1, we further analyzed the upregulated DEGs. The top 40 upregulated DEGs are shown in Figure 4A. We subjected the complete list of upregulated DEGs through Gene Ontology (GO) enrichment analysis. The top enriched GO BP (p < 10–10) and MF, are depicted in Figure 4B. The majority of the top enriched BPs were related to cytokine production and function (response to cytokine, cytokine-mediated signaling pathway, cytokine production, and regulation of cytokine production), immunoregulation (immune response, immune system process, innate immune response, and regulation of immune system process) and host defense response (defense response, response to other organism, 2'-5'-oligoadenylate synthetase activity and dsRNA binding).

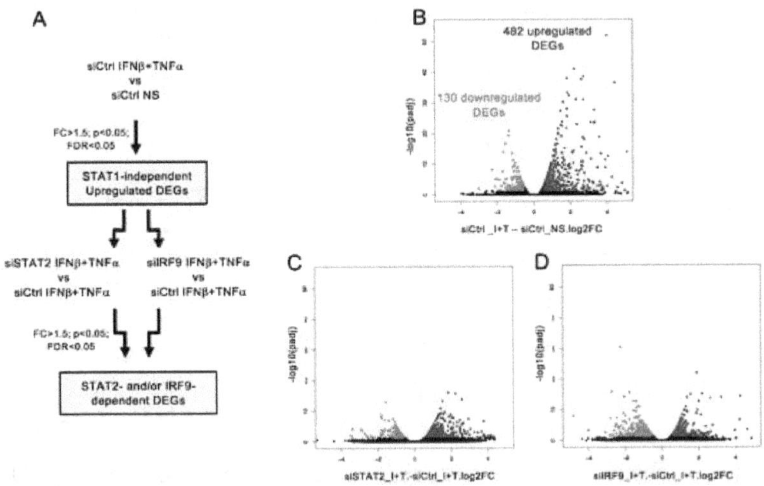

Figure 3. Analysis of STAT1-independent IFNβ+ TNF-induced DEGs. (A) Diagram describing the bioinformatics analysis strategy used to determine STAT1-independent differentially expressed genes (DEGs) and their regulation by STAT2 and IRF9. (B) Volcano plots of the fold-change (FC) vs. adjusted p-value of IFNβ + TNF (I +T) vs. non-stimulated (NS) siCtrl conditions. (C) Volcano plots of the fold-change vs. adjusted

p-value of siSTAT2 IFNβ + TNF vs. siCTRL IFNβ + TNF (I + T) conditions. (D) Volcano plots of the fold-change vs. adjusted p-value of siIRF9 IFNβ + TNF vs. siCTRL IFNβ + TNF conditions.

Fourteen MF categories were found enriched in IFNβ + TNF. The top enriched MFs were related to cytokine and chemokine functions (cytokine activity, cytokine receptor binding, chemokine activity, and Interleukin 1-receptor binding). Other enriched MF included peptidase related functions (endopeptidase inhibitor activity, peptidase regulator activity, and serine-type endopeptidase activity).

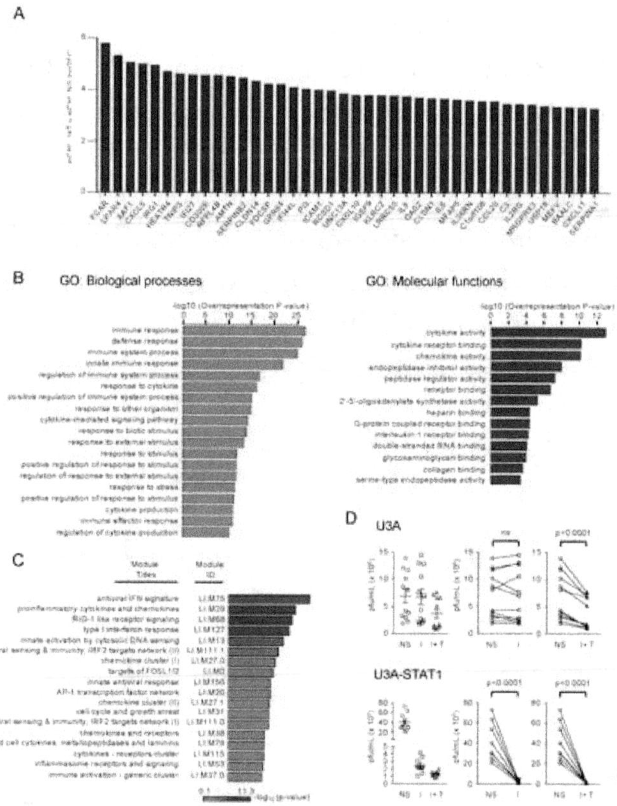

Figure 4. Functional characterization of STAT1-independent IFNβ + TNF-induced DEGs. (A) Top forty IFNβ + TNF- upregulated DEGs. (B) Gene ontology (GO) enrichment analysis of the differentially upregulated genes in IFNβ + TNF vs. non-stimulated siCtrl conditions based on the Biological Processes and Molecular Function categories. (C) Modular transcription analysis of upregulated DEGs. Eighteen enriched modules are shown. (D) U3A and STAT1-rescued U3A-STAT1 cells were stimulated with IFNβ (I) or

IFNβ + TNF (I + T) for 30 h before infection with VSV at a MOI of 5 for 12 h. The release of infectious viral particles was quantified by plaque forming unit (pfu) assay. The left graphs show dot-plots of all stimulations. Statistical comparisons were performed on the "before and after" plots (displayed on the right of dot-plot graphs) using ratio paired t-tests.

To gain deeper insight into the relevance of the STAT1-independent IFNβ + TNF-induced transcripts towards immunological and host defense responses, we conducted a modular transcription analysis of upregulated DEGs against 606 immune-related functional modules. These modules were previously defined from co-clustered gene sets built via an unbiased data-driven approach as detailed in the material and methods section [30,31]. STAT1-independent IFNβ + TNF-induced DEGs showed significant enrichment in 37 modules. Six of these modules were associated with virus sensing/Interferon antiviral response, including LI.M75 (antiviral IFN signature), LI.M68 (RIG-I-like receptor signaling), LI.M127 (type I interferon response), LI.M111.0 and LI.M111.1 (IRF2 target network), and LI.M150 (innate antiviral response) (Figure 4C). Additionally, six modules were associated with immunoregulatory functions, including LI.M29 (proinflammatory cytokines and chemokines), LI.M27.0 and LI.M27.1 (chemokine cluster I and II), LI.M38 (chemokines and receptors), LI.M115 (cytokines receptors cluster), and LI.M37.0 (immune activation - generic cluster) (Figure 4C). Module analysis also showed enriched AP-1 transcription factor-related network modules, LI.M20 and LI.M0, and cell cycle and growth arrest LI.M31 module.

GO enrichment and modular transcription analyses were suggestive of an antiviral response being induced by IFNβ + TNF. To demonstrate the physiological relevance of STAT1-independent gene expression, we evaluated the capacity of IFNβ + TNF to restrict virus replication in the absence of STAT1. Previous study has shown that infection by Vesicular Stomatitis Virus (VSV) is sensitive to IFNβ, but not to TNF, in 2ftGH-derived cells [35]. Thus, U3A cells were stimulated with IFNβ alone or IFNβ + TNF and further infected with (VSV). While VSV replicated similarly in untreated U3A cells and in cells treated with IFNβ, treatment with IFNβ + TNF significantly limited VSV replication (Figure 4D). As a control of the efficiency of IFNβ alone treatment, we observed a significant antiviral response when STAT1 expression was stably restored to endogenous levels in U3A cells (U3A-STAT1) (Figure 2B and Figure 4D). Collectively, these data demonstrate that the combination of IFNβ + TNF induces an efficient antiviral and immunoregulatory transcriptional response in the absence of STAT1.

3.4. Clustering of STAT1-Independent IFNβ + TNF Induced DEGs According to their Regulation by STAT2 and IRF9

Having shown that IFNβ + TNF induce a broad antiviral and immunoregulatory transcriptional response independently of STAT1, we next sought to gain insight into the role of STAT2 and IRF9. To do so, we compared transcripts levels in siSTAT2_IFNβ + TNF vs. siCTRL_IFNβ + TNF and siIRF9_IFNβ + TNF vs. siCTRL_IFNβ + TNF conditions (Figure 3A). Volcano plots revealed that a fraction of IFNβ + TNF-induced DEGs were significantly ($FC > 1.5$, $p < 0.05$, $FDR < 0.05$) downregulated or upregulated upon silencing of STAT2 (Figure 3C) or IRF9 (Figure 3D). Nine distinct theoretical categories of DEGs can be defined based on their potential individual behavior across siSTAT2 and siIRF9 groups (Categories A–I, Figure 5A): a gene can either be downregulated upon STAT2 and IRF9 silencing, indicative of a positive regulation by STAT2 and IRF9 (Category A); conversely, a gene negatively regulated by STAT2 and IRF9 would exhibit upregulation upon STAT2 and IRF9 silencing (Category B); genes that do not exhibit significant differential expression in siSTAT2 and siIRF9 groups would be classified as STAT2 and IRF9 independent (Category C); IRF9-independent genes could exhibit positive (Category D) or negative (Category E) regulation by STAT2; conversely, STAT2-independent genes might be positively (Category F) or negatively (Category G) regulated by IRF9; lastly, STAT2 and IRF9 could have opposite effects on a specific gene regulation (Category H and I). Based on a priori clustering of RNASeq data we found that in the absence of STAT1 IFNβ + TNF-induced DEGs clustered into only seven of the nine possible categories (Figure 5B). The top 15 upregulated DEGs of each category are shown in Figure 5C. Amongst the 482 DEGs, 163 genes exhibited either inhibition or upregulation following silencing of

STAT2 and/or IRF9 (Categories B–G). A large majority of upregulated DEGs, i.e., 319 out of the 482 DEGs, were not significantly affected by either STAT2 or IRF9 silencing, and were therefore classified as STAT2/IRF9-independent (Figure 5B). No genes were found in Category H and only one gene was found in Category I.

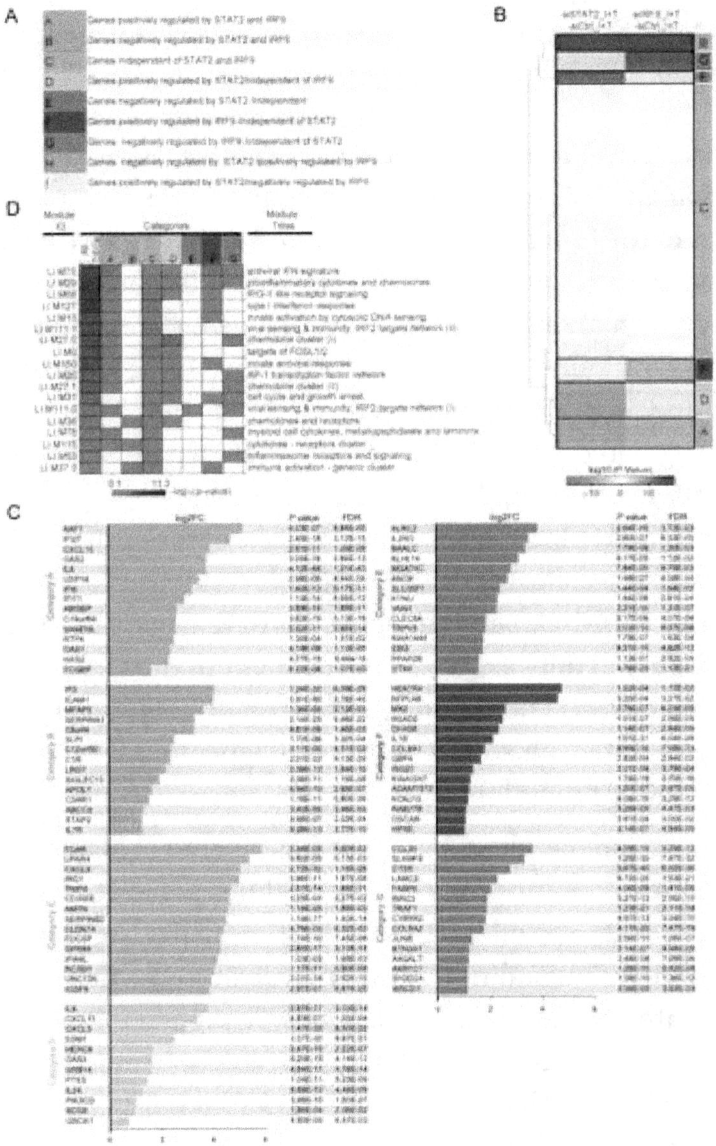

Figure 5. Clustering of IFNβ + TNF-induced DEGs according to their regulation by STAT2 and IRF9. (A) Theoretical categories in which IFNβ +

TNF-induced DEGs can be segregated based on their potential individual regulation by STAT2 and IRF9. (B) Hierarchical clustering of the categories of DEG responses according to their regulation by STAT2 and IRF9. Euclidean distance metric is used for the construction of distance matrix and the categories are used as a priori input into clustering algorithm as detailed in Materials and Methods. (C) Top 15 induced DEGs (Log2FC, siCTRL NS vs. siCTRL IFNβ + TNF) in each category identified in (B) are displayed. Note that category D contains only twelve genes, so all genes are shown. (D) Diagram showing enriched transcription modules in each gene category.

To functionally interpret these clusters, we applied the modular transcription analysis to each of the categories to assess for the specific enrichment of the functional modules found associated with IFNβ + TNF-upregulated DEGs (Figure 5D). First, most modules, except LI.M31 (cell cycle and growth arrest), LI.M38 (chemokines and receptors), LI.M37.0 (immune activation - generic cluster), and LI.M53 (inflammasome receptors and signaling), were enriched in the category of DEGs positively regulated by STAT2 and IRF9 (Category A). Conversely, the cluster negatively regulated by STAT2 and IRF9 (Category B) exclusively contains enriched LI.M38 (chemokines and receptors), LI.M37.0 (immune activation - generic cluster) and LI.M115 (cytokines receptors cluster). The cluster of IRF9-independent genes that are negatively regulated by STAT2 (Category E) only exhibited enrichment in the virus sensing/IRF2 target network LI.M111.0 module, while the IRF9-independent/STAT2-positively regulated cluster (Category D) encompasses antiviral and immunoregulatory functions.

The STAT2-independent but IRF9 positively regulated transcripts (Category F) were mainly enriched in modules associated with the IFN antiviral response, including LI.M75 (antiviral IFN signature), LI.M68 (RIG-I-like receptor signaling), LI.M127 (type I interferon response), and LI.M150 (innate antiviral response). Finally, the STAT2-independent but IRF9 negatively regulated cluster (Category G) was mostly enriched in modules associated with immunoregulatory functions, including LI.M29 (proinflammatory cytokines and chemokines), LI.M27.0 and LI.M27.1 (chemokine cluster I and II), LI.M78 (myeloid cell cytokines), but also with cell cycle and growth arrest (LI.M31) and inflammasome receptors and signaling (LI.M53). Of note, all modules were found enriched in the cluster of genes induced independently of STAT2 and IRF9 (Category C), pointing to a broad function of this pathway(s) in the regulation of the antiviral and immunoregulatory program elicited by IFNβ + TNF. Altogether, these observations reveal that STAT2 and IRF9 are involved in the regulation of a subset of the genes induced in response to the co-stimulation by IFNβ and TNF in the absence of STAT1. Importantly, our results also reveal that STAT2 and IRF9 act in part in a concerted manner, but also independently in distinct non-canonical pathways, to regulate specific subsets of the IFNβ + TNF-induced DEGs.

3.5. Differential Regulation of CXCL10 in Response to IFNβ and IFNβ+TNF

The canonical ISGF3 complex mediates ISGs transcriptional regulation through binding to ISRE consensus sequences [7]. Identification of DEGs upregulated by IFNβ+TNF in a STAT1-independent, but STAT2 and IRF9-dependent, manner raised the question of the ISRE site usage compared to the ISGF3 pathway. To address this question, we studied the regulation of the CXCL10 gene promoter that was found induced by STAT2 and IRF9 in the absence of STAT1 in response to IFNβ and TNF, but is also inducible by IFNβ alone in the presence of STAT1 (Figure 1 and Figure 2A). The CXCL10 promoter contains three ISRE sites. We used full-length wild-type (972bp containing the three ISRE sites), truncated (376bp containing only the ISRE (3) site) or mutated (972bp containing a mutated ISRE (3) site) CXCL10 promoter luciferase (CXCL10prom-Luc) reporter constructs (Figure 6A) to determine the ISRE consensus site(s) requirement. U3A and STAT1-rescued U3A-STAT1 cells were transfected with the CXCL10prom-Luc constructs and further stimulated with IFNβ or IFNβ+TNF to monitor the canonical ISGF3-dependent and the STAT1-independent responses. In STAT1-deficient U3A cells, IFNβ was unable to activate the CXCL10prom reflecting the dependence on the ISFG3 pathway. In contrast, induction of CXCL10prom was significantly induced when cells were stimulated with IFNβ +TNF. This induction in the absence of STAT1 was not altered by the deletion of ISRE (1) and (2) sites, but was significantly impaired when the ISRE (3) site was mutated (Figure 6B). STAT1 expression rescue led to a higher induction of CXCL10prom by IFNβ + TNF. Additionally, IFNβ-mediated induction of CXCL10prom was restored, albeit to a much lower extent than IFNβ +TNF. The activation of the CXCL10prom in the presence of STAT1 involved both the distal ISRE (1) and/or ISRE (2) sites and the proximal ISRE (3) site (Figure 6B). Altogether, this shows that the STAT1-independent, but STAT2/IRF9-dependent, pathway mediates gene expression through a restricted ISRE site usage compared to the ISGF3-dependent regulation.

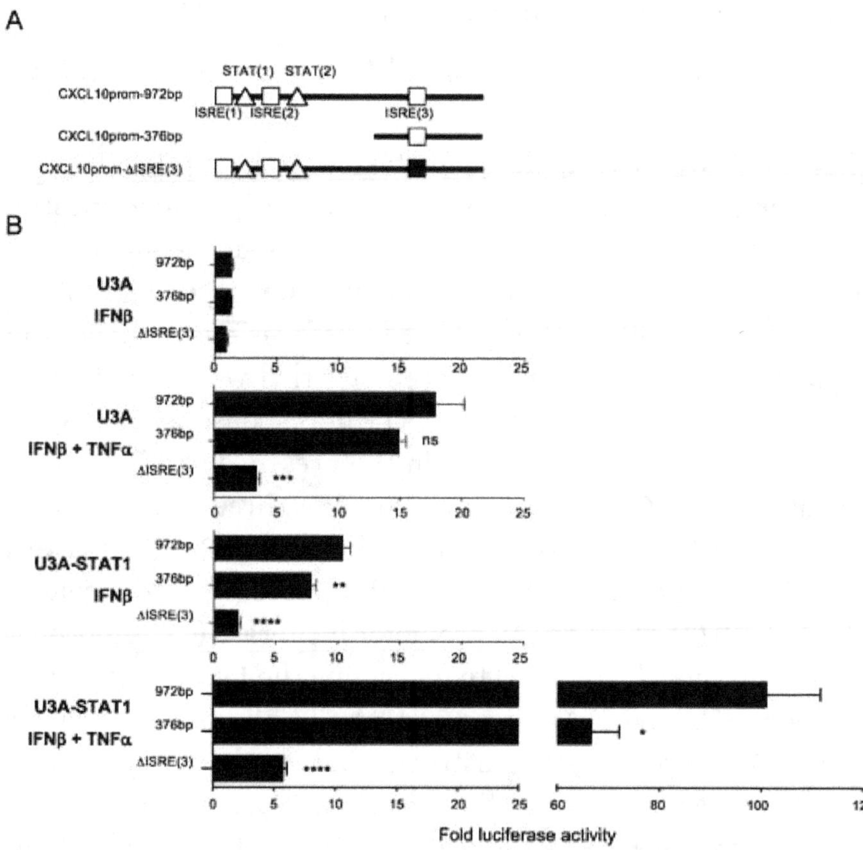

Figure 6. Analysis of the CXCL10 promoter regulation in response to IFNβ + TNF vs. IFNβ. (A) Schematic representation of the CXCL10 promoter (CXCL10prom) luciferase constructs used in this study indicating the main transcription factors consensus binding sites. (B) U3A and U3A-

STAT1 cells were transfected with the indicated CXCL10prom-luciferase reporter constructs and either left untreated or stimulated with IFNβ or IFNβ + TNF. Relative luciferase activities were measured at 16 h post-stimulation and expressed as fold over the corresponding unstimulated condition. Mean +/- SEM, n = 6. Statistical analyses were performed using an unpaired t-test comparing each promoter to the CXCL10prom-972bp construct. $p < 0.01$ (**), $p < 0.001$ (***), and $p < 0.0001$ (****).

4. DISCUSSION

Previous studies have described that IFNβ and TNF synergize to elicit a specific delayed transcriptional program that differs from the one induced by either cytokine alone [13,36]. The mechanisms underlying the transcriptional induction of genes specifically regulated by IFNβ and TNF remain poorly defined. The present study was specifically designed to document the functional relevance of a previously observed delayed gene expression induced by IFNβ in the presence of TNF in the absence of STAT1 [14] and to document the role of STAT2 and IRF9 in this response.

The observation that STAT2 and IRF9 activation in response to IFNβ + TNF is reduced in STAT1-deficient U3A cells compared to the wild-type 2ftGH parental cells and that IFNβ + TNF-mediated induction of the STAT2- and IRF9-dependent CXCL10 promoter exhibits partial dependence on STAT1 support a model in which a canonical ISGF3 pathway is engaged downstream of the costimulation. Importantly, it also implied the existence of a STAT2- and/or IRF9-dependent transcriptional response occurring in the absence of STAT1. The human STAT1-deficient U3A cell model offered a unique opportunity to specifically pinpoint this STAT1-independent response. In this model, genome wide RNA sequencing highlighted that the transcriptional program induced by IFNβ + TNF in the absence of STAT1 encompasses a wide range of immunoregulatory and antiviral functions. The functional relevance of this response was confirmed by the observation that the treatment with IFNβ + TNF induced an antiviral state capable of restricting VSV replication in the absence of STAT1. This points to a significant role of the STAT1-independent pathway in the establishment of the antiviral state induced by the synergistic action of IFNβ and TNF that enhances the restriction of VSV (Figure 4D and [12]), Myxoma virus [13], and paramyxoviruses [14]. Although previous reports have shown that type I IFNs, mostly IFNα, alone can trigger STAT1-independent responses [8], we neither observed establishment of an IFNβ-induced antiviral state against VSV, nor activation of the CXCL10 promoter in the absence of STAT1 in our model (Figure 4D and Figure 6).

We previously reported that IFNβ + TNF induces the DUOX2 gene via a STAT2- and IRF9-dependent pathway in the absence of STAT1 [14]. To what extent this pathway contributes to the STAT1-independent transcriptional response elicited by IFNβ + TNF remained to be addressed. Here, we demonstrate that IFNβ + TNFα-induced DEGs segregate into seven categories that reflect distinct contributions of STAT2 and/or IRF9, thereby highlighting an unexpected heterogeneity of the STAT1-independent pathways engaged downstream of IFNβ + TNF. Importantly, only one anecdotic gene was found in categories implying inverse regulation by STAT2 and IRF9 (categories H and I) pointing to convergent functions of STAT2 and IRF9 when both are engaged in gene regulation. We can rule out that these distinct regulation mechanisms reflect specific induction profiles by IFNβ + TNF as CXCL10, IL33, CCL20, and ISG20 all exhibit synergistic induction by IFNβ + TNF, but are differentially regulated by STAT2 and/or IRF9; while CXCL10 is dependent on STAT2 and IRF9, IL33 is independent on STAT2 and IRF9, and CCL20 and ISG20 are STAT2-independent but IRF9-dependent (Figure 5). Consistent with our previous observation [14], we found several STAT1-independent genes positively regulated by STAT2 and IRF9 (Category A). DEGs in this category encompass most of the functions induced in response to IFNβ + TNF, with the notable exception of cell cycle and growth arrest and inflammasome and receptor signaling functions. Genes negatively regulated by STAT2 and IRF9 were also identified (Category B). Formation of an alternative STAT2/IRF9-containing complex mediating gene expression in the absence of STAT1 [37,38,39,40,41] has been reported, but with limited DNA-binding affinity for the typical ISRE sequence [37]. The existence of a STAT2/IRF9 complex is also supported by our recent observation of a high affinity of IRF9 for STAT2 with an equilibrium dissociation constant (Kd) of 10 nM [42]. A recent report of experiments, performed in murine bone marrow-derived macrophages proposes a model in which murine STAT2/IRF9 complex drives basal expression of ISGs, while IFNβ-inducible expression of ISGs depends on a switch to the ISGF3 complex [43]. This differs from our results as silencing of either STAT2 or IRF9 did not alter basal gene expression. Further analysis of the CXCL10 promoter demonstrates a restricted usage of ISRE sites by the STAT2/IRF9 pathway compared to the ISGF3 pathway. Further large-scale studies will be needed to identify the parameters allowing binding of ISGF3, but not STAT2/IRF9, to specific ISRE sequences upon IFNβ + TNF.

The observation that some IFNβ + TNF-induced genes were solely dependent on STAT2 (either positively or negatively) but not on IRF9, (Categories D and E) is a rare genome wide demonstration of gene regulation by STAT2 independently of STAT1 and IRF9. Previous reports have identified ISGF3-independent, STAT2-dependent genes but the association with IRF9 was not formally excluded [44,45,46,47]. STAT2 was shown to associate with STAT3 and STAT6, but it is not clear whether IRF9 is also part of these alternative complexes [44,46]. Transcriptional module analyses demonstrated that the functional distribution of genes negatively regulated by STAT2 is very limited compared to other categories; only a virus-sensing module was enriched in this category. In contrary, IRF9-independent genes positively regulated by STAT2 mediate broader antiviral and immunoregulatory functions.

ISGF3-independent functions of IRF9 have been proposed based on the study of IRF9 deficiencies [11,48]. However, IRF9 target genes in these contexts have been barely documented. Intriguingly, Li et al. [49] studied IFNα-induced genes and their dependency on the ISGF3 subunits. While they confirmed previous studies showing that IFNα can trigger a delayed and sustained ISG response via an ISGF3-independent pathway, it is very striking that they did not find STAT1- and STAT2-independent but IRF9-dependent genes. All identified IRF9-dependent genes were either STAT2- or STAT1-dependent. This result greatly differs with our study. Here, we found several IFNβ + TNF-induced DEGs independent of STAT1 and STAT2, but positively or negatively regulated by IRF9 (Categories F and G). Typically, IRF9 is considered a positive regulator of gene transcription. However, our findings are consistent with recent reports documenting the role of IRF9 in the negative regulation of the TRIF/NF-κB transcriptional response [50] or the expression of SIRT1 in acute myeloid leukemia cells [51]. The molecular mechanisms underlying gene regulation by IRF9 without association with either STAT1 or STAT2 remain to be elucidated. To the best of our knowledge, no alternative IRF9-containing complex has yet been described.

Our analysis showed that a large number of genes were induced by IFNβ + TNF independently of STAT2 and IRF9 (Category C). All transcriptional modules were enriched in this category pointing to a major role of this pathway in the establishment of a host defense and immunoregulatory response. The STAT2 and IRF9 independent genes does not solely reflects induction by TNF alone. For instance, APOBEC3G that is amongst the STAT2- and IRF9- independent genes is not induced by TNF alone (Figure 2A). While NF-κB, a downstream effector of the TNF receptor, is an obvious candidate for the regulation of these DEGs, this might fall short in explaining the synergistic action of IFNβ + TNF as we did not observe enhanced NF-κB activation compared to TNF alone [14]. Alternatively, the potential role of AP-1 is supported by the finding that the AP-1 transcription network module is enriched amongst IFNβ + TNF-induced DEGs. However, this module is not restricted to genes regulated independently of STAT2 and IRF9. It is also worth noting that two modules of IRF2-target genes were enriched, although again not specifically in the STAT2- and IRF9-independent category. A similar crosstalk was reported between IFNα and TNF in macrophages resulting in increased colocalized recruitment of IRF1 and p65 to the promoter of a subset of genes [52]. However, while IRF1 was found synergistically induced by IFNβ + TNF at early stages (Figure 1), we did not observe significant induction of IRF1 in the absence of STAT1 by RNASeq or qRT-PCR (data not shown), suggesting that IRF1 is unlikely to be involved in our system. Further studies will be required to uncover these STAT2- and IRF9-independent pathways.

This study provides novel insight into the molecular pathways leading to delayed antiviral and immunoregulatory gene expression in conditions where elevated levels of both IFNβ and TNF are present. Altogether our results demonstrate that in addition to the engagement of an ISGF3-dependent canonical response, a broad transcriptional program is elicited independently of STAT1, and support a model in which STAT2 and IRF9 contribute to the regulation of this response through non-canonical parallel pathways involving their concerted or independent action (Figure 7).

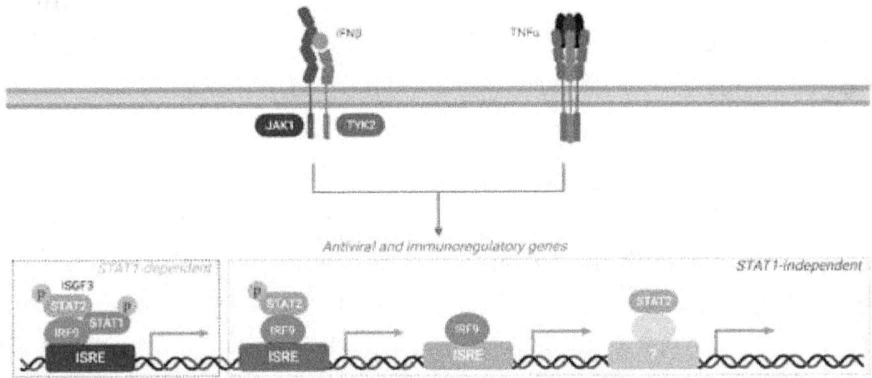

Figure 7. Role of distinct STAT2 and/or IRF9-dependent pathways in the regulation of distinct subset of antiviral and immunoregulatory genes in response to IFNβ and TNF. Our data supports a model in which multiple pathways participate to the synergistic action of IFNβ + TNF. While the STAT1-dependent pathway, likely ISFG3, is engaged downstream of IFNβ and TNF, STAT1-independent pathways are also involved in the control of the delayed gene expression. STAT2 and IRF9 act not only in a concerted fashion, likely as a complex, but also independently. IRF9 is known to act as the DNA-binding subunit of the ISGF3 complex and therefore likely mediates binding of STAT2/IRF9 complexes and of alternative complexes devoid of STAT2. The mechanisms of STAT2-dependent regulation of gene expression remains to be characterized.

Based on the increasing evidence [8], it is highly probable that these separate activities of STAT2 and IRF9 occur because of the creation of unique complexes that exist alongside ISGF3 during the stimulation of IFNβ and TNF. Ongoing research is being conducted to biochemically unravel the intricate and precise mechanisms via which the alternative

STAT2 and/or IRF9-containing complexes are activated in a normal cell environment. The aim is to gain a deeper understanding of the transcriptional response triggered by IFNβ and TNF.

CONCISE SUMMARY

Interferon (IFN) β and Tumor Necrosis Factor (TNF) play crucial roles in the immune response against viral infections. Conclusive evidence has demonstrated that the antiviral and inflammatory transcriptional response triggered by IFNβ is altered through communication with TNF. IFNβ primarily triggers the activation of interferon-stimulated genes through the Janus kinase (JAK)/signal transducer and activator of transcription (STAT) pathway. This pathway involves the classical ISGF3 transcriptional complex, which consists of STAT1, STAT2, and IRF9. The specific signaling pathways activated by the combination of IFNβ and TNF are not yet fully understood, however prior findings have indicated the presence of a response that is not dependent on STAT1. In this study, we utilized genome-wide transcriptional profiling by RNASeq to observe a comprehensive antiviral and immunoregulatory response that was triggered in the absence of STAT1 following IFNβ and TNF costimulation. Further analysis of this transcriptional response demonstrated that STAT2 and IRF9 are responsible for regulating the expression of a diverse range of genes. While a certain group of genes was controlled by the combined influence of STAT2 and IRF9, other sets of genes were regulated separately by either STAT2 or IRF9. Our evidence collectively indicates that STAT2 and IRF9 function through non-canonical parallel pathways to control specific sets of antiviral and immunoregulatory genes in situations where both IFNβ and TNF levels are high.

REFERENCES

1. McNab, F.; Mayer-Barber, K.; Sher, A.; Wack, A.; O'Garra, A. Type I interferons in infectious disease. Nat. Rev. Immunol. 2015, 15, 87–103.

2. Tomasello, E.; Pollet, E.; Vu Manh, T.P.; Uze, G.; Dalod, M. Harnessing Mechanistic Knowledge on Beneficial Versus Deleterious IFN-I Effects to Design Innovative Immunotherapies Targeting Cytokine Activity to Specific Cell Types. Front. Immunol. 2014, 5, 526.
3. Lee-Kirsch, M.A. The Type I Interferonopathies. Annu. Rev. Med. 2017, 68, 297–315.
4. Yarilina, A.; Park-Min, K.H.; Antoniv, T.; Hu, X.; Ivashkiv, L.B. TNF activates an IRF1-dependent autocrine loop leading to sustained expression of chemokines and STAT1-dependent type I interferon-response genes. Nat. Immunol. 2008, 9, 378–387.
5. Palucka, A.K.; Blanck, J.P.; Bennett, L.; Pascual, V.; Banchereau, J. Cross-regulation of TNF and IFN-alpha in autoimmune diseases. Proc. Natl. Acad. Sci. USA 2005, 102, 3372–3377.
6. Tliba, O.; Tliba, S.; Da Huang, C.; Hoffman, R.K.; DeLong, P.; Panettieri, R.A., Jr.; Amrani, Y. Tumor necrosis factor alpha modulates airway smooth muscle function via the autocrine action of interferon beta. J. Biol. Chem. 2003, 278, 50615–50623.
7. Au-Yeung, N.; Mandhana, R.; Horvath, C.M. Transcriptional regulation by STAT1 and STAT2 in the interferon JAK-STAT pathway. JAKSTAT 2013, 2, e23931.
8. Fink, K.; Grandvaux, N. STAT2 and IRF9: Beyond ISGF3. JAKSTAT 2013, 2, e27521.
9. Schneider, W.M.; Chevillotte, M.D.; Rice, C.M. Interferon-stimulated genes: A complex web of host defenses. Annu. Rev. Immunol. 2014, 32, 513–545.
10. Levy, D.E.; Marie, I.J.; Durbin, J.E. Induction and function of type I and III interferon in response to viral infection. Curr. Opin. Virol. 2011, 1, 476–486.
11. Majoros, A.; Platanitis, E.; Kernbauer-Holzl, E.; Rosebrock, F.; Muller, M.; Decker, T. Canonical and Non-Canonical Aspects of JAK-STAT Signaling: Lessons from Interferons for Cytokine Responses. Front. Immunol. 2017, 8, 29.

12. Mestan, J.; Brockhaus, M.; Kirchner, H.; Jacobsen, H. Antiviral activity of tumour necrosis factor. Synergism with interferons and induction of oligo-2′, 5′-adenylate synthetase. J. Gen. Virol. 1988, 69, 3113–3120.

13. Bartee, E.; Mohamed, M.R.; Lopez, M.C.; Baker, H.V.; McFadden, G. The addition of tumor necrosis factor plus beta interferon induces a novel synergistic antiviral state against poxviruses in primary human fibroblasts. J. Virol. 2009, 83, 498–511.

14. Fink, K.; Martin, L.; Mukawera, E.; Chartier, S.; De Deken, X.; Brochiero, E.; Miot, F.; Grandvaux, N. IFNbeta/TNFalpha synergism induces a non-canonical STAT2/IRF9-dependent pathway triggering a novel DUOX2 NADPH oxidase-mediated airway antiviral response. Cell Res. 2013, 23, 673–690.

15. McKendry, R.; John, J.; Flavell, D.; Muller, M.; Kerr, I.M.; Stark, G.R. High-frequency mutagenesis of human cells and characterization of a mutant unresponsive to both alpha and gamma interferons. Proc. Natl. Acad. Sci. USA 1991, 88, 11455–11459.

16. Horvath, C.M.; Wen, Z.; Darnell, J.E., Jr. A STAT protein domain that determines DNA sequence recognition suggests a novel DNA-binding domain. Genes Dev. 1995, 9, 984–994.

17. Schindler, C.; Fu, X.Y.; Improta, T.; Aebersold, R.; Darnell, J.E., Jr. Proteins of transcription factor ISGF-3: One gene encodes the 91- and 84-kDa ISGF-3 proteins that are activated by interferon alpha. Proc. Natl. Acad. Sci. USA 1992, 89, 7836–7839.

18. Robitaille, A.C.; Mariani, M.K.; Fortin, A.; Grandvaux, N. A High Resolution Method to Monitor Phosphorylation-dependent Activation of IRF3. J. Vis. Exp. 2016, 107, e53723.

19. Dussault, A.A.; Pouliot, M. Rapid and simple comparison of messenger RNA levels using real-time PCR. Biol. Proced. Online 2006, 8, 1–10.

20. Bolger, A.M.; Lohse, M.; Usadel, B. Trimmomatic: A flexible trimmer for Illumina sequence data. Bioinformatics 2014, 30, 2114–2120.

21. Dobin, A.; Davis, C.A.; Schlesinger, F.; Drenkow, J.; Zaleski, C.; Jha, S.; Batut, P.; Chaisson, M.; Gingeras, T.R. STAR: Ultrafast universal RNA-seq aligner. Bioinformatics 2013, 29, 15–21.

22. Roberts, A.; Pimentel, H.; Trapnell, C.; Pachter, L. Identification of novel transcripts in annotated genomes using RNA-Seq. Bioinformatics 2011, 27, 2325–2329.

23. Trapnell, C.; Hendrickson, D.G.; Sauvageau, M.; Goff, L.; Rinn, J.L.; Pachter, L. Differential analysis of gene regulation at transcript resolution with RNA-seq. Nat. Biotechnol. 2013, 31, 46–53.

24. Anders, S.; Huber, W. Differential expression analysis for sequence count data. Genome Biol. 2010, 11, R106.

25. Robinson, M.D.; McCarthy, D.J.; Smyth, G.K. edgeR: A Bioconductor package for differential expression analysis of digital gene expression data. Bioinformatics 2010, 26, 139–140.

26. Young, M.D.; Wakefield, M.J.; Smyth, G.K.; Oshlack, A. Gene ontology analysis for RNA-seq: Accounting for selection bias. Genome Biol. 2010, 11, R14.

27. Weiner, J.; Domaszewska, T. tmod: An R package for general and multivariate enrichment analysis. PeerJ Prepr. 2016, 4, e2420v1.

28. Bar-Joseph, Z.; Gerber, G.K.; Lee, T.I.; Rinaldi, N.J.; Yoo, J.Y.; Robert, F.; Gordon, D.B.; Fraenkel, E.; Jaakkola, T.S.; Young, R.A.; et al. Computational discovery of gene modules and regulatory networks. Nat. Biotechnol. 2003, 21, 1337–1342.

29. Chaussabel, D.; Baldwin, N. Democratizing systems immunology with modular transcriptional repertoire analyses. Nat. Rev. Immunol. 2014, 14, 271–280.

30. Chaussabel, D.; Quinn, C.; Shen, J.; Patel, P.; Glaser, C.; Baldwin, N.; Stichweh, D.; Blankenship, D.; Li, L.; Munagala, I.; et al. A modular analysis framework for blood genomics studies: Application to systemic lupus erythematosus. Immunity 2008, 29, 150–164.

31. Li, S.; Rouphael, N.; Duraisingham, S.; Romero-Steiner, S.; Presnell, S.; Davis, C.; Schmidt, D.S.; Johnson, S.E.; Milton, A.; Rajam, G.; et

al. Molecular signatures of antibody responses derived from a systems biology study of five human vaccines. Nat. Immunol. 2014, 15, 195–204.

32. Benjamini, Y.; Hochberg, Y. Controlling the False Discovery Rate: A Practical and Powerful Approach to Multiple Testing. J. R. Stat. Soc. Ser. B (Methodol.) 1995, 57, 289–300.
33. Ross, I.; Gentleman, R. R: A language for data analysis and graphics. J. Comput. Graph. Stat. 1996, 5, 299–314.
34. Zaheer, R.S.; Koetzler, R.; Holden, N.S.; Wiehler, S.; Proud, D. Selective transcriptional down-regulation of human rhinovirus-induced production of CXCL10 from airway epithelial cells via the MEK1 pathway. J. Immunol. 2009, 182, 4854–4864.
35. Davis, A.M.; Hagan, K.A.; Matthews, L.A.; Bajwa, G.; Gill, M.A.; Gale, M., Jr.; Farrar, J.D. Blockade of virus infection by human CD4+ T cells via a cytokine relay network. J. Immunol. 2008, 180, 6923–6932.
36. Bartee, E.; McFadden, G. Human cancer cells have specifically lost the ability to induce the synergistic state caused by tumor necrosis factor plus interferon-beta. Cytokine 2009, 47, 199–205.
37. Bluyssen, H.A.; Levy, D.E. Stat2 is a transcriptional activator that requires sequence-specific contacts provided by stat1 and p48 for stable interaction with DNA. J. Biol. Chem. 1997, 272, 4600–4605.
38. Martinez-Moczygemba, M.; Gutch, M.J.; French, D.L.; Reich, N.C. Distinct STAT structure promotes interaction of STAT2 with the p48 subunit of the interferon-alpha-stimulated transcription factor ISGF3. J. Biol. Chem. 1997, 272, 20070–20076.
39. Gupta, S.; Jiang, M.; Pernis, A.B. IFN-alpha activates Stat6 and leads to the formation of Stat2:Stat6 complexes in B cells. J. Immunol. 1999, 163, 3834–3841.
40. Lou, Y.J.; Pan, X.R.; Jia, P.M.; Li, D.; Xiao, S.; Zhang, Z.L.; Chen, S.J.; Chen, Z.; Tong, J.H. IRF-9/STAT2 [corrected] functional interaction drives retinoic acid-induced gene G expression independently of STAT1. Cancer Res. 2009, 69, 3673–3680.

41. Abdul-Sater, A.A.; Majoros, A.; Plumlee, C.R.; Perry, S.; Gu, A.D.; Lee, C.; Shresta, S.; Decker, T.; Schindler, C. Different STAT Transcription Complexes Drive Early and Delayed Responses to Type I IFNs. J. Immunol. 2015, 195, 210–216.
42. Rengachari, S.; Groiss, S.; Devos, J.; Caron, E.; Grandvaux, N.; Panne, D. Structural basis of STAT2 recognition by IRF9 reveals molecular insights into ISGF3 function. Proc. Natl. Acad. Sci. USA 2018, 115, E601–E609.
43. Platanitis, E.; Demiroz, D.; Schneller, A.; Fischer, K.; Capelle, C.; Hartl, M.; Gossenreiter, T.; Muller, M.; Novatchkova, M.; Decker, T. A molecular switch from STAT2-IRF9 to ISGF3 underlies interferon-induced gene transcription. Nat. Commun. 2019, 10, 2921.
44. Ghislain, J.J.; Fish, E.N. Application of genomic DNA affinity chromatography identifies multiple interferon-alpha-regulated Stat2 complexes. J. Biol. Chem. 1996, 271, 12408–12413.
45. Brierley, M.M.; Marchington, K.L.; Jurisica, I.; Fish, E.N. Identification of GAS-dependent interferon-sensitive target genes whose transcription is STAT2-dependent but ISGF3-independent. FEBS J. 2006, 273, 1569–1581.
46. Wan, L.; Lin, C.W.; Lin, Y.J.; Sheu, J.J.; Chen, B.H.; Liao, C.C.; Tsai, Y.; Lin, W.Y.; Lai, C.H.; Tsai, F.J. Type I IFN induced IL1-Ra expression in hepatocytes is mediated by activating STAT6 through the formation of STAT2: STAT6 heterodimer. J. Cell Mol. Med. 2008, 12, 876–888.
47. Perry, S.T.; Buck, M.D.; Lada, S.M.; Schindler, C.; Shresta, S. STAT2 mediates innate immunity to Dengue virus in the absence of STAT1 via the type I interferon receptor. PLoS Pathog. 2011, 7, e1001297.
48. Suprunenko, T.; Hofer, M.J. The emerging role of interferon regulatory factor 9 in the antiviral host response and beyond. Cytokine Growth Factor Rev. 2016, 29, 35–43.
49. Li, W.; Hofer, M.J.; Songkhunawej, P.; Jung, S.R.; Hancock, D.; Denyer, G.; Campbell, I.L. Type I interferon-regulated gene expression and

signaling in murine mixed glial cells lacking signal transducers and activators of transcription 1 or 2 or interferon regulatory factor 9. J. Biol. Chem. 2017, 292, 5845–5859.

50. Zhao, X.; Chu, Q.; Cui, J.; Huo, R.; Xu, T. IRF9 as a negative regulator involved in TRIF-mediated NF-kappaB pathway in a teleost fish, Miichthys miiuy. Mol. Immunol. 2017, 85, 123–129.

51. Tian, W.L.; Guo, R.; Wang, F.; Jiang, Z.X.; Tang, P.; Huang, Y.M.; Sun, L. IRF9 inhibits human acute myeloid leukemia through the SIRT1-p53 signaling pathway. FEBS Lett. 2017, 591, 2951.

52. Park, S.H.; Kang, K.; Giannopoulou, E.; Qiao, Y.; Kang, K.; Kim, G.; Park-Min, K.H.; Ivashkiv, L.B. Type I interferons and the cytokine TNF cooperatively reprogram the macrophage epigenome to promote inflammatory activation. Nat. Immunol. 2017, 18, 1104–1116.

11 FLUORESCENT TAP AS A PLATFORM FOR VIRUS-INDUCED DEGRADATION OF THE ANTIGENIC PEPTIDE TRANSPORTER

1. INTRODUCTION: AN OVERVIEW

The coexistence of a host and a virus is contingent upon a delicate equilibrium between the replication of the pathogen and the immunological response of the host. Peptides produced from viruses, primarily by proteasomal degradation, are presented by major histocompatibility complex class I (MHC I) molecules. This presentation allows cytotoxic CD8+ T lymphocytes (CTLs) to recognize infected cells. (Reviewed in [1]). The transporter associated with antigen processing (TAP) is crucial in facilitating MHC I-restricted antigen presentation, making it an appealing target for viruses seeking to evade the immune system.

The TAP protein is a heterodimer that is classified as a member of the ATP-binding cassette (ABC) family of transporters. The structure comprises of two subunits, namely TAP1 (ABCB2) and TAP2 (ABCB3) [2]. The central component of each subunit consists of a transmembrane domain (TMD) found at the N-terminus. This domain is made up of six transmembrane helices (TMs) and is responsible for recognizing and binding peptides [3]. Additionally, there is a C-terminal nucleotide-binding domain (NDB) that is highly conserved. This domain has the ability to bind and break down ATP [4]. The acquisition of both substrates, ATP and the peptide, happens autonomously [3]. This causes structural changes in TAP, leading to a transition from a conformation that is open inward to one that is open outward, allowing the peptide to be released into the lumen of the endoplasmic reticulum (ER). Subsequently, the process of ATP hydrolysis initiates the liberation of phosphate and reestablishes the original state of TAP [5]. The core-flanking TMD0 domains, consisting of four transmembrane domains in TAP1 and three transmembrane domains in TAP2, are not essential for peptide transport. However, they play a critical role in the formation of the peptide-loading complex (PLC) and the subsequent presentation of antigenic peptides to cytotoxic T lymphocytes (CTLs) [6]. Throughout their co-evolution with their hosts, multiple herpesviruses and one currently identified poxvirus have developed specific mechanisms to suppress TAP (Transporter Associated with Antigen Processing) by various means, as discussed in a previous review [7]. The herpes simplex virus 1 and 2 (HSV-1 and HSV-2) have the ICP47 protein, which competes for the peptide-binding site. The ICP47 protein has a unique structure that keeps the TAP-ICP47 complex in an inward-facing position [8,9,10]. On the other hand, the US6 protein of human cytomegalovirus (HCMV) and the cowpox virus (CPXV) strain Brighton Red-encoded CPXV012 protein have the ability to prevent ATP from binding to NDBs, without affecting peptide binding. The precise mechanisms by which the UL49.5 protein family, encoded by members of the Varicellovirus genus, inhibits TAP (Transporter associated with Antigen Processing) are not yet well comprehended. Furthermore, it appears that these mechanisms vary in specific details across different virus species. The majority of UL49.5 orthologs impede conformational rearrangements inside TAP [17]. Bovine herpesvirus 1 (BoHV-1) UL49.5 has the distinct ability to specifically degrade bovine, human, and murine TAP through proteasomal degradation after conformational arrest [7,18,19]. The UL49.5 protein encoded by the Varicella-zoster virus (VZV) has the ability to bind to TAP, however it does not have any inhibitory effects [20]. The murine gammaherpesvirus-68 MK3 protein [21] and the rodent herpesvirus Peru pK3 ortholog [22] have been reported to possess TAP degrading activity. These proteins have a cytoplasmic RING (really intriguing novel gene) finger domain and are capable of targeting the murine transporter. The MC80 protein of the molluscum contagiosum virus has been found to destabilize the human TAP protein. However, unlike the UL49.5 protein of the BoHV-1 virus, the inhibitor does not primarily target the transporter. This information was reported in a recent study [23]. In recent times, the use of fluorescent tags and gene fusion technologies has become essential in several biochemical and cell biology applications. However, there are still situations when creating a functional fluorescent fusion protein poses a challenge. Multiple studies have demonstrated that the selection of a linker can greatly influence the correct folding, production efficiency, and functional properties of the fusion protein, as well as its interaction with other proteins. Flexible linkers are commonly used to allow for a specific range of motion, whereas rigid linkers are preferred for spatially separating two bioactive domains [24]. In order to study how TAP is inhibited or removed, researchers used a TAP-GFP fusion protein that emits green fluorescence. However, they noticed that when TAP was tagged with GFP, it prevented TAP from being degraded by the BoHV-1-encoded UL49.5 [18]. In this study, we describe the creation of a range of complete TAP1 and TAP2 variants that contain either N-terminal or C-terminal GFP, connected by various types of linkers. We then assess how the design of the TAP-GFP fusion affects their fluorescence and functionality, as well as their susceptibility to inhibition and degradation caused by viruses. A fluorescent TAP platform has the ability to elucidate the molecular mechanism of UL49.5 activity and enhance the characterisation of the transporter.

2. MATERIALS AND METHODS

2.1. Cells and Viruses

Madin-Darby bovine kidney (MDBK) cells (ATCC, Manassas, VA, USA, CCL-22), human melanoma Mel JuSo (MJS) cells, MJS TAP1 CRISPR/Cas9 knock-out (TAP1 KO), MJS TAP2 CRISPR/Cas9 knock-out (TAP2 KO) [25], and U937 (ATCC, CRL-1593) were cultured in RPMI 1640 (Corning, Corning, NY, USA) supplemented with 10% fetal bovine serum (FBS, Thermo Scientific (Thermo Scientific, Waltham, MA, USA)) and Antibiotic Antimycotic Solution (Thermo Scientific). Lenti-X HEK293T and GP2-293 cells (both from Takara/Clontech, Kusatsu, Japan) used for lentivirus and retrovirus production, respectively, were cultured in Iscove's modified Dulbecco's medium (IMDM, Lonza, Basel, Switzerland) supplemented as above. HEK293T (ATCC, CRL-3216) cells were cultured in Dulbecco's modified Eagle's medium (DMEM, high glucose, Lonza) supplemented as above. BoHV-1 field strain Lam (Institute for Animal Health and Science, Lelystad, The Netherlands) was propagated and titrated on MDBK cells.

2.2. DNA Constructs

All TAP constructs were cloned in lentiviral vectors downstream of an EF1α promoter. For unmodified (wild-type, wt) TAP1 or TAP2 reconstitution, dual promoter lentiviral vectors described in [25] (pPuroR-GFP-TAP1 and pZeoR-mAmetrine-TAP2) were used. mAmetrine and marker GFP genes were removed from these vectors. Fragments of TAP1 and TAP2 sequences were ordered as synthetic genes designed for cloning in pEGFP-N3 or pEGFP-C1 (Takara/Clontech). For TAP1-N-GFP (TAP1 with the N-terminal GFP, random linker), TAP1-C-GFP (TAP1 with the C-terminal GFP, random linker), TAP2-N-GFP (TAP2 with the N-terminal GFP, random linker), and TAP2-C-GFP (TAP2 with the C-terminal GFP, random linker), fusion genes were re-cloned in the original lentiviral vectors. The amino acid sequences of random linkers resulting from the cloning procedure are depicted in Figure 1A. Fragments coding for TAP1 with helical linker sequences were ordered as synthetic genes designed for cloning in pEGFP-N3 or pEGFP-C1. TAP1-HN-GFP (TAP1 with the N-terminal GFP, helical linker) or TAP1-HC-GFP (TAP1 with the C-terminal GFP, helical linker) were re-cloned in the lentiviral vector pCDH-EF1α-MCS-(PGK-Puro) (System Biosciences, Palo Alto, CA, USA).

Genes coding for viral TAP inhibitors were cloned in retroviral vectors downstream of a retroviral promoter. The BoHV-1 UL49.5 gene was cloned from pLZRS-BoHV-1 UL49.5-IRES-GFP [18] in BamHI-EcoRI sites of pLZRS-IRES-ΔNGFR [26]. The VZV UL49.5 gene was amplified from the pLZRS-VZV UL49.5-IRES-GFP vector [20] using KOD Hot Start DNA polymerase (Merck, Darmstadt, Germany) and the following primers: forward 5’-CGGGATCCCACCATGGGATCAATTACC-3’ and reverse 5’-CCGGAATTCTTACCACGTGCTGCGTAATAC-3’. The PCR product was verified by DNA sequencing and introduced into BamHI and EcoRI sites of the pBABEpuro vector [27]. Synthetic genes encoding: myc-tagged HSV-1 ICP47 (Gene ID: 2703441), myc-tagged HCMV US6 (Gene ID: 3077555), or the FLAG-N-CPXV012 (Gene ID: 1485887) variantlacking six N-terminal amino acid residues [28] were introduced into BamHI and EcoRI restriction sites of pBABEpuro.

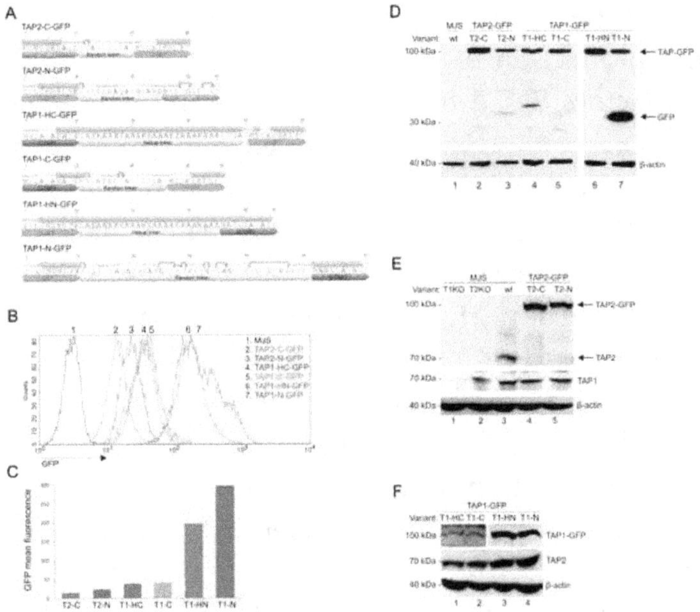

Figure 1. Construction and characterization of fluorescent transporter associated with antigen processing (TAP)-green fluorescent protein (GFP) variants. Mel JuSo (MJS) cells with CRISPR/Cas9 TAP1 or TAP2 knockouts (T1KO, T2KO) were stably reconstituted with fluorescent TAP1 or TAP2 constructs using lentivirus vectors and cell sorting. (A) Schematic representation of TAP-GFP constructs. Secondary structures of linkers flanked by ten amino acid residues of fused proteins were determined by the Geneious software; α-helices are depicted in pink, coiled regions in gray, β-strands in yellow, and turns in blue. (B) Representative histograms of TAP-GFP fluorescence intensity. (C) Comparative TAP-GFP analysis by flow cytometry. The mean fluorescence intensity of three independent measurements is represented as bars with standard deviations. The statistical significance was assessed by t-test; $p \leq 0.001$. (D–F) Expression

of TAP-GFP variants in stable cell lines was determined by SDS-PAGE and immunoblotting using: (D) anti-GFP monoclonal antibody (Mab) (E) anti-TAP2 MAb (F) anti-TAP1 MAb. β-actin was used as a loading control. Abbreviations: T2-C: TAP2-C-GFP (TAP2 with the C-terminal GFP, random linker); T2-N: TAP2-N-GFP (TAP2 with the N-terminal GFP, random linker); T1-C: TAP1-C-GFP (TAP1 with the C-terminal GFP, random linker); T1-HC: TAP1-HC-GFP (TAP1 with the C-terminal GFP, helical linker); T1-N: TAP1-N-GFP (TAP1 with the N-terminal GFP, random linker); T1-HN: TAP1-HN-GFP (TAP1 with the N-terminal GFP, helical linker).

2.3. Retroviral and Lentiviral Transduction

For the production of recombinant lentiviruses, third-generation packaging vectors based on the pRSV-Rev and pCgpV plasmids (Cell Biolabs, San Diego, CA, USA), the obtained lentiviral expression vectors, and pCMV-VSV-G (Cell Biolabs) for pseudotyping were co-transfected into Lenti-X HEK293T cells using CalPhos mammalian transfection kit (Takara/Clontech). For recombinant retroviruses, a transfer plasmid (pBABEpuro-based or pLZRS-IRES-ΔNGFR-based) and pCMV-VSV-G were co-transfected into GP2-293 packaging cells as above. Twenty-four hours after transfection the medium was refreshed; for lentiviruses it was supplemented with 1 mM sodium butyrate (Sigma-Aldrich, Saint Louis, MO, USA). Virus-containing supernatants were collected after 48 h, concentrated with PEGit (System Biosciences), and used for transduction in the presence of 0.01 mg mL−1 polybrene (Sigma-Aldrich). MJS cells with TAP1 or TAP2 knock-outs were stably reconstituted with the wt or fluorescent TAP1 or TAP2 constructs using lentivirus vectors and cell-sorting for GFP- and MHC I-positive cells. The cells were subsequently transduced with a retrovirus coding for BoHV-1 UL49.5 and sorted for nerve growth receptor (NGFR)-positive cells or with a retrovirus coding for HSV-1 ICP47, HCMV US6, VZV UL49.5, or CPXV012, and selected with puromycin (2 μg mL−1) (Sigma-Aldrich).

2.4. Plasmid Transfection

HEK293T cells were transfected with plasmids encoding fluorescent TAP variants using JetPRIME (Polyplus-transfection, Illkirch, France) according to the manufacturer's protocol and analyzed after 24 h by flow cytometry.

2.5. Generation of A TAP1/TAP2 Double Knock-Out U937 Cells for Reconstitution with Fluorescent TAP Variants

U937 TAP1/TAP2 KO cells were generated with a strategy described for MJS TAP1/TAP2 KO in [25] Briefly, U937 cells were transfected with pSico-CRISPR-PuroR containing the TAP2-targeting crRNA sequence 5'-GGAAGAAGAAGGCGGCAACG-3'. The cells were selected with puromycin (4 μg mL−1), and cloned by limiting dilution. Individual clones were analyzed by flow cytometry to identify clones with low cell surface MHC I expression, followed by immunoblotting and DNA sequencing of the genomic target site. A clone lacking TAP2 was subsequently transfected with a pSicoR-CRISPR-PuroR vector containing the TAP1-targeting crRNA sequence 5'-GGGGTCCTCAGGGCAACGGT-3'. After selection with puromycin and cell cloning, the clones were analyzed for TAP1 expression by immunoblotting and DNA sequencing of the genomic target site. Genomic DNA sequence analysis revealed a 16-bp deletion around the TAP2 gRNA target site and multiple short deletions altering the whole TAP1 gene sequence downstream of the target site. A monoclonal cell line lacking TAP1 and TAP2 was used for reconstitution with a combination of unmodified and fluorescent TAP-encoding sequences delivered by lentivirus vectors. Reconstituted U937 cells were sorted for GFP and high MHC I expression. The cells were subsequently transduced with the BoHV-1 UL49.5-encoding retrovirus and sorted for NGFR.

2.6. Antibodies

Antibodies used for immunoblotting: mouse anti-TAP1 monoclonal antibody MAb 143.5 (kindly provided by R. Tampé, Institute of Biochemistry, The Johann Wolfgang Goethe University, Frankfurt, Germany); mouse anti-TAP2 MAb 435.3 (a kind gift from P. van Endert, INSERM U25, Institute Necker, Paris, France); rabbit anti-TAP1 (Enzo Life Sciences, Farmingdale, NY, USA); rat anti-GFP 3H9 (Chromotek, Planegg, Germany); mouse anti-myc tag 9B11 (Cell Signaling, Danvers, MA, USA); rabbit anti-β-actin (Novus Biologicals, Centennial, CO, USA); rabbit anti-β-catenin (Santa Cruz Biotechnology, Dallas, TX, USA); rabbit antibodies (H11) against a synthetic peptide derived from the N-terminal domain of BoHV-1 UL49.5 [26] and mouse anti-OctA (FLAG) G-8 (Santa Cruz Biotechnology); and mouse anti-HC10 [19] and rabbit anti-ERp57 H-220 (Santa Cruz Biotechnology). Probes used for immunofluorescence: Alexa 633-conjugated concanavalin A (ConA) (Thermo Scientific). Antibodies used for flow cytometry: mouse anti-MHC I W6/32 (Novus Biologicals); mouse anti-NGFR (Sigma-Aldrich); and Alexa 633-conjugated goat anti-mouse IgG (Thermo Scientific).

2.7. Flow Cytometry

Cell surface expression of MHC I was determined by indirect immunofluorescence using primary anti-MHC I antibodies (1:1000) and secondary antibodies (1:1000), all in phosphate buffered saline (PBS) buffer containing 1% bovine serum albumin and 0.02% sodium azide. For cell sorting, anti-NGFR antibodies (1:1000) and secondary antibodies were used. Cells were analyzed using a FACS Calibur flow cytometer (Becton Dickinson, Franklin Lakes, NJ, USA) and CellQuest software (version 5.2.1, Becton Dickinson)); for cell sorting, the sorting option of FACS Calibur was applied.

2.8. Immunoblotting and Immunoprecipitation

For immunoblotting, the cells were lysed in Cell Lytic M buffer (Sigma-Aldrich); for immunoprecipitation, the cells were lysed in a buffer containing 1% digitonin (Merck), 50 mM Tris-HCl, pH 7.5, 5 mM MgCl2, and 150 mM NaCl. The buffers were supplemented with the cOmplete mini protease inhibitor cocktail (Roche, Basel, Switzerland). Cell lysates were analyzed by SDS-PAGE and immunoblotting as previously described [26] or incubated with GFP-Trap (Chromotek) according to the manufacturer's protocol to isolate protein complexes.

2.9. Peptide Transport Assay

The peptide transport assay was performed as described before [26]. Briefly, the cells were permeabilized with 2 IU mL−1 of Streptolysin O (Sigma-Aldrich) at 37 °C for 15 min. The cells (2 × 106 cells/sample) were subsequently incubated with 600 pmol of the fluorescein-conjugated synthetic peptide CVNKTERAY (JPT Peptide Technologies, Berlin,

Germany) in the presence or absence of ATP (10 mM final concentration) at 37 °C for 10 min. Peptide translocation was terminated by adding 1 mL of ice-cold lysis buffer containing 1% Triton X-100. After 20 min of lysis, cell debris was removed by centrifugation, and supernatants were collected and incubated with 100 μL of concanavalin A (ConA)-Sepharose (Sigma-Aldrich) at 4 °C for 1 h to isolate the glycosylated peptides. After extensive washing of the beads, the peptides were eluted with elution buffer (500 mM mannopyranoside, 10 mM EDTA, 50 mM Tris-HCl, pH 8.0) by rigorous shaking for 1 h at 25 °C. Eluted peptides were separated from ConA by centrifugation at 12,000× g. The fluorescence intensity was measured using a fluorescence plate reader (EnVision, PerkinElmer, Waltham, MA, USA) with excitation and emission wavelengths of 485 nm and 530 nm, respectively.

2.10. Immunofluorescence

MJS cells were grown on microcover glass, fixed with 4% paraformaldehyde in PBS, permeabilized with 0.2% Triton X-100 in PBS, and stained with Alexa 633-ConA (1:1000), prepared in PBS containing 1% bovine serum albumin (Sigma-Aldrich). GFP booster (1:100, Chromotek) was used for MJS-TAP2-C-GFP to enhance the green fluorescence. The blue signal was electronically converted into the red during the analysis of images using Leica TCS SP8X confocal laser scanning microscope (Leica, Wetzlar, Germany).

3. RESULTS

3.1. Construction and Characterization of Fluorescent TAP-GFP Variants

In order to develop a universal fluorescent platform for virus-induced TAP degradation, we constructed six versions of TAP-GFP fusion: four with two types of linkers, a random linker or a helical one, placed at the N- or C-terminus of TAP1, and two with a random linker at the N- or C-terminus of TAP2 (Figure 1A). A number of studies regarding fusion protein linker design have suggested that the most important features of a proper linker are its hydrophilicity and flexibility [24]. The random linkers used to join TAP and GFP have resulted from the cloning procedure into one of the pEGFP plasmid series. The analysis of their amino acid sequence revealed they were unstructured; thus, they should be more flexible. In some cases, flexible linkers may result in undesired interactions or interference between the fusion partners. In such cases, rigid linkers are preferable to separate two independently active domains spatially. An example of a rigid linker is the α helix-forming peptide AEAAAKEAAAKEAAAKA, stabilized by salt bridges between glutamate and lysine residues [29]. The distance between two separated domains can be regulated by changing the number of EAAAK motif repetitions. By using fluorescence resonance energy transfer (FRET) measurement, the helical linker with four repetitions of this motif has been demonstrated as the most efficient in separating two fluorescent proteins [30]. Therefore, this linker was selected for our studies to generate N- and C-terminal fusion of TAP1 and GFP. To introduce fluorescent TAP-GFP into human melanoma MJS cells, we first generated, by using CRISPR/Cas9-based technology, TAP1 or TAP2 knock-outs (KO). This enabled stable reconstitution of the cells with a fluorescent TAP1 or TAP2 using lentiviral vectors. The cells were subsequently sorted based on GFP to high purity (>98%). MJS cells have been shown to be permissive for BoHV-1 infection and are widely used in the research on modulation of antigen presentation by viruses [31,32,33].

First, we analyzed the GFP fluorescence intensity of our constructs (Figure 1B,C). Flow cytometry analysis revealed that N-terminal fusions of GFP to TAP1 exhibited the highest fluorescence, followed by TAP1 C-terminal fusion constructs. The type of linker seemed to have no crucial impact on the fluorescence intensity, although, for TAP1-N-GFP, a population of brighter green fluorescent cells could be observed when compared to TAP1-HN-GFP. TAP2 constructs exposed the lowest fluorescence, with a similar tendency of N-terminal fusion outperforming the C-terminal one. In addition, for TAP1-N-GFP, we could observe a more heterogeneous distribution of GFP fluorescence than for the other variants.

During the transient expression of fluorescent constructs in plasmid-transfected HEK293T cells, we could observe a similar range of

GFP fluorescence, which indicates that differences in GFP fluorescence depend on the properties of individual fusion proteins rather than result from random incorporation of a lentivirus into a host genome. TAP2-C-GFP was performing noticeably better in transfected HEK293T than in the stable MJS cell line, and this time TAP1-HN-GFP slightly outperformed TAP1-N-GFP.

Next, we characterized the expression of TAP-GFP constructs by SDS-PAGE and immunoblotting. TAP1 and TAP2 have a similar apparent molecular mass of approximately 75 kDa. The fusion to GFP should yield a single protein at 100 kDa. All constructs were detected in cell lysates with anti-GFP antibodies at the predicted molecular weight of 100 kDa; only for TAP1-N-GFP did we observe an additional 30-kDa band corresponding to, most probably, cleaved free GFP or cleaved TAP-GFP fragment (Figure 1D). Fluorescent TAP could also be detected with anti-TAP1 or anti-TAP2 antibodies (Figure 1E,F). The reconstitution of TAP1KO cells with a fluorescent variant of TAP1 resulted in increased stability of endogenous TAP2 as it could be detected in higher amounts than in TAP1KO cells (compare lane 1 in Figure 1E with lanes 1–4 in Figure 1F). This is in agreement with a previous study reporting that unlike TAP1, TAP2 is unstable when expressed without the other half of the transporter [34]. The higher sensitivity of TAP2 can most probably explain why no band corresponding to TAP2 could be identified in MJS TAP1KO cells (Figure 1E, lane 1), while TAP1 could be easily detected in MJS TAP2KO cells (Figure 1E, lane 2).

3.2. TAP-GFP Localizes in the ER and Forms a Functional Transporter

TAP localizes in the cells predominantly in the ER [35]. To assess if the fluorescent TAP constructs acquired their proper localization, we stained the cells with Alexa 633-conjugated concanavalin A (ConA), an ER-cis Golgi marker, and analyzed fluorescence distribution by confocal laser scanning microscopy (Figure 2). In all cell lines, TAP-GFP (green) localized predominantly in the ER (red), as it co-localized with the ER marker (yellow). No granular localization patterns that would indicate TAP aggregates could be visualized.

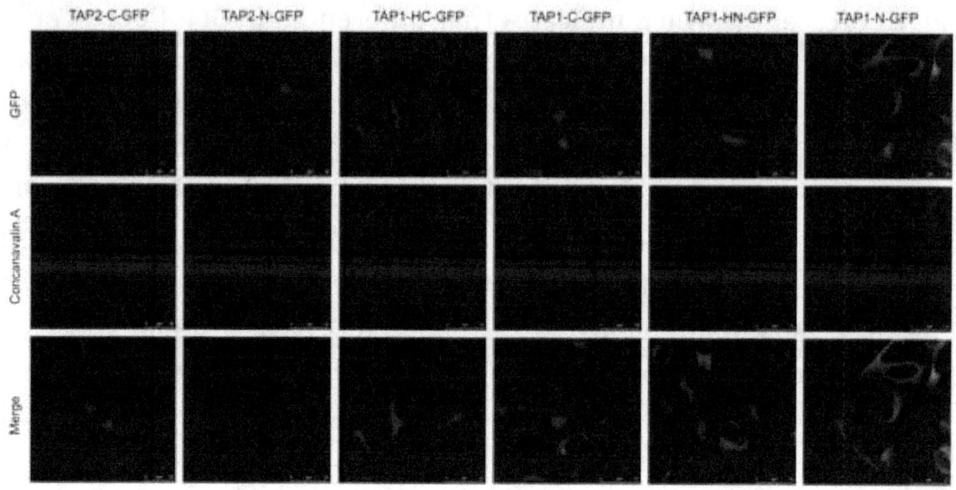

Figure 2. TAP-GFP variants demonstrate endoplasmic reticulum (ER) localization. MJS cells reconstituted with TAP-GFP variants were fixed, permeabilized, and stained with Alexa 633-conjugated concanavalin A (ER marker). Colocalization with GFP was analyzed using fluorescent confocal microscopy. For MJS TAP2-C-GFP, the GFP booster was used to enhance very weak green fluorescence.

Based on the currently available evidence, a lack of at least one fully functional TAP subunit results in a suppressed peptide translocation and production of empty or suboptimally loaded unstable MHC I molecules, mostly retained in the ER [36]. GFP fusion could potentially affect TAP structure and activity. Therefore, the functionality of fluorescent TAP constructs was tested by measuring cell surface expression of MHC I by flow cytometry (Figure 3). As expected, MJS cells with TAP1KO or TAP2KO exhibited strongly reduced levels of MHC I. To have a proper control for testing the effect of fluorescent tagging on TAP performance, especially during overexpression of a TAP subunit, we generated control cell lines reconstituted with unmodified TAP subunits (MJS TAP1KO + TAP1 and MJS TAP2KO + TAP2). Overexpression of individual TAP subunits might result in the formation of their homodimers, affecting the interpretation of our further experiments [19,37]. Therefore, as an additional control, we transduced TAPKO cells with vectors enabling overexpression of the existing half-transporter (MJS TAP1KO + TAP2 and MJS TAP2KO + TAP1). As expected, MHC I surface levels were reduced in those cells to a similar extent as in TAP1KO or TAP2KO cells. On the other hand, reconstitution of the missing TAP subunit with its fluorescent variant rescued surface MHC I to levels slightly higher than on parental ("wild-type") MJS cells, but similar to the ones observed on the cells reconstituted with the non-fluorescent TAP.

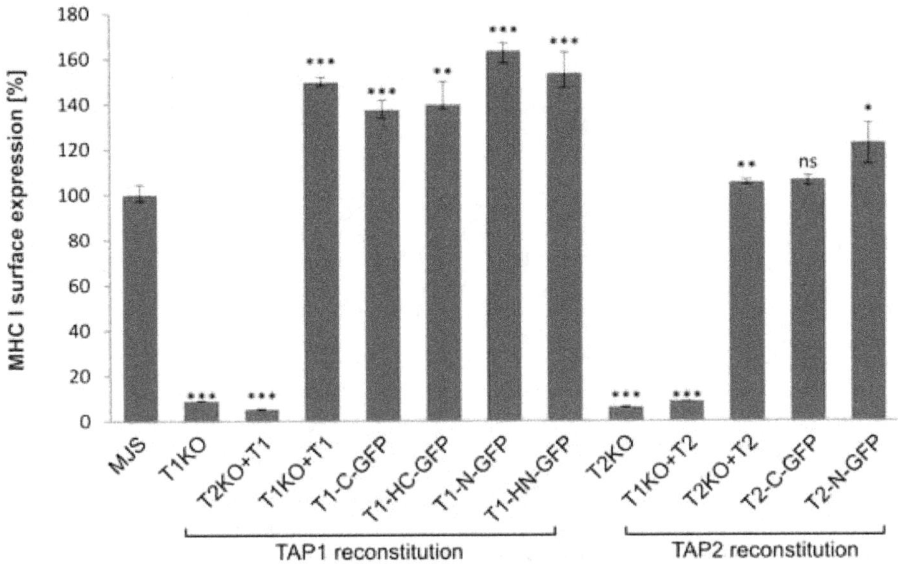

Figure 3. TAP-GFP forms a functional transporter. MJS cells with CRISPR/Cas9 TAP1 or TAP2 knockouts (T1KO or T2KO) were stably reconstituted with wild-type or fluorescent TAP1 or TAP2 variants. Surface expression of major histocompatibility complex class I (MHC I) was assessed by flow cytometry using specific antibodies (W6/32). MHC I expression on MJS cells with TAP reconstitution is presented as the percentage of MHC I mean fluorescence intensity on MJS cells (set as 100%). The analysis was performed in triplicates. The statistical significance of differences between MHC I on MJS cell with TAP reconstitution and MJS wild-type (wt) cells was estimated by t-test; *** $p \leq 0.001$, ** $p \leq 0.01$, * $p \leq 0.05$, ns: not significant.

3.3. The Sensitivity of TAP-GFP Variant to UL49.5-Mediated MHC I Downregulation and TAP Degradation

To examine the sensitivity of fluorescent TAP to BoHV-1-encoded UL49.5-mediated inhibition, UL49.5 was introduced to stable cell lines with TAP-GFP variants using a retrovirus vector. The cells were subsequently sorted for high purity based on the expression of truncated NGFR as a marker, and analyzed by flow cytometry for surface expression of MHC I

(Figure 4A). According to the obtained data, all fluorescent transporters were prone to UL49.5-induced inhibition, which was illustrated as the downregulation of MHC I, to a similar extent for all the tested variants.

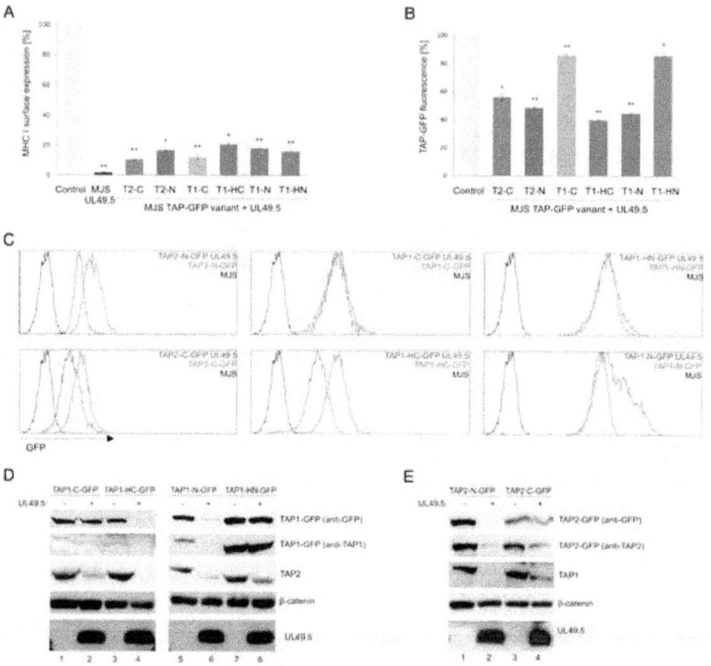

Figure 4. TAP-GFP variants differ in their sensitivity to UL49.5-mediated inhibition and degradation. MJS cells with fluorescent TAP1 or TAP2 variants were transduced with a retrovirus encoding bovine herpesvirus 1 (BoHV-1) UL49.5. (A) Surface expression of MHC I was assessed by flow cytometry using specific antibodies (W6/32). MHC I expression is presented as the percentage of mean fluorescence intensity; fluorescence of parental cells without UL49.5 was set as 100%. The analysis was performed in triplicates. The statistical significance of differences between MJS TAP-GFP and MJS TAP-GFP UL49.5 cell lines was estimated by t-test; ** $p \leq 0.001$ * $p \leq 0.005$. (B) GFP mean fluorescence intensity is presented as the percentage of GFP fluorescence of parental cells (set as 100%). The analysis was performed in triplicates. The statistical significance of differences between MJS TAP-GFP and MJS TAP-GFP UL49.5 cell lines was estimated by t-test; ** $p \leq 0.001$ * $p \leq 0.005$. (C) The effect of UL49.5 on GFP level in MJS TAP-GFP cells was assessed by flow cytometry. (D,E) Degradation of TAP-GFP variants in the presence of BoHV-1 UL49.5 in stable cell lines was determined by SDS-PAGE and immunoblotting using: anti-GFP, anti-TAP1, anti-TAP2, or anti-UL49.5 antibodies. β-catenin was used as a loading control.

Next, we analyzed the susceptibility of TAP-GFP variants to UL49.5-triggered degradation (Figure 4B,C). Flow cytometry analysis revealed a reduction of GFP mean fluorescence intensity by approximately 50% in the cells co-expressing UL49.5 with TAP1-N-GFP, TAP1-HC-GFP, and both TAP2 variants, in comparison to the control cells expressing the fluorescent transporter without UL49.5. There were no significant changes in GFP fluorescence observed in the cells expressing TAP1-C-GFP or TAP1-HN-GFP with UL49.5.

Susceptibility of TAP1-HC-GFP and TAP2-N-GFP constructs to UL49.5-dependent TAP inhibition and degradation was also confirmed in a reconstituted U937 cell line, where downregulation of surface MHC I, as well as a decrease of mean GFP fluorescence to the similar extent as in MJS cells, could be denoted.

To verify if changes in GFP fluorescence correspond with the decreased TAP protein level, immunoblotting analysis of cell lysates was performed (Figure 4D,E). A similar level of BoHV-1 UL49.5 protein could be observed in all cell lines. Decreased amounts of TAP1-N-GFP, TAP1-HC-GFP, and both TAP2-GFP fusion proteins could be detected in the presence of UL49.5, with the use of both anti-GFP and specific anti-TAP antibodies. Of note, in the case of non-degradable fluorescent TAP1 constructs (TAP1-HN-GFP and TAP1-C-GFP), the level of endogenous TAP2 was decreased (compare TAP2 in Figure 4D in lanes 1–2 and 5–6), what may suggest partial degradation of the untagged TAP subunit only. Degradation in TAP1-HC-GFP and TAP2-N-GFP cell lines was observed for both, exogenous fluorescent and endogenous TAP subunits. TAP1-N-GFP and TAP2-C-GFP characterized with only partial degradation.

3.4. Interaction of TAP-GFP with UL49.5 and the Peptide-Loading Complex

To assess the interaction of a non-degradable TAP with BoHV-1-encoded UL49.5 and selected components of PLC, we chose the TAP1-HN-GFP construct as demonstrating high expression. Among proteins co-immunoprecipitating with the fluorescent transporter (using GFP-Trap), we could identify UL49.5 and endogenous TAP2, as well as the known components of PLC: MHC I heavy chain and ERp57 (Figure 5). These results highlight the fact that UL49.5 is still capable of interacting with the non-degradable TAP1-HN-GFP. Unspecific binding of UL49.5 was excluded by immunoprecipitation from MJS wt cell lysate (with unmodified TAP) expressing UL49.5.

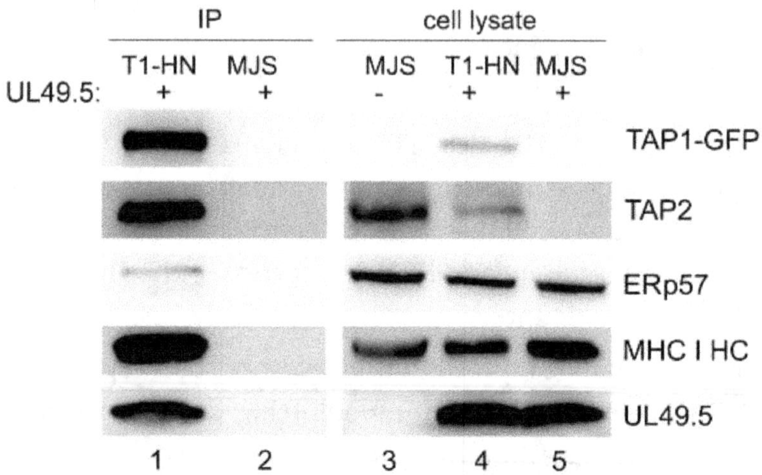

Figure 5. TAP1-GFP interacts with UL49.5 and the peptide-loading complex. TAP1-HN-GFP (T1-HN) was immunoprecipitated by GFP-Trap from lysates of MJS cells expressing TAP1-HN-GFP and UL49.5 or wt MJS with UL49.5 only. Co-precipitating proteins were analyzed by SDS-PAGE and immunoblotting using antibodies against GFP, TAP2, ERp57, MHC I HC, and UL49.5. Right panel: cell lysates were loaded on SDS-PAGE directly and analyzed by immunoblotting.

3.5. Application of the TAP2-N-GFP Variant as a Platform to Study BoHV-1 UL49.5 Activity in Virus-Infected Cells

In the more detailed studies on our fluorescent TAP platform, we focused on a single TAP-GFP variant, selecting TAP2-N-GFP. TAP1-N-GFP was excluded based on the presence of a free form of GFP. The constructs resistant to UL49.5-triggered degradation were eliminated as well, as was TAP2-C-GFP due to its very week, nearly undetectable basic green fluorescence and only partial degradation in the presence of UL49.5.

First, we confirmed that UL49.5-mediated MHC I downregulation relies on the inhibition of antigenic peptide translocation. TAP transport assay was performed in TAP2-N-GFP MJS cells and TAP2-N-GFP cells expressing UL49.5 or (as controls) in MJS wt, MJS TAP2KO cells, and MJS TAP2KO cells reconstituted with a non-fluorescent TAP2. The assay was based on cytoplasm-to-ER translocation of fluorescein-conjugated substrate peptides, in the presence of ATP or EDTA, as a passive diffusion control (Figure 6). Reconstitution with fluorescent TAP restored peptide transport when compared to the parental TAP2KO cells to a level of almost 50% higher than for MJS wt cells. This might result from high expression of exogenous TAP2 as a very similar transport activity was denoted for the non-fluorescent TAP2 reconstitution. The presence of UL49.5 inhibited peptide translocation to the level of TAP2KO cells.

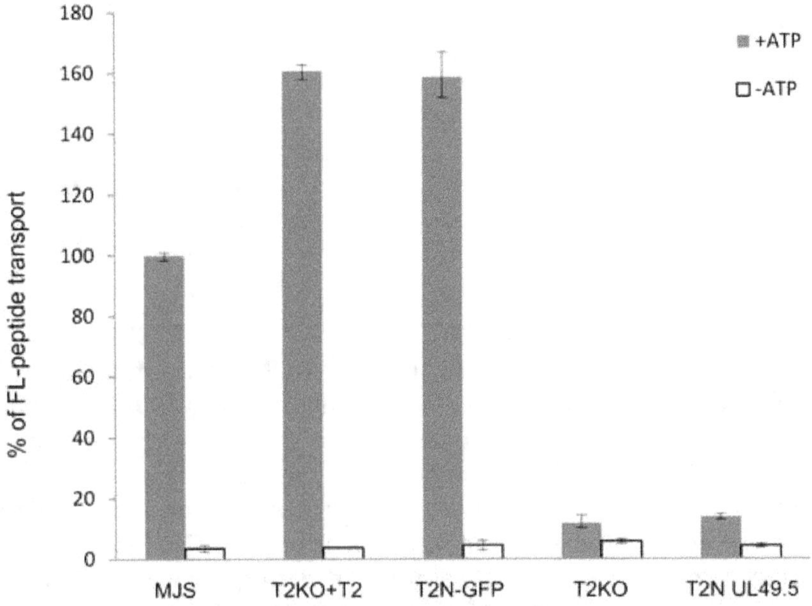

Figure 6. UL49.5-induced inhibition of peptide transport. MJS cells with CRISPR/Cas9 TAP2 knockout (T2KO) were stably reconstituted with wild-type TAP2 (T2KO+T2) or fluorescent TAP2 (T2N-GFP), subsequently transduced with a retrovirus encoding BoHV-1 UL49.5, and sorted (T2N UL49.5). Transport activity of TAP was analyzed using fluorescein-labeled peptide CVNKTERAY in the presence of ATP (gray bars) or EDTA (white bars). Peptide transport is expressed as a percentage of translocation, relative to the translocation observed in control MJS cells (set at 100%). The experiment was performed in triplicates. The statistical significance of differences between MJS controls and reconstituted or KO variants was estimated by t-test; for all the samples $p \leq 0.005$.

Finally, to test whether we can apply our fluorescent TAP platform to quantify the TAP level during virus infection, we infected MJS TAP2-N-GFP cells with BoHV-1 and analyzed TAP-derived GFP fluorescence by flow cytometry (Figure 7A,B). TAP2-N-GFP and endogenous TAP1 levels were assessed in infected cell lysates by immunoblotting (Figure 7C,D). The viral infection resulted in a 30% decrease of mean GFP fluorescence intensity, while approximately 70% and 50% reduction of TAP2-N-GFP and TAP1 protein levels, respectively, could be demonstrated by immunoblotting.

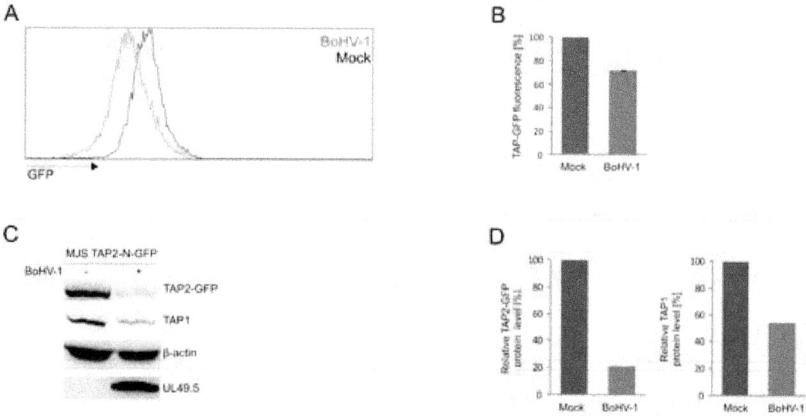

Figure 7. BoHV-1 infection results in TAP2-GFP degradation. MJS TAP2-N-GFP cells were infected with BoHV-1 at a multiplicity of infection (moi) = 10. Twenty-four hours post-infection, cells were collected and analyzed. (A) TAP2-GFP fluorescence was assessed by flow cytometry; histograms from a representative analysis are shown, and (B) depicted as the percentage of fluorescence in mock-infected MJS TAP2-N-GFP cells (set as 100%). The analysis was performed in triplicates. The statistical significance was assessed by t-test; $p \leq 0.0005$ (C) TAP2-GFP degradation was determined by SDS-PAGE and immunoblotting using anti-GFP, anti-TAP1, or anti-UL49.5 antibodies; β-actin was used as a loading control. (D) The relative amount of TAP2-GFP detected by immunoblotting was normalized to β-actin.

3.6. Only UL49.5 among Different Viral TAP Inhibitors Can Induce Human TAP Degradation

The effect of viral TAP inhibitors on TAP stability was reported, usually, in separate studies. TAP levels were analyzed in those reports by immunoblotting. To compare the sensitivity of fluorescent TAP to different viral inhibitors, we selected several representatives with distinct modes of action. Those were: competition for peptide (HSV-1 ICP47) or ATP (HCMV US6 and CPXV012) binding, as well as conformational arrest and degradation (BoHV-1 UL49.5) or a TAP-binding protein with no activity towards the transporter (VZV-encodedUL49.5). MJS TAP2-N-GFP cells were transduced with a retrovirus vector encoding a viral TAP inhibitor and subsequently selected to high purity. The presence of inhibitors was determined by immunoblotting (Figure 8A). Downregulation of MHC I surface expression could be observed during flow cytometry analysis, as expected, for all but the VZV-encoded protein (Figure 8B). Mean GFP fluorescence intensity measurement illustrated that BoHV-1 UL49.5 was unique in causing TAP-GFP degradation. ICP47 even seemed to slightly stabilize TAP-GFP (Figure 8C).

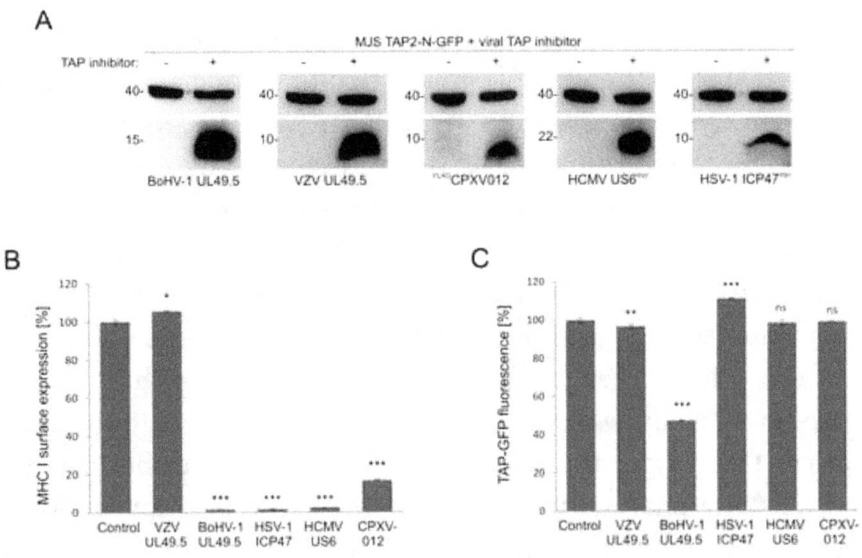

Figure 8. BoHV-1 UL49.5-mediated TAP degradation is unique among viral inhibitors of human TAP. MJS TAP2-N-GFP cells were transduced with a retrovirus encoding BoHV-1 UL49.5, varicella-zoster virus (VZV) UL49.5, human cytomegalovirus (HCMV) US6, herpes simplex virus 1 (HSV-1) ICP47, and cowpox virus CPXV012 and selected with puromycin. (A) The presence of viral TAP inhibitors was confirmed by SDS-PAGE and immunoblotting using anti-β, anti-BHV-1 UL49.5, anti-VZV UL49.5, anti-c-myc for HSV-1 ICP47, and HCMV US6 or anti-FLAG antibodies for CPXV012;

β-actin (upper panels) was used as a loading control. Size markers are in kDa. (B) Surface expression of MHC I was assessed by flow cytometry using specific antibodies (W6/32). MHC I expression is presented as the percentage of MHC I on MJS TAP2-N-GFP cells (set as 100%). The analysis was performed in triplicates. The statistical significance of differences between MJS TAP2-N-GFP cells and cells with viral inhibitor was assessed by t-test; *** $p \leq 0.001$ * $p \leq 0.05$. (C) The mean fluorescence intensity of GFP was analyzed by flow cytometry and presented as the percentage of GFP fluorescence of MJS TAP2-N-GFP cells (set as 100%). The analysis was performed in triplicates. The statistical significance of differences between MJS TAP2-N-GFP cells and cells with a viral inhibitor was assessed by t-test; *** $p \leq 0.001$ ** $p \leq 0.01$ ns: not significant.

3.7. UL49.5-Induced TAP-GFP Degradation Is p97-Dependent

The fluorescent TAP platform can be potentially applied to search for cellular proteins involved in the activity of BoHV-1 UL49.5. According to previous studies, UL49.5-induced TAP degradation is proteasome-dependent [18]. Since TAP is an ER-resident protein, it is presumed to be removed via one of the endoplasmic reticulum-associated degradation (ERAD) pathways. To verify this hypothesis, we used NMS-873 (NMS), an allosteric inhibitor of p97/VCP (valosin-containing protein) a major AAA ATPase belonging to ERAD [38]. MJS TAP2-N-GFP cells with or without UL49.5 were treated with a 2 μM concentration of NMS for 24 h and analyzed by flow cytometry. Inhibition of p97 drastically rescued the mean fluorescence intensity of TAP2 in the presence of UL49.5 to 170% of mean fluorescence intensity in cells treated with DMSO, while in TAP2-N-GFP cells without UL49.5 we could observe only minimal increase to 115% (Figure 9A,B). Compatible results were obtained during immunoblotting analysis of cell lysates (Figure 9C, compare TAP2-GFP in lanes 1–2 and 3–4). Interestingly, inhibition of p97 seemed to stabilize also the expression of UL49.5 (Figure 9C, lanes 1–2). However, apparently, this did not result in a more pronounced TAP-GFP degradation when p97 was blocked.

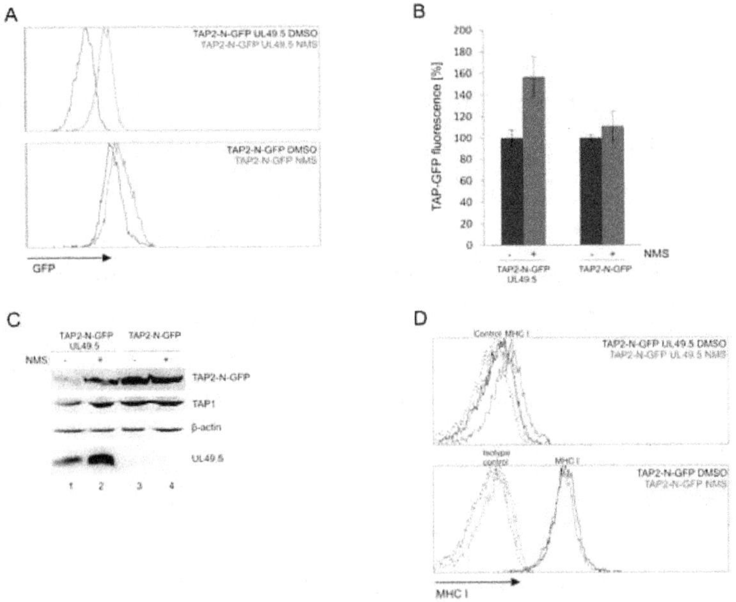

Figure 9. UL49.5 induced TAP-GFP degradation is p97-dependent. MJS TAP2-N-GFP UL49.5 cells were treated with p97 inhibitor NMS-873 (NMS) at 2 μM concentration for 24 hours. (A) Flow cytometry analysis of GFP fluorescence in NMS-treated and control (DMSO-treated) cells. (B) Relative GFP fluorescence in NMS-treated cells calculated as a percentage of GFP fluorescence in the control cells. The analysis was performed in triplicates. (C) The level of TAP-GFP in the presence of p97 inhibitor was determined by SDS-PAGE and immunoblotting using anti-GFP, anti-TAP1 or anti-UL49.5 antibodies; β-actin was used as a loading control. (D) Surface expression of MHC I assessed by flow cytometry using specific antibodies (W6/32).

Next, by using flow cytometry, we determined how inhibition of p97 influences MHC I cell surface expression in the presence (and also absence) of UL49.5. NMS-873-treated cells had only slightly improved surface MHC I levels (Figure 9D). This effect was reminiscent of MHC I levels in the presence of some UL49.5 point mutants, which lost the ability to induce TAP degradation, like, for instance, mutants in the C-terminal RGRG motif [32].

4. DISCUSSION

Quantitative studies on protein stability and degradation may require proper tools and platforms that grant full functionality of the protein of interest and, at the same time, assess the protein levels accurately. Here, we tested the application potential of various full-length TAP to GFP fusion constructs for the studies on TAP stability in the presence of four inhibiting proteins encoded by viruses, to obtain the most suitable fluorescent TAP platform. All the tested TAP-GFP variants were functional. Nevertheless, our results point toward a critical role of GFP localization on fluorescence intensity of the tagged transporter, which in concert with the type a linker used to separate TAP and GFP may regulate its susceptibility to virally induced degradation. By using this platform, we also provide evidence that BoHV-1 UL49.5-induced TAP degradation is p97-dependent.

In a study on HSV-1 ICP47, a truncated fluorescent TAP complex (the so-called 6+6 transmembrane TAP core C-terminally fused to mVenus or mCerulean) was used to determine the effect of the viral protein on TAP thermostability [39]. However, for the studies on UL49.5, which was our primary protein of interest, the full-length TAP should constitute a better platform, as N-terminal TMD0s are required for maximum efficiency of UL49.5 binding and inhibition [19]. Full-length fluorescent TAP has been successfully used in several basic studies, some of which were to elucidate the association of H2Ld molecules with the TAP complex [40], follow lateral mobility of TAP in living cells [41], or illustrate its cellular distribution [42,43]. In those reports, the addition of a relatively large GFP tag to a much larger multiple membrane-spanning partner protein was tolerated to grant proper localization and functionality of the transporter. However, when exploited in a study on varicellovirus immune evasion, GFP-tagged TAP (C-terminal fusion using a random linker) failed to be degraded by BoHV-1 UL49.5, contrary to non-fluorescent wt TAP [18]. Since BoHV-1-encoded UL45.9 has been, so far, the only known viral inhibitor which can cause human TAP degradation apart from its inhibition, further investigation into this mechanism seemed very intriguing, and for this purpose, construction of fluorescent TAP was instrumental.

Designing an optimal fluorescent TAP construct was hampered by the lack of complete structural information about TAP-UL49.5 interaction, and thus, it required an experimental evaluation of different TAP-GFP variants. The latest structural study on BoHV-1 UL49.5 revealed its 3D structure, while subsequent molecular docking experiments proposed three different possible orientations of TAP-UL49.5 complex in which UL49.5 was suggested to interact simultaneously with both TAP subunits [44]. However, these models were predicted based on the structure of ICP47-arrested TAP conformation [10], and therefore the actual UL45.9-TAP binding model needs to be further confirmed.

Fluorescence analysis of constructed TAP-GFP variants in stable MJS cell lines provides evidence that the tag location, rather than the type of a linker used to separate TAP and GFP, has a pivotal impact on fluorescence intensity. N-terminal fusions generally granted stronger fluorescence (Figure 1B,C). It is worth mentioning here that both ends of TAP1 are present in the cytoplasm, while TAP2 incorporates its C terminus in the cytoplasm, and the N terminus localizes to the ER lumen [45,46]. Together with the fact that fluorescence of both TAP1 fusions was more intense than of TAP2, our results lead to speculations that it is the structure of both TMD0 and C-terminal NBDs that determines the fluorescent potential of the tagged constructs. For some constructs, especially for TAP1-N-GFP, we could observe additional protein products reacting with GFP-specific antibodies (Figure 1D), which might correspond to cleaved GFP and could also, most probably, explain higher and heterogeneous GFP signal of this construct observed by flow cytometry. The reason for the presence of free GFP in the case of TAP1-N-GFP is not fully understood. The length of this linker exceeds the size of other tested linkers, so it has a higher chance of affecting the stability of the protein. Another explanation might be the presence of a sequence recognized by cellular proteases, but the ExPASy PeptideCutter software analysis (https://web.expasy.org/peptide_cutter) did not reveal any significant candidates.

Fluorescent tagging did not affect the subcellular localization and function of the transporter, even upon overexpression of only one TAP subunit (Figure 2 and Figure 3). Both TAP1 and TAP2 lack an N-terminal signal sequence for ER targeting [47], and the exact ER-targeting or ER-retention signals have not been identified to date. This encouraged us to design N-terminal GFP fusions with no additional signaling sequences preceding the tag. The localization of our constructs resembles patterns previously described for other recombinant fluorescent TAP proteins [38,39,40,41]. Our results stay in line with the studies on truncated TAP1/TAP2 [43] or functional dissection of transmembrane regions of TAP [6], which have indicated that the transmembrane segments themselves determine ER-localization. It is interesting that even genetically separated TMD0 and the core domains of TAP1 and TAP2 were previously found in the ER (TMD0 additionally localizing to the ER-Golgi intermediate compartment (ERGIC)), when co-expressed [6].

Replacing endogenous TAP1 or TAP2 with TAP-GFP or the untagged subunit restored MHC I on a cell surface equally well and to a level higher than on MJS wt cells (especially in the case of TAP1 constructs). One possible explanation could be the stronger stabilization of endogenous TAP2 by overexpression of TAP1. This effect might be especially noticeable in MJS cells since many melanoma-derived lines have lower endogenous expression of TAP, normally limiting MHC I surface levels [48]. Transduction of MJS TAP1KO or TAP2KO cells with the endogenously present subunit of the transporter did not increase MHC I level, which stays in line with the current view that although TAP1 and TAP2 can form homodimers under certain conditions, they are not functional in antigen presentation [19,35].

One of the most important results of this work provides evidence that all TAP-GFP variants were susceptible to UL49.5-induced inhibition to a similar extent, as assessed by surface MHC I downregulation (Figure 4A). However, only some of them were prone to degradation (both TAP2 fusions, TAP1-N-GFP and TAP1-HC-GFP, Figure 4B,D,E). TAP1-C-GFP remained resistant to UL49.5 what stays in agreement with the previous report [18]. As an interpretation of these data, we can suggest that the helical linker, in contrast to the random one, located at the C terminus of TAP1, effectively separates TAP from GFP to enable undisturbed TAP-UL49.5 interaction, resulting eventually in TAP degradation. Alternatively, it may also permit better access to ERAD components. In the case of the fluorescent TAP2 subunit, the location of GFP, despite the presence of random linkers, did not affect degradation, which could arise from structural differences between the TAP subunits. The TAP2 construct with GFP located in the ER lumen (TAP2-N-GFP) manifested more prominent degradation than the one with GFP in the cytoplasm. An additional observation from this experiment demonstrates that even in the case of a non-degradable fluorescent TAP variant, the second untagged endogenous TAP subunit seems to be sensitive to UL49.5-induced degradation (Figure 4D,E). This, in our opinion, supports the idea of reduced access to ERAD components in TAP1-C-GFP, whereas the access of the second untagged destabilized subunit remains, in this case, undisturbed. It is still unsolved whether UL49.5 can bind single TAP subunits, and the current mechanism points out at the heterodimer as the primary target [19]. As in MJS cells with non-degradable TAP variants, we could observe very efficient MHC I reduction, and the PLC composition in those cells seemed to be intact (Figure 5), at least with regard to the interaction of TAP with ERp57 and MHC I; our data confirm the previous report by [18], demonstrating that abolished degradation does not exclude inhibition. It even seems that TAP degradation might be only an auxiliary event, a "finish-off" effect, in the mechanism of UL49.5 action.

For further studies, we selected and validated TAP2-N-GFP as the most promising variant. TAP transport assay performed on this cell line confirmed that changes in MHC I surface levels reflect TAP transport efficiency (Figure 6). TAP transport in reconstituted cell lines, either with wt or a fluorescent version of the TAP subunit, was higher than in wt MJS. Then we demonstrated that results obtained in a stable cell line model system reflect a situation that occurs upon BoHV-1 infection, which was illustrated as loss of GFP fluorescence observed by flow cytometry and reduction of protein level shown by immunoblotting (Figure 7).

A former pulse-chase experiment with the use of proteasome inhibitor postulated co-degradation of TAP with UL49.5 [18]. In line with this working model, our data show that inhibition of p97 increases levels of both TAP and UL49.5, and demonstrate for the first time that UL49.5-induced TAP degradation requires functional p97. Most of known ER-resident substrates of this ATPase, which retrotranslocates proteins back to the cytoplasm, are ubiquitinated and targeted for proteasomal degradation [38], indicating that UL49.5 mediated TAP degradation occurs via ERAD.

Finally, the TAP2-N-GFP construct was verified as a platform for different viral TAP inhibitors, representing distinct mechanisms of transport inhibition and, most probably, binding another TAP conformation [28,39,49]. BoHV-1, HSV-1, and HCMV-encoded proteins were capable of drastic reduction of surface MHC I; CPXV012 contributed to a slightly weaker but still significant downregulation of MHC I, whereas VZV UL49.5, as expected, did not cause any changes. In terms of degradation, only BoHV-1 UL49.5 was able to decrease TAP-GFP levels, while ICP47 seemed even to stabilize TAP, which is in accordance with its reported effect on TAP thermostability. We believe that the fluorescent TAP platform provides more quantitative data in this respect when compared to previous immunoblotting analyses, which generally are more technically error-prone.

5. CONCISE SUMMARY

The Transporter associated with antigen processing (TAP) is a crucial component in the presentation of antigens restricted to major histocompatibility complex class I. This makes it a desirable target for viruses seeking to evade the immune system. The mechanisms by which TAP inhibition occurs differ across different virus species. Bovine herpesvirus 1 (BoHV-1) has the special capacity to specifically target the transporter associated with antigen processing (TAP) for degradation by the proteasome. This degradation is triggered by the UL49.5 gene product after it causes a conformational arrest in TAP. The precise technique by which TAP is removed still has to be clarified. To achieve this objective, the use of a TAP-GFP fusion protein is crucial. However, it is important to note that GFP-tagging could potentially impact the UL49.5-induced degradation process. Consequently, we created a range of TAP-GFP variations by employing different linkers in order to achieve an ideal cellular fluorescent TAP platform. Mel JuSo (MJS) cells, which had TAP1 or TAP2 genes knocked out using CRISPR/Cas9, were restored by introducing TAP-GFP constructs. The results indicate that the location of GFP plays a crucial role in determining the luminous characteristics of the fusion proteins. Additionally, the sensitivity to virally-induced inhibition and degradation is influenced by both the kind of linker and the GFP localization. The fluorescent TAP platform was utilized to reassess the stability of TAP in the presence of various established viral TAP inhibitors. Among these inhibitors, only UL49.5 shown the capability to decrease TAP levels. Ultimately, we present proof that the elimination of TAP triggered by BoHV-1 UL49.5 is reliant on p97, suggesting that it undergoes degradation through endoplasmic reticulum-associated degradation (ERAD).

In this study, we were able to validate the application potential of fluorescent TAP as a platform for viral immune evasion studies. Our results indicate TAP-GFP variants susceptible to BoHV-1 UL49.5-induced degradation, demonstrate that this degradation is p97-dependent, and emphasize the importance of linker design in fusion protein construction. The fluorescent TAP platform can be now applied in further research on BoHV-1 UL49.5, for instance in the genome-wide search for cellular proteins responsible for UL49.5-induced degradation, where the fluorescent signal can be measured and indicate even small changes in TAP levels. TAP-GFP could be also exploited to identify the active motifs or amino acid residues of UL49.5 affecting TAP stability. The same platform with viral inhibitors can be applied, in a similar way as in the study by [28], to identify TAP conformation recognized by UL49.5.

REFERENCES

1. Jensen, P.E. Recent advances in antigen processing and presentation. Nat. Immunol. 2007, 8, 1041–1048.
2. Dean, M.; Annilo, T. Evolution of the ATP-binding cassette (ABC) transporter superfamily in vertebrates. Annu. Rev. Genom. Hum. Genet. 2005, 6, 123–142.

3. van Endert, P.M.; Tampé, R.; Meyer, T.H.; Tisch, R.; Bach, J.-F.; McDevitt, H.O. A sequential model for peptide binding and transport by the transporters associated with antigen processing. Immunity 1994, 1, 491–500.

4. Neefjes, J.J.; Momburg, F.; Hämmerling, G.J. Selective and ATP-dependent translocation of peptides by the MHC-encoded transporter. Science 1993, 261, 769–771.

5. Geng, J.; Sivaramakrishnan, S.; Raghavan, M. Analyses of conformational states of the transporter associated with antigen processing (TAP) protein in a native cellular membrane environment. J. Biol. Chem. 2013, 288, 37039–37047.

6. Koch, J.; Guntrum, R.; Heintke, S.; Kyritsis, C.; Tampé, R. Functional dissection of the transmembrane domains of the transporter associated with antigen processing (TAP). J. Biol. Chem. 2004, 279, 10142–10147.

7. Verweij, M.C.; Horst, D.; Griffin, B.D.; Luteijn, R.D.; Davison, A.J.; Ressing, M.E.; Wiertz, E.J.H.J. Viral Inhibition of the transporter associated with antigen processing (TAP): A striking example of functional convergent evolution. PLoS Pathog. 2015, 11, e1004743.

8. Früh, K.; Ahn, K.; Djaballah, H.; Sempé, P.; Peter, M.; van Endert, P.M.; Tampé, R.; Peterson, P.A.; Yang, Y. A viral inhibitor of peptide transporters for antigen presentation. Nature 1995, 375, 415–418.

9. Tomazin, R.; Hill, A.B.; Jugovic, P.; York, I.; van Endert, P.; Ploegh, H.L.; Andrews, D.W.; Johnson, D.C. Stable binding of the herpes simplex virus ICP47 protein to the peptide binding site of TAP. EMBO J. 1996, 15, 3256–3266.

10. Oldham, M.L.; Grigorieff, N.; Chen, J. Structure of the transporter associated with antigen processing trapped by herpes simplex virus. eLife 2016, 5, e21829.

11. Ahn, K.; Gruhler, A.; Galocha, B.; Jones, T.R.; Wiertz, E.J.H.J.; Ploegh, H.L.; Peterson, P.A.; Yang, Y.; Früh, K. The ER-luminal domain of the HCMV glycoprotein US6 inhibits peptide translocation by TAP. Immunity 1997, 6, 613–621.

12. Kyritsis, C.; Gorbulev, S.; Hutschenreiter, S.; Pawlitschko, K.; Abele, R.; Tampé, R. Molecular mechanism and structural aspects of transporter associated with antigen processing inhibition by the cytomegalovirus protein US6. J. Biol. Chem. 2001, 276, 48031–48039.

13. Halenius, A.; Momburg, F.; Reinhard, H.; Bauer, D.; Lobigs, M.; Hengel, H. Physical and functional interactions of the cytomegalovirus US6 glycoprotein with the transporter associated with antigen processing. J. Biol. Chem. 2006, 281, 5383–5390.

14. Alzhanova, D.; Edwards, D.M.; Hammarlund, E.; Scholz, I.G.; Horst, D.; Wagner, M.J.; Upton, C.; Wiertz, E.J.; Slifka, M.K.; Früh, K. Cowpox virus inhibits the transporter associated with antigen processing to evade T cell recognition. Cell Host Microbe 2009, 6, 433–445.

15. Byun, M.; Verweij, M.C.; Pickup, D.J.; Wiertz, E.J.H.J.; Hansen, T.H.; Yokoyama, W.M. Two mechanistically distinct immune evasion proteins of cowpox virus combine to avoid antiviral CD8 T Cells. Cell Host Microbe 2009, 6, 422–432.

16. Luteijn, R.D.; Hoelen, H.; Kruse, E.; van Leeuwen, W.F.; Grootens, J.; Horst, D.; Koorengevel, M.; Drijfhout, J.W.; Kremmer, E.; Früh, K.; et al. Cowpox virus protein CPXV012 eludes CTLs by blocking ATP binding to TAP. J. Immunol. 2014, 193, 1578–1589.

17. Verweij, M.C.; Lipinska, A.D.; Koppers-Lalic, D.; van Leeuwen, W.F.; Cohen, J.I.; Kinchington, P.R.; Messaoudi, I.; Bienkowska-Szewczyk, K.; Ressing, M.E.; Rijsewijk, F.A.M.; et al. The capacity of UL49.5 proteins to inhibit TAP is widely distributed among members of the genus varicellovirus. J. Virol. 2011, 85, 2351–2363.

18. Koppers-Lalic, D.; Reits, E.A.J.; Ressing, M.E.; Lipinska, A.D.; Abele, R.; Koch, J.; Rezende, M.M.; Admiraal, P.; van Leeuwen, D.; Bienkowska-Szewczyk, K.; et al. Varicelloviruses avoid T cell recognition by UL49.5-mediated inactivation of the transporter associated with antigen processing. Proc. Natl. Acad. Sci. USA 2005, 102, 5144–5149.

19. Verweij, M.C.; Koppers-Lalic, D.; Loch, S.; Klauschies, F.; de la Salle, H.; Quinten, E.; Lehner, P.J.; Mulder, A.; Knittler, M.R.; Tampé, R.;

et al. The varicellovirus UL49.5 protein blocks the transporter associated with antigen processing (TAP) by inhibiting essential conformational transitions in the 6+6 transmembrane TAP core complex. J. Immunol. 2008, 181, 4894–4907.

20. Koppers-Lalic, D.; Verweij, M.C.; Lipińska, A.D.; Wang, Y.; Quinten, E.; Reits, E.A.; Koch, J.; Loch, S.; Rezende, M.M.; Daus, F.; et al. Varicellovirus UL49.5 proteins differentially affect the function of the transporter associated with antigen processing, TAP. PLoS Pathog. 2008, 4, e1000080.

21. Boname, J.; May, J.; Stevenson, P. The murine gamma-herpesvirus-68 MK3 protein causes TAP degradation independent of MHC class I heavy chain degradation. Eur. J. Immunol. 2005, 35, 171–179.

22. Herr, R.A.; Wang, X.; Loh, J.; Virgin, H.W.; Hansen, T.H. Newly discovered viral E3 ligase pK3 induces endoplasmic reticulum-associated degradation of class I major histocompatibility proteins and their membrane-bound chaperones. J. Biol. Chem. 2012, 287, 14467–14479.

23. Harvey, I.B.; Wang, X.; Fremont, D.H. Molluscum contagiosum virus MC80 sabotages MHC-I antigen presentation by targeting tapasin for ER-associated degradation. PLoS Pathog. 2019, 15, e1007711.

24. Chen, X.; Zaro, J.L.; Shen, W.-C. Fusion protein linkers: Property, design and functionality. Adv. Drug Deliver. Rev. 2013, 65, 1357–1369.

25. Praest, P.; Luteijn, R.D.; Brak-Boer, I.G.J.; Lanfermeijer, J.; Hoelen, H.; Ijgosse, L.; Costa, A.I.; Gorham, R.D.; Lebbink, R.J.; Wiertz, E.J.H.J. The influence of TAP1 and TAP2 gene polymorphisms on TAP function and its inhibition by viral immune evasion proteins. Mol. Immunol. 2018, 101, 55–64.

26. Lipińska, A.D.; Koppers-Lalic, D.; Rychlowski, M.; Admiraal, P.; Rijsewijk, F.A.M.; Bienkowska-Szewczyk, K.; Wiertz, E.J.H.J. Bovine herpesvirus 1 UL49.5 protein inhibits the transporter associated with antigen processing despite complex formation with glycoprotein M. J. Virol. 2006, 80, 5822–5832.

27. Morgenstern, J.P.; Land, H. Advanced mammalian gene transfer: High titre retroviral vectors with multiple drug selection markers and a complementary helper-free packaging cell line. Nucleic Acid Res. 1990, 18, 3587–3596.

28. Lin, J.; Eggensperger, S.; Hank, S.; Wycisk, A.I.; Wieneke, R.; Mayerhofer, P.U.; Tampé, R. A negative feedback modulator of antigen processing evolved from a frameshift in the cowpox virus genome. PLoS Pathog. 2014, 10, e1004554.

29. Marqusee, S.; Baldwin, R.L. Helix stabilization by Glu-Lys+ salt bridges in short peptides of de novo design. Proc. Natl. Acad. Sci. USA 1987, 84, 8898–8902.

30. Arai, R.; Ueda, H.; Kitayama, A.; Kamiya, N.; Nagamune, T. Design of the linkers which effectively separate domains of a bifunctional fusion protein. Protein Eng. Des. Sel. 2001, 14, 529–532.

31. Koppers-Lalic, D.; Rychlowski, M.; van Leeuwen, D.; Rijsewijk, F.A.M.; Ressing, M.E.; Neefjes, J.J.; Bienkowska-Szewczyk, K.; Wiertz, E.J.H.J. Bovine herpesvirus 1 interferes with TAP-dependent peptide transport and intracellular trafficking of MHC class I molecules in human cells. Arch. Virol. 2003, 148, 2023–2037.

32. Verweij, M.C.; Lipińska, A.D.; Koppers-Lalic, D.; Quinten, E.; Funke, J.; van Leeuwen, H.C.; Bieńkowska-Szewczyk, K.; Koch, J.; Ressing, M.E.; Wiertz, E.J.H.J. Structural and functional analysis of the TAP-inhibiting UL49.5 proteins of varicelloviruses. Mol. Immunol. 2011, 48, 2038–2051.

33. Praest, P.; de Buhr, H.; Wiertz, E.J.H.J. A flow cytometry-based approach to unravel viral interference with the MHC class I antigen processing and presentation pathway. In Antigen Processing; van Endert, P., Ed.; Springer: New York, NY, USA, 2019; Volume 1988, pp. 187–198.

34. Keusekotten, K.; Leonhardt, R.M.; Ehses, S.; Knittler, M.R. Biogenesis of functional antigenic peptide transporter TAP requires assembly of pre-existing TAP1 with newly synthesized TAP2. J. Biol. Chem. 2006, 281, 17545–17551.

35. Kleijmeer, M.J.; Kelly, A.; Geuze, H.J.; Slot, J.W.; Townsend, A.; Trowsdale, J. Location of MHC-encoded transporters in the endoplasmic reticulum and cis-Golgi. Nature 1992, 357, 342–344.

36. Van Kaer, L.; Ashton-Rickardt, P.G.; Ploegh, H.L.; Tonegawa, S. TAP1 mutant mice are deficient in antigen presentation, surface class I molecules, and CD4−8+ T cells. Cell 1992, 71, 1205–1214.

37. Antoniou, A.N.; Ford, S.; Pilley, E.S.; Blake, N.; Powis, S.J. Interactions formed by individually expressed TAP1 and TAP2 polypeptide subunits. Immunology 2002, 106, 182–189.

38. Xia, D.; Tang, W.K.; Ye, Y. Structure and function of the AAA+ ATPase p97/Cdc48p. Gene 2016, 583, 64–77.

39. Herbring, V.; Bäucker, A.; Trowitzsch, S.; Tampé, R. A dual inhibition mechanism of herpesviral ICP47 arresting a conformationally thermostable TAP complex. Sci. Rep. 2016, 6, 36907.

40. Marguet, D.; Spiliotis, E.T.; Pentcheva, T.; Lebowitz, M.; Schneck, J.; Edidin, M. Lateral diffusion of GFP-tagged H2Ld molecules and of GFP-TAP1 reports on the assembly and retention of these molecules in the endoplasmic reticulum. Immunity 1999, 11, 231–240.

41. Reits, E.A.J.; Vos, J.C.; Grommé, M.; Neefjes, J. The major substrates for TAP in vivo are derived from newly synthesized proteins. Nature 2000, 404, 774–778.

42. Kobayashi, A.; Maeda, T.; Maeda, M. Membrane localization of transporter associated with antigen processing (TAP)-like (ABCB9) visualized in vivo with a fluorescence protein-fusion technique. Biol. Pharm. Bull. 2004, 27, 1916–1922.

43. Ghanem, E.; Fritzsche, S.; Al-Balushi, M.; Hashem, J.; Ghuneim, L.; Thomer, L.; Kalbacher, H.; van Endert, P.; Wiertz, E.; Tampe, R.; et al. The transporter associated with antigen processing (TAP) is active in a post-ER compartment. J. Cell Sci. 2010, 123, 4271–4279.

44. Karska, N.; Graul, M.; Sikorska, E.; Zhukov, I.; Ślusarz, M.J.; Kasprzykowski, F.; Lipińska, A.D.; Rodziewicz-Motowidło, S. Structure determination of UL49.5 transmembrane protein

from bovine herpesvirus 1 by NMR spectroscopy and molecular dynamics. BBA Biomembr. 2019, 1861, 926–938.

45. Vos, J.C.; Spee, P.; Momburg, F.; Neefjes, J. Membrane topology and dimerization of the two subunits of the transporter associated with antigen processing reveal a three-domain structure. J. Immunol. 1999, 163, 6679–6685.

46. Schrodt, S.; Koch, J.; Tampé, R. Membrane topology of the transporter associated with antigen processing (TAP1) within an assembled functional peptide-loading complex. J. Biol. Chem. 2006, 281, 6455–6462.

47. Lankat-Buttgereit, B.; Tampé, R. The transporter associated with antigen processing (TAP): A peptide transport and loading complex essential for cellular immune response. In ABC Proteins: From Bacteria to Man; Holland, B., Cole, S.P.C., Kuchler, K., Higgins, C.F., Eds.; Academic Press: Cambridge, MA, USA, 2003; pp. 533–550.

48. Kageshita, T.; Hirai, S.; Ono, T.; Hicklin, D.J.; Ferrone, S. Down-regulation of HLA class I antigen-processing molecules in malignant melanoma. Am. J. Pathol. 1999, 154, 745–754.

49. Matschulla, T.; Berry, R.; Gerke, C.; Döring, M.; Busch, J.; Paijo, J.; Kalinke, U.; Momburg, F.; Hengel, H.; Halenius, A. A highly conserved sequence of the viral TAP inhibitor ICP47 is required for freezing of the peptide transport cycle. Sci. Rep. 2017, 7, 2933.

12 Human Cells Exhibit The Expression Of A Novel Erap2/Iso3 Isoform IN RESPONSE TO VARIOUS MICROBIAL STIMULI CAN IT CONTRIBUTE TO THE REGULATION OF SARS-COV-2 INFECTION?

1. INTRODUCTION: AN OVERVIEW

ERAP1 and ERAP2, also known as endoplasmic reticulum aminopeptidases 1 and 2, are two human enzymes that are expressed in all tissues and are induced by IFNγ and TNFα. These enzymes belong to the M1 family of zinc aminopeptidases [1]. Within the endoplasmic reticulum (ER), ERAPs modify the collection of antigens by removing a portion of the N-terminus from precursor peptides that were initially produced in the cytoplasm by the proteasome. ERAPs synthesize peptides of the ideal length to be loaded onto the MHC class I groove and presented to CD8+ T cells [2,3]. Although these two enzymes have distinct enzymatic specialization, they can work together in a coordinated manner by forming homo- or heterodimers. This collaboration enables the production of a diverse and more immunogenic range of antigens [4]. Specifically, the production of ERAP1-ERAP2 heterodimers has been shown to enhance the formation of peptides that are capable of binding to MHC class I molecules [5]. ERAPs are encoded by two genes, with approximately 49% similarity in their genetic sequences. These genes are located on chromosome 5q15, facing opposing directions. They are known to exhibit a significant degree of genetic variation. Given their prominent role in the antigen processing pathway, numerous studies have explored the potential connection between ERAP polymorphic variants and changes in their functionality. These alterations could potentially lead to the onset of MHC-I-associated disorders [2,7,8] and variations in susceptibility or progression of microbial infections [9]. Regarding ERAP2, the primary significant single nucleotide polymorphism (SNP) is the non-coding rs2248374 (A/G), which distinguishes two haplotypes known as HapA (A allele for rs2248374) and HapB (G allele for rs2248374). Within HapB, the G allele of this SNP initiates the transcription of a modified form of ERAP2 (ERAP2/Iso2), which includes an elongated exon 10 consisting of 56 more nucleotides and two in-frame TAG stop codons. Consequently, this modified form undergoes nonsense-mediated decay (NMD) [10]. On the other hand, HapA is converted into a protein consisting of 965 amino acids. It is linked to several conditions such as Crohn's disease, birdshot uveitis associated with HLA-A29, ankylosing spondylitis, juvenile idiopathic arthritis, and natural immunity to HIV infection. Due to a balanced selection, these variants are maintained at a frequency of around 50% (HapB: 53% and HapA: 47%). As a result, around 25% of the population does not produce the ERAP2 protein [18]. This discovery prompts a logical inquiry: in what specific environment does balancing selection occur to preserve the seemingly non-functional HapB and the disease-causing HapA in the human population? Recently, Ye and his colleagues offered a comprehensive explanation for this seeming inconsistency [19]. For the first time, researchers recorded the transcription of two new short isoforms (ERAP2/Iso3, ERAP2/Iso4) from flu-infected monocyte-derived dendritic cells that were taken from individuals expressing the homozygous HapB gene variant [19].

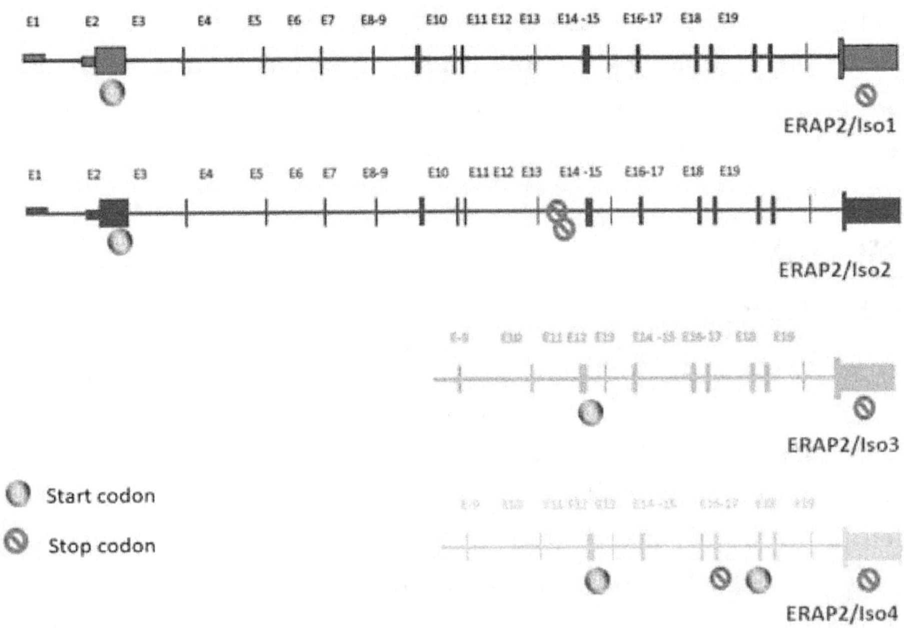

Figure 1. Genetics of ERAP2 isoforms regulation. Structures of transcripts derived from each ERAP2 isoform are represented. Start and stop codons for each isoform are reported.

The two abbreviated isoforms, derived from HapB, exhibit distinct characteristics compared to the full-length isoform ERAP2/Iso1, which is transcribed from HapA. The transcription of the abbreviated isoforms initiates at exon 9 and involves alternative splicing of an elongated exon 10. In addition, they differ from each other by alternative splicing at a secondary splice site located at exon 15 (Figure 1). It is important to mention that ERAP2/Iso4 is likely to have a premature termination codon that may trigger nonsense-mediated decay (NMD). On the other hand, ERAP2/Iso3 is anticipated to be translated into a protein. This protein lacks the catalytic domain [19], but it can nevertheless play a crucial role in shaping the cellular immune-peptidome by forming dimers with ERAP1 and perhaps ERAP2-wild type (wt) [20]. Based on these premises, the aim of our study was to investigate if the transcription of ERAP2/Iso3 is either flu-specific or if it can be induced by other kind of stimuli, such as other viruses, bacteria, or inflammatory triggers. Given the foremost role played by ERAPs in the field of both acquired and innate immunity, the characterization of the different isoforms produced in a particular pathological setting, such as the one caused by microbial infections, could lead to the identification of new molecular targets to be exploited in the setting up of innovative therapeutically or vaccinal approaches.

2. METHODS

2.1. Study Population

Twenty healthy controls (HC) were enrolled in the study in order to investigate the induction of ERAP isoform transcription in response to common recall antigens such as influenza-antigens (flu), Lipopolisaccaride (LPS), Citomegalovirus (CMV), in addition to Aldrithiol-2 (AT-2)-inactivated R5-tropic human immunodeficiency virus-1Ba-L (HIV-AT-2) and acute respiratory syndrome coronavirus 2 (SARS-CoV-2) inactivated virus (i-SARS-CoV-2). The Ethical Committee of the Fondazione IRCCS Ca' Granda Ospedale Maggiore Policlinico approved the study (Prot. N°0028257). All the donors signed an informed consent form, in agreement with the Declaration of Helsinki of 1975, revised in 2013.

2.2. Viruses

The laboratory-adapted HIV-1 strain used in the experiments was the R5 tropic HIV-1BaL (courtesy of Drs. S. Gartner, M. Popovic, and R. Gallo; NIH AIDS Research and Reference Reagent Program) provided through the EU program EVA Centre for AIDS Reagents (NIBSC, Potter Bars, UK). The virus was inactivated with AT-2, which is able to change the zinc finger nucleocapsid proteins of HIV-1, therefore deactivating the viral infectivity as previously described [21].

SARS-CoV-2 (Virus Human 2019-nCoV strain 2019-nCoV/Italy-INMI1, Rome, Italy) was expanded on Calu-3 cells (ATCC® HTB-55™) and TCID50 was calculated as previously reported [22]. SARS-CoV-2 inactivation (i-SARS-CoV-2) was obtained by incubation at 65 °C for 30 min [23].

2.3. ERAP2 Genotyping Analyses

Whole blood was collected by all the subjects enrolled in the study by venipuncture in Vacutainer tubes containing EDTA (BD Vacutainer, San Diego, CA, USA). Total DNA was extracted by DNA purification Maxwell® RSC Instrument (Promega, Fitchburg, WI, USA) and quantified using the Nanodrop 2000 Instrument (Thermo Scientific, Waltham, MA, USA). Two-hundred ng of DNA were used to perform an SNP genotyping assay for ERAP2 rs2549782 (G/T) (TaqMan SNP Genotyping Assay; Applied Biosystems, Foster City, CA, USA), which is in linkage disequilibrium with the non-coding rs2248374 (A/G). Analyses were performed on Peripheral blood mononuclear cells (PBMCs) from all the subjects recruited in the study as well as on lung adenocarcinoma cells (Calu3; ATCC® HTB-55™). Allelic discrimination real-time PCR method was used to analyze the results.

2.4. Isolation of PBMCs and Monocyte-Derived Macrophages (MDMs) Differentiation

PBMCs, obtained from density gradient centrifugation on Ficoll (Cedarlane Laboratories Limited, Hornby, ON, Canada), were counted by automated cell counter ADAM-MC (NanoEnTek Inc., Seoul, Korea), which distinguishes viable from non-viable cells. Flow cytometer analyses was used to quantify the percentage of CD14+ monocytes in PBMCs isolated from 3 HeteroAB HC. MDMs were obtained as previously described [24]. Briefly, 5 × 105 adherent monocytes were incubated for 5 days in RPMI with 20% of fetal bovine serum (FBS) (Euroclone, Milan, Italy) and 100 ng/mL macrophage-colony stimulating factor (M-CSF) (R&D Systems, Minneapolis, MN, USA). Optical microscope (ZOE™ Fluorescent Cell Imager, Bio-Rad, Hercules, CA, USA) observation allows to verify MDM differentiation.

2.5. Cell Cultures for Microbial Antigen Stimulation

PBMCs were resuspended at the concentration of 1 × 106 PBMCs/mL in RPMI 1640 medium (Euroclone, Milan, Italy) containing 10% fetal bovine serum (FBS), 1% levo-glutammin LG and 2% penstreptomicin. Subsequently, PBMCs and MDMs from HC were stimulated with antigens from different pathogens: 32 μg/mL of CMV grade 2 Antigen (Microbix Biosystem, Mississauga, ON, Canada), 1 μg/mL of LPS, 1 ng/mL of HIV-AT-2 equivalents and 5 multiplicity of infection (MOI) of i-SARS-CoV-2 inactivated virus. Two live UV-inactivated influenza viruses (flu) were used as well: an influenza A virus (A/RX73 and A/Puerto Rico/8/34 strains; 1:800) and the 1998–1999 formula of flu vaccine (1:5000; Wyeth Laboratories Inc., Marietta, PA, USA). Cells were stimulated even with non-microbial stimuli: 100U of IFNα and 1 μg/mL of IL-1β (Sigma, Saint Louis, MO, USA). Unstimulated PBMCs were cultured as control as well. Cells were harvested 10 (PBMCs) and 36 (MDMs) h post-treatment for RNA and protein analyses, respectively.

2.6. In Vitro Infection of PBMCs and Calu3 Cells with SARS-CoV-2

2.5 × 105 Calu3 cells (ATCC® HTB-55™) were cultured in DMEM medium (Euroclone, Milan, Italy) supplemented with 2% FBS in a 24-well plate. DMEM containing 100 U/mL penicillin and 100 μg/mL streptomycin was used as inoculum in the mock-infected cells. Cell cultures were incubated with serial dilutions of virus supernatant in duplicate, (1000 MOI, 5 MOI, 0.5 MOI) for three h at 37 °C and 5% CO2. Cells were washed two times with lukewarm PBS and refilled with the proper growth medium (10%FBS). Optical microscope observation (ZOE™ Fluorescent Cell Imager, Bio-Rad, Hercules, CA, USA) was performed daily to investigate the cytopathic effect. The infected cells were harvested for SARS-CoV-2 detection in the supernatant and mRNA collection at 48 h. Each culture condition was run in triplicate. All the procedures were performed in agreement with the GLP guidelines adopted in our laboratory.

Maxwell® RSC Viral Total Nucleic Acid Purification Kit (Promega, Fitchburg, WI, USA) was used to extract RNA from Calu3 cell culture supernatants by the Maxwell® RSC Instrument (Promega, Fitchburg, WI, USA). Viral RNA was quantified, by single-step RT PCR -time PCR (GoTaq® 1-Step RT-qPCR) (Promega, Fitchburg, WI, USA) on a CFX96 (Bio-Rad, Hercules, CA, USA) by using TaqMan probes which target two portions of SARS-CoV-2 nucleocapsid (N) gene (N1 and N2). Specifically, we used the 2019-nCoV CDC qPCR Probe Assay emergency kit (IDT, Coralville,

IA, USA), which allows also to amplify the human RNase P gene. Viral copy number quantification was performed by generating a standard curve from the quantified 2019-nCoV_N positive Plasmid Control (IDT, Coralville, IA, USA).

2.7. In Vitro HIV-Infection Assay

3 × 106 PBMCs from all the subjects included in the study were in vitro HIV-infected as previously described [25] with 10, 1, and 0.1 ng p24 HIV-1Bal/1 × 106 PBMCs. After 5 days, supernatants were collected for p24 antigen ELISA (Cell Biolabs, San Diego, CA, USA), whereas PBMCs collected at 2 days post-infection were used for RNA extraction and gene expression analyses.

2.8. Gene Expression Analysis

RNA extracted from 1 × 106 PBMCs, and Calu3 cell lines were retrotranscribed as previously described [16]. cDNA quantification for ERAPs was performed on antigen-stimulated and HIV-infected PBMCs as well as on SARS-CoV-2 infected Calu3 cells through a real-time PCR (CFX96 connect, Bio-Rad, Hercules, CA, USA) and an SYBR Green PCR mix (Bio-Rad, Hercules, CA, USA); all the reactions were carried out in duplicate. Results are shown as the media of the relative expression units to the glyceraldehyde-3-phosphate dehydrogenase (GAPDH) and β-actin housekeeping genes calculated by the 2-ΔΔCt equation. The following thermal protocol was used: initial denaturation (95 °C, 15 min) followed by 40 cycles of 15 s at 95 °C (denaturation), 20 s at 60 °C (annealing) and 20 s at 72 °C (extension). Furthermore, a melting curve analysis was assessed for amplicon characterization. Ct values of 35 or higher were let off the analyses.

2.9. Quantigene Plex Gene Expression Assay

Gene expression of 8 × 105 PBMCs was performed by quantiGene Plex assay (Thermo Scientific, Waltham, MA, USA) which provides a fast and high-throughput solution for multiplexed gene expression quantitation, allowing the simultaneous measurement of 70 custom selected genes of interest in a single well of a 96-well plate. The QuantiGene Plex assay is hybridization-based and incorporates branched DNA (bDNA) technology, which uses signal amplification for direct measurement of RNA transcripts. The assay does not require RNA purification.

2.10. Western Blot Analyses

Cultured MDMs were removed by non-enzymatic cell dissociation solution (Sigma, Saint Louis, Missouri, USA), counted by the automated cell counter ADAM-MC (Digital Bio) and used for protein extraction by RIPA buffer (Sigma, Saint Louis, MO, USA). Extracted proteins were stored at -80 °C for further analyses. For WB analyses, samples from 3 HeteroAB subjects were sub-pooled (50 µg per pool). Equal amounts of proteins were separated by 4–20% SDS-polyacrylamide gel electrophoresis (Criterion TGX Stain-free precast gels and Criterion Cell system; Bio-Rad) and transferred onto nitrocellulose membrane using a Bio-Rad Trans-Blot Turbo System. Membranes were probed using a 1:1000 dilution of primary antibody goat anti-ERAP1 (AF2334; R&D Systems, Minneapolis, MN, USA) goat anti-ERAP2 polyclonal antibody (AF3830; R&D Systems, Minneapolis, MN, USA), rabbit anti-GAPDH polyclonal (VPA00187); Bio-Rad, Hercules, CA, USA] and a 1:10,000 dilution of secondary antibody conjugated with alkaline phosphatase anti-goat IgG (A4187; Sigma; goat anti-rabbit (STAR208P); Bio-Rad]. Membranes were incubated with the appropriate antibody and, after being excited using the Clarity Western ECL substrate, bands were visualized with a ChemiDoc MP imaging system (Bio-Rad) and quantified for densitometry with the Bio-Rad Image Lab software.

2.11. Statistical Analyses

Data are shown as mean and standard deviation. Analysis and figures were performed by GRAPHPAD PRISM version 5 (Graphpad software, La Jolla, CA, USA) and SPSS Statistics, version 25 (IBM software, Armonk, NY, USA). Gene expressions of ERAP2/iso1, ERAP2/iso1, and ERAP1 in PBMCs, monocyte-derived macrophages, and Calu-3 cells upon stimulation were compared to that of untreated cells by Wilcoxon test and Mann–Whitney test, as appropriate. P-values were corrected for false discovery rate (FDR) by using Microsoft R software, p-values ≤ 0.005 were considered to be significant.

3. RESULTS

3.1. ERAP2 Allelic Variants Analyses

Analysis of ERAP2 SNP prevalence was aligned with European population distribution reported in the U.S. National Library of Medicine Database [https://www.ncbi.nlm.nih.gov/snp/rs2549782?fbclid=IwAR1ZdwC747PDWvtAzt6hZBV5j7oFZiPkLjY-JdSee1Plzvym7fhJVQc1Aks (data not shown). Among the 20 genotyped HC: 6 were HomoA, 8 heterozygous, 6 HomoB. Subsequent analyses were performed only on PBMCs and MDMs isolated from HomoB and heteroAB donors to exclude confounding results. Indeed, HomoA individuals express negligible levels of ERAP2/Iso2 and Iso3. Calu3 cell lines were heterozygous for rs2248374 ERAP2 genotype.

3.2. mRNA Expression of ERAPs in PBMCs from Subjects Carrying Different ERAP2 Genotypes Following Microbial Stimulation

To verify whether the expression of ERAP2/Iso3 isoform is exclusively flu-specific or it may be triggered by other stimuli, we analyzed its expression on PBMCs isolated from HomoB and HeteroAB HC following HIV-AT-2, i-SARS-CoV-2, CMV, LPS, flu, IFNα, and IL-1β. As expected, the expression of the newly identified ERAP2/Iso3 was significantly augmented in PBMCs from all the subjects included in the study in response to flu ($p < 0.002$). However, even following CMV ($p < 0.004$), LPS ($p < 0.002$), HIV-AT-2 ($p < 0.003$), i-SARS-CoV-2 ($p < 0.002$) and IFNα ($p < 0.003$) stimulation, we observed a significant increase of its expression (Figure 2A). Conversely, IL-1β addition to cell culture did not result in ERAP2/Iso3 induction (Figure 2A). ERAP2/Iso1 expression was observed only in PBMCs from HeteroAB subjects and was induced following all the microbial-stimuli employed plus IFNα but not in response to IL-1β (Figure 2B). However, statistical significance was observed exclusively following HIV-AT-2 ($p < 0.025$), i-SARS-CoV-2 ($p < 0.03$) and CMV ($p < 0.02$). Likewise, ERAP1 expression was induced by all the stimuli employed excepting IL-1β and reached statistical significance following flu ($p < 0.002$), HIV-AT-2 ($p < 0.03$) CMV ($p < 0.002$), i-SARS-CoV-2 ($p <$

0.008) LPS ($p < 0.05$). We also observed a reduction following IL-1β ($p <$ 0,001) stimulations (Figure 2C).

Notably, the microbial-dependent genetic control of ERAP2 isoform usage is sustained by further evidence. Indeed, there was a significant correlation between whole microbial transcript quantity and ERAP2/Iso3 transcript abundances in heterozygotes compared to HapB homozygotes (Figure 2D).

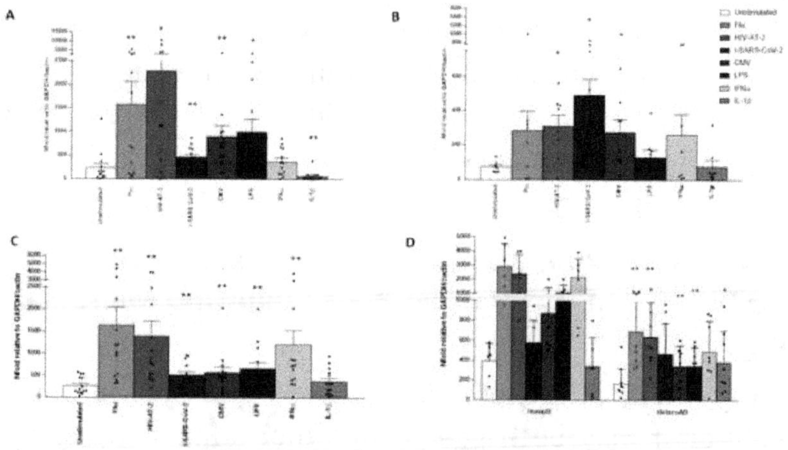

Figure 2. ERAP1, ERAP2/Iso1 and ERAP2/Iso3 mrna expression is increased following microbial stimulation. pbmcs isolated from 8 heteroab and 6 homob individuals were in vitro stimulated with microbial antigens (flu, CMV) inactivated viruses (i-SARS-CoV-2, HIV-AT-2) bacterial by-products (LPS) or inflammatory stimuli (IFNα, IL-1β) for ten h. mRNA expression for ERAP2/Iso3 (A), ERAP2/Iso1 (B) and ERAP1 (C) were assessed by RT-Real-Time PCR. Results are shown as the media of the relative expression units to the glyceraldehyde-3-phosphate dehydrogenase (GAPDH) and β-actin reference genes calculated by the 2-ΔΔCt equation. (D) The microbial-dependent genetic control of ERAP2/Iso3 expression is underlined by the observation that its abundances are nearly doubled in HapB homozygotes compared to heterozygotes. Results are expressed as mean ± ES. * = $p <$ 0.05; ** = $p < 0.01$.

3.3. Gene Expression of Immune Selected Effectors in PBMCs Following Microbial Stimulation

To verify if the increased expression of ERAPs in response to microbial stimulation could be extended to other factors involved in the orchestration of the immune response, mRNA expression of 70 selected effectors was investigated by the innovative Quantigene Plex Gene expression technology. The genes whose mRNA expression was upregulated are implicated in almost all phases of immune response, including: chemokines, cytokines and cytokine receptors, pathogen recognition receptor, inflammasome, cholesterol metabolism, interferon stimulated genes, adhesion molecules, activation/inhibition markers, antigen presentation factors (Figure 3). Notably, the gene expression pattern was partially shared by all the stimuli employed and partially pathogen-specific as summarized in Figure 3. In particular, following i-SARS-CoV-2-stimulation a significantly higher transcription rate was observed for: CCL2 ($p < 0.05$); CCL5 ($p < 0.035$), HMGCS1 ($p < 0.002$), PYCARD ($p < 0.002$), CASP1 ($p < 0.001$), CD44 ($p < 0.016$), CD274 ($p > 0.008$), IL-8 ($p < 0.003$); IL-1β ($p < 0.006$), ABCA1 ($p < 0.02$), IL-6R ($p < 0.004$), CCL3 ($p < 0.01$), IFNγ ($p < 0.05$); TAP1 ($p < 0.05$).

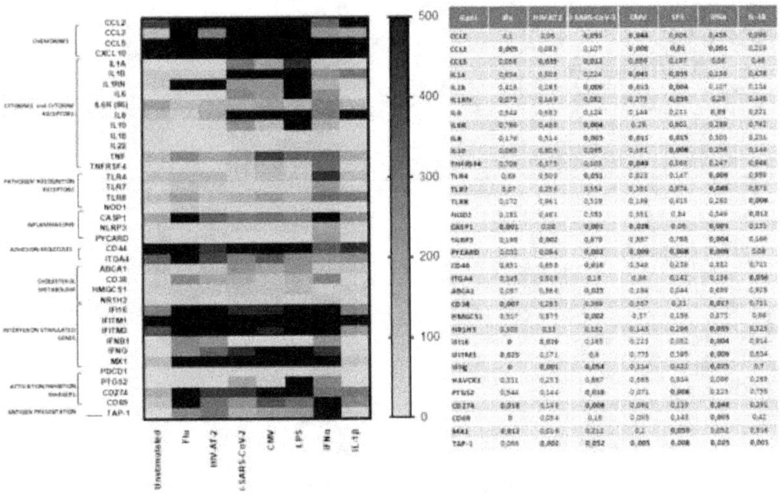

Figure 3. mRNA expression of genes involved in the anti-microbial immune response was modulated in response to different pathogens. Quantigene Plex Gene expression technology was applied to quantify gene expression on PBMCs isolated from 8 HeteroAB and 6 HomoB individuals and stimulated with microbial antigens (flu, CMV) inactivated viruses (i-SARS-CoV-2, HIV-AT-2) bacterial by-products (LPS) or inflammatory stimuli (IFNα, IL-1β). Gene expression (mean values) is shown as a color scale from white to blue (Heatmap). Only statistically significant p values from T-test comparison between unstimulated and stimulated PBMCs are shown in table.

Unlike ERAP2 expression, no differences were observed in mRNA expression levels of all the analyzed genes in PBMCs from HomoB and HeteroAB subjects in response to all the stimuli engaged (data not shown).

3.4. mRNA Expression of ERAPs in In Vitro SARS-CoV-2 Infected Calu3 Cell Lines and HIV-Infected PBMCs

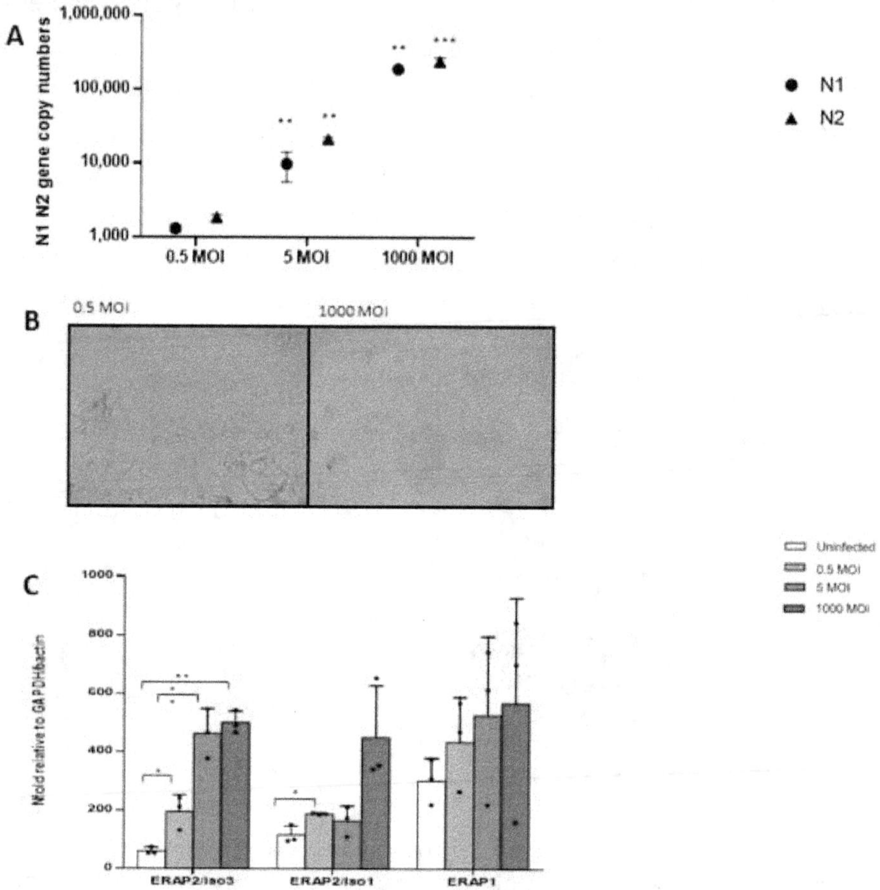

Figure 4. In vitro SARS-CoV-2 infection assay on Calu3 cells. (A) SARS-CoV-2 replication was assessed on Calu3 cells infected at 0.5, 5, and 1000 MOI 48 h post-infection. Viral copy number quantification was performed by generating a standard curve from the quantified 2019-nCoV_N positive plasmid control for Nucleocapsid (N) region 1 and 2, showing a significant increase in response to the increased viral input. (B) SARS-CoV-2-induced cytopathic effects were assessed in Calu3 infected cells. Representative images of SARS-CoV-2 infected-cells at 0.5 and 1000 MOI are reported. At

48 h post-infection, typical cytopathic effects, including cell rounding, detachment, degeneration, and syncytium formation were seen only in cells infected at 1000 MOI. Cells were imaged by optical microscope observation (ZOE™ Fluorescent Cell Imager, Bio-Rad, Hercules, CA, USA). (C) ERAP2/Iso3, Iso1, and ERAP1 mRNA expression by in vitro SARS-CoV-2 infected Calu3 cells increased according to the rising viral input. Mean values ± ES are reported. * = $p < 0.05$; ** = $p < 0.01$; *** = $p < 0.001$. MOI = multiplicity of infection.

To verify whether ERAP2/Iso3 expression varies in response to growing viral concentrations we adopted two in vitro model of infection. Thus, Calu3 cell lines were infected with different SARS-CoV-2 viral input and after 48 hours' viral replication as well as ERAP mRNA expression were assessed. As expected, SARS-CoV-2 replication increased according to the rising viral input as assessed analyzing both N1 (MOI 0.5 vs. 5: $p < 0.01$; MOI 0.5 vs. 1000: $p < 0.001$) and N2 (MOI 0.5 vs. 5: $p < 0.01$; MOI 0.5 vs. 1000: $p < 0.001$) (Figure 4A). Images of cellular cytopathic effect on SARS-CoV-2 infected cells at 48 h showed that despite robust SARS-CoV-2 replication in Calu3 cells, substantial cell death was detected only in cells infected with 1000 MOI (Figure 4B). Notably, this increase was coupled with a progressive rise of ERAP2/Iso3 expression compared to the uninfected condition (MOI 0.5: $p < 0.04$; MOI 5: $p < 0.01$; MOI 1000: $p < 0.01$). Likewise, ERAP2/Iso1 and ERAP1 expression was induced in a viral-dose dependent manner, although statistical significance was observed only for ERAP2/Iso1 (MOI 0.5: $p < 0.05$) (Figure 4C).

In vitro HIV-1 infection of PBMCs isolated from HapB HC produced similar results. Indeed, as the viral input raised, viral replication quantified through p24 concentration analyses at 5 days post infection increased (0.1 vs. 1 ng p24 HIV-1Bal/1 × 106 PBMCs: $p < 0.02$; 0.1 vs. 10 ng p24 HIV-1Bal/1 × 106 PBMCs: $p < 0.001$) (Figure 5A). Likewise, ERAP2/Iso3 gene expression increased compared to the uninfected condition (0.1 ng p24 HIV-1Bal/1 × 106 PBMCs: $p < 0.003$; 0.1 ng p24 HIV-1Bal/1 × 106 PBMCs: $p < 0.04$; 10 ng p24 HIV-1Bal/1 × 106 PBMCs: $p < 0.03$) (Figure 5B). ERAP2/Iso1 (0.1 ng p24 HIV-1Bal/1 × 106 PBMCs: $p < 0.05$) and ERAP1 (0.1 ng p24 HIV-1Bal/1 × 106 PBMCs: $p < 0.05$) mRNA levels showed a similar trend (Figure 5B).

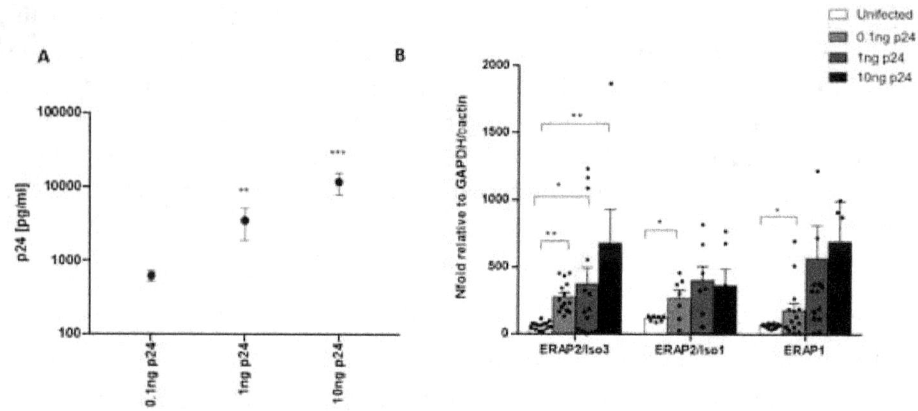

Figure 5. In vitro HIV-1 infection assay on PBMCs. (A) HIV-1 replication was assessed by p24 quantification on PBMCs isolated from 8 HeteroAB and 6 HomoB individuals infected with 0.1, 1, and 10 ng p24 HIV-1Bal/1 × 106 PBMCs 5 days post infection. Results showed a significant increase in viral replication according to the increased viral input. (B) ERAP2/Iso3, Iso1, and ERAP1 mRNA expression by in vitro HIV-1-infected PBMCs increased according to the rising viral input. Mean values ± ES are reported. * = $p < 0.05$; ** = $p < 0.01$; *** = $p < 0.001$.

3.5. ERAP2/Iso3 Protein Production by MDMs from HeteroAB Subjects Following Microbial Specific Stimulation

To verify if the short ERAP2/Iso3 isoform would function as an RNA or is translated into a protein product we performed a western blot assay on MDMs from 3 HeteroAB subjects triggered with different microbial antigens. Remarkably, the antibody which recognizes the full-length ERAP2 (Iso1) was able to detect also one short protein isoform (~50 kDa) in CMV, flu, HIV-AT-2, i-SARS-CoV-2, LPS, IFNα stimulated cells from HeteroAB subjects, suggesting the translation of the short microbial-specific ERAP2/Iso3 (Figure 6). Conversely, following IL-1β stimulation, only ERAP2/Iso1 isoform was detected. The production of ERAP1 protein was observed in all the stimulated conditions except IL-1β (Figure 6). As the proteins extracted from the 3 HeteroAB subjects were sub-pooled for WB analyses, statistical evaluation of the results was not possible.

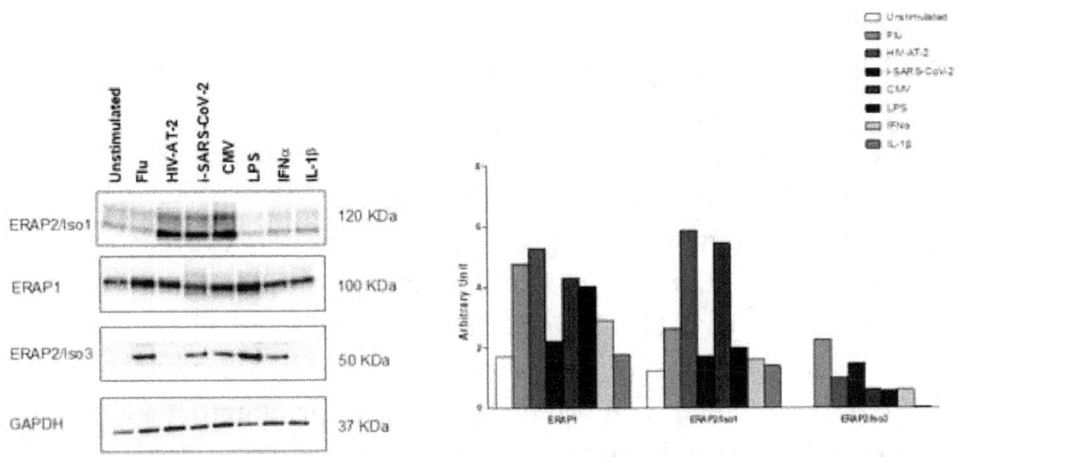

Figure 6. ERAP1, ERAP2/Iso1, and ERAP2/Iso3 production in pathogen stimulated monocyte-derived macrophages (MDMs). MDMs differentiated from 3 HeteroAB participants stimulated for 36 h with microbial antigens (flu, CMV) inactivated viruses (i-SARS-CoV-2, HIV-AT-2) bacterial by-products (LPS) or inflammatory stimuli (IFNα, IL-1β) were tested for protein using primary antibodies specific to a ERAP1 (goat polyclonal), ERAP2 and β-actin. Proteins extracted from the 3 HeteroAB subjects were sub-pooled for WB analyses. Histograms representing ERAP1, ERAP2/Iso1, and ERAP2/Iso3 densitometric quantification. Quantification was performed by Quantity One 4.6.6 software (Bio-Rad) and normalization was permed on GAPDH.

4. DISCUSSION

Given the documented role of ERAP2 in antigen presentation [20] and viral infections [21], we examined the genetic control of ERAP2 transcripts in the human antimicrobial response. In particular, based on the results recently reported by Ye and co-workers [19], we investigated if the expression of the recently characterized ERAP2/Iso3 is flu-specific or if it can be triggered by other microbial stimuli. Our results suggest that: (1) ERAP2/Iso3 mRNA expression is not restricted to flu-infection but it can be prompted by other pathogens including HIV, SARS-CoV-2, CMV, Bacteria (LPS); (2) ERAP2/Iso3 mRNA is translated into a protein following microbial induction; (3) ERAP2/Iso3 mRNA expression is sensitive to viral concentration.

Remarkably, Ye et al. did not detect ERAP2/Iso3 expression in IFNB1 stimulated cells leading to the conclusion that the transcription of novel ERAP2 isoforms is likely initiated by viral sensing pathways upstream of type 1 interferon signaling [19]. Conversely, in our cell culture condition, IFNα-stimulation was able to induce ERAP2/Iso3 expression in a consistent way, suggesting the participation of type 1 interferon cascade to the induction of ERAP2/Iso3 transcription, as already documented for the wild type forms of ERAP1 and 2 [26]. The different cell kinds and IFN subtypes adopted in the two experimental settings could at least partially justify the discrepancies reported in the two studies, but further analyses are needed to clarify the pathways and molecules induced by microbial exposure, directly responsible for ERAP2/Iso3 synthesis. For example, in our study IL-1β stimulus was not able to trigger the expression of any ERAP variants, further strengthening the assumption of a specific response of ERAP gene transcription controlled by pathogen exposure. Furthermore, as IFNα is also strongly associated to various type of I IFN conditions such as Sjogren's disease [27], systemic lupus erythematosus [28] and Scleroderma [29] it would be valuable to verify if ERAP2/Iso3 expression varies in patients affected by these pathologies and/or following administration of IFNα therapy.

The discovery of these new ERAP2 isoforms is of great importance and gives a plausible explanation to the maintenance of one of the major gene expression quantitative trait loci (eQTL) and alternate isoform usage in most tissues and cell types [19,30]. Indeed, until last year, the preservation of HapB at intermediate frequency in human population was almost unexplained, as its transcript was believed to be addressed to NMD [9,10]. The identification of this previously uncharacterized short isoform, ERAP2/Iso3, transcribed from HapB, results in the partial rescue of ERAP2 expression, suggesting its involvement in the anti-microbial response.

The mechanism of action and functional impact of this new genetic variant to host defense; however, is still indefinite. The lack of the aminopeptidase domain proves that it does not directly participate in the shaping of antigenic peptides to be presented to CD8+ T cells. However, in 2005 Saveanu et al. conducted an immunoblot analysis detecting the presence of ERAP1-ERAP2 heterodimers and possibly homodimers of each enzyme [4]. This crystallographic dimer is described as mediated by domains I and II of the enzyme that is missing from ERAP2/Iso3 [31]. However, this observation has been reiterated by Ye et al. who hypothesized that ERAP2 isoforms could have a dominant-negative effect on either ERAP1 and possibly ERAP2 wt, through the formation of hetero- or homodimeric complexes [15]. Despite still speculative the fact that ERAP2/Iso3 may exert such an effect cannot be ruled out: this, in turn, could lead to an altered peptide processing, which could confer an advantage/disadvantage against infections by presenting a more/less immunogenic antigen repertoire. In relation to this aspect, previous studies have demonstrated that the functional skills of ERAP monomers, homo, or heterodimers may significantly differ in terms of both substrate specificity and trimming efficiency [32]. In particular, ERAP1/2 dimerization creates complexes with superior peptide-trimming efficacy and a higher affinity towards ERAP1 preferential substrates. This is allowed by the adoption of a modified physical conformation by ERAP1, caused by its interaction with ERAP2, which mainly works as an enhancer of ERAP1 role upon dimer assembling [33]. Further studies are needed to verify if also ERAP2/Iso3 physical interaction with the wt ERAP variants prompts an allosteric effect able to modify basic enzymatic parameters and to improve their substrate-binding affinity. However, the observation that ERAP2/Iso3 mRNA is translated into a protein allows to speculate on the generation of a new ERAP member which further contributes to enrich the non-redundant, yet a complete and potent system of aminopeptidases, warranting an efficient trimming of various kinds of precursors.

The results obtained in this study definitely establish a link between invading pathogens and ERAP2/Iso3 expression which further strengthens the significance of the results reported by Ye and collaborators. Supporting this hypothesis, we observed that ERAP2/Iso3 expression progressively increased in response to growing doses of viral input in both SARS-CoV-2 and HIV-1 in vitro infection assay, as if the production of this genetic variant, as well as one of the other elements within the ERAP family, were directly dependent on the viral dose of exposure. Additionally, the observation that the increased expression of ERAP2/Iso3 in response to pathogen exposure is accompanied by the modulation of many other determinants (chemokines, cytokines, pathogen recognition receptor, inflammasome, interferon-stimulated genes, adhesion molecules, activation/inhibition markers, antigen presentation elements) orchestrating the anti-microbial immune response, further supports its direct intervention in this defensive pathway. Notably, as its expression is increased by a wide range of microbial stimuli including viral antigens, inactivated virus as well as bacterial by-products, it is possible to assume that ERAP2/Iso3 expression is not pathogen-specific, but it's secondary to the activation of an antimicrobial cascade commonly shared by different pathogens. Meanwhile, we cannot exclude that ERAPs responses observed following microbial stimulations and in vitro viral infections result from an erroneous transcription and translation due to pathogen-induced cellular stress. Indeed, as ERAP1 and ERAP2 expression has been demonstrated to be prompted by IFNγ stimulation, ERAPs production could be secondary to the innate immune response of the cells to infections, rather than to an immune response or an immune evasion mechanism. Further analyses will be necessary to verify this hypothesis.

The involvement of ERAPs in modulating viral infections is widely recognized as recently reviewed in [34]. Several studies, indeed, have demonstrated the intervention of ERAP genetic variants in the life cycle of HCV, flu, CMV, HPV, HIV, and other pathogens at different levels. In particular, studies performed by our research group have established an association between ERAP2/Iso1 and HIV-infection in terms of both susceptibility [17,35] and progression [36]. However, to our knowledge ERAP expression and/or genetic variants have been correlated to the recent coronavirus disease 2019 (COVID-19), provoked by SARS-CoV-2, only by two recent studies [34,37,38]. In the first one by Stamatakis et al. ERAP2 trimming ability has been investigated in SARS-CoV-2 infection together with ERAP1 and IRAP, and it has been proved as the most stable of the enzymes generating optimal length antigenic peptides for HLA binding [38]. In the second one by Lu et al. by examining 193 deaths from 1,412 confirmed infections in a group of 5,871 UK Biobank participants tested for the virus, rs150892504 variant in ERAP2 gene came up as potentially being implicated in risk from SARS-CoV-2 infection. Although rs150892504 variant is not in linkage disequilibrium with rs2248374, this finding suggests the involvement of ERAP2 in the modulation of SARS-CoV-2 infection. Such an assumption is further supported by other intriguing observations. This virus enters the cells by spike protein binding to ACE2 (angiotensin-converting enzyme 2), which is responsible for the conversion of angiotensin I to angiotensin I-9 and of angiotensin II to angiotensin I-7, an effective vasodilator, thus working as a negative regulator of the renin-angiotensin system (RAS) Our results, for the first time demonstrate that SARS-CoV-2 exposure triggers the expression of ERAP1, ERAP2/Iso1 and also the recently detected ERAP2/Iso3 in a dose-dependent mode, suggesting its participation in the control of the anti-SARS-CoV-2 response. This observation is far more important, considering that besides their involvement in the antigen presentation pathway, ERAPs display several key anti-SARS-CoV-2 functions. Indeed, they intervene in the RAS, where ERAP1 efficiently cleaves angiotensin II to angiotensin III and IV, and ERAP2 cuts angiotensin III to angiotensin IV, thus influencing both ACE2 virus receptor bio-availability and blood pressure levels [39]. Furthermore, ERAPs modulate the proteolytic cleavage of IL-6 receptor (IL-6Rα) [40] a function which can improve the clinical conditions in COVID-19 patients, as recently documented following its pharmacological inhibition by Tocilizumab [41]. Last but not least, Ranjit and colleagues recently demonstrated that sex-specific differences in ERAP1 modulation influence blood pressure and RAS responses [42], and a male bias in mortality has emerged in the COVID-19 pandemic since the very beginning. As ERAPs genetic variants have been demonstrated to orchestrate and condition the result of infection of coronavirus in other animal species [43,44] detailed studies investigating SARS-CoV-2-host interplay is absolutely mandatory.

Another query which needs to be addressed in the near future concerns the cellular localization of ERAP2/Iso3 isoform. In particular, it would be interesting to verify if, as with the wt ERAP variants [17], even ERAP2/Iso3 may be secreted into the extracellular milieu following inflammatory stimulation, with which substrates it may interact, which functions may eventually exert in this environment and if its administration can interfere with viral replication.

Considering the key role played by ERAPs in antigen processing and presentation, it is plausible that these aminopeptidases may be potential targets and controllers of the pathogenicity of infectious diseases, shaping the susceptibility and response to microbial infections. The recent acquisition of ERAP intervention even in the modulation of innate immunity further reinforces this assumption.

Given the growing number of viral epidemics, the identification of molecular mechanisms driven by factors such as ERAPs that can interfere, control or modulate viral replication is unequivocally needed as they could be widely exploited for the inception of future, still unknown viral infections.

CONCISE SUMMARY

After being infected with influenza, the genetic variant rs2248374-G is present. Cells that express ERAP2 have the ability to transcribe an alternate spliced isoform known as ERAP2/Iso3. This mutant, in contrast to ERAP2-wt, lacks the ability to cleave peptides for loading onto MHC class I molecules. However, it is still capable of forming dimers with both ERAP2-wt and ERAP1-wt, so aiding in the characterization of an alternative cellular immune-peptidome. To determine if the expression of ERAP2/Iso3 can be triggered by other pathogens, peripheral blood mononuclear cells (PBMCs) and monocyte-derived macrophages (MDMs) obtained from 20 healthy individuals were exposed to influenza virus (flu), lipopolysaccharide (LPS), cytomegalovirus (CMV), human immunodeficiency virus (HIV-AT-2), and severe acute respiratory syndrome coronavirus 2 (SARS-CoV-2) antigens. The mRNA and protein expression of ERAP2/Iso3 were then analyzed. Simultaneously,

Calu3 cell lines and PBMCs were infected in a laboratory setting with increasing amounts of SARS-CoV-2 (0.5, 5, 1000 MOI) and HIV-1BAL (0.1, 1, and 10 ng p24 HIV-1Bal/1 × 106 PBMCs) viruses, respectively. The results indicate that the expression of ERAP2/Iso3 mRNA can be induced by various pathogens and is associated with the regulation of several factors (cytokines, interferon-stimulated genes, activation/inhibition markers, antigen-presentation elements) that coordinate the immune response against microbial infections (Quantigene). Additionally, the ERAP2/Iso3 mRNA is translated into a protein (western blot). Furthermore, the expression of ERAP2/Iso3 mRNA is influenced by the concentration of SARS-CoV-2 and HIV-1. Given the significant role that ERAPs play in the processing and presentation of antigens, it is possible that these enzymes could serve as targets and regulators of the severity of infectious diseases. Further investigations are necessary to determine the specific role played by each isoform.

REFERENCES

1. Neefjes, J.; Jongsma, M.L.M.; Paul, P.; Bakke, O. Towards a systems understanding of MHC class I and MHC class II antigen presentation. Nat. Rev. Immunol. 2011, 11, 823–836.

2. Cifaldi, L.; Romania, P.; Lorenzi, S.; Locatelli, F.; Fruci, D. Role of Endoplasmic Reticulum Aminopeptidases in Health and Disease: From Infection to Cancer. Int. J. Mol. Sci. 2012, 13, 8338–8352.

3. Compagnone, M.; Fruci, D. Peptide Trimming for MHC Class I Presentation by Endoplasmic Reticulum Aminopeptidases. Methods Mol. Biol. Clifton NJ 2019, 1988, 45–57.

4. Saveanu, L.; Carroll, O.; Lindo, V.; Del Val, M.; Lopez, D.; Lepelletier, Y.; Greer, F.; Schomburg, L.; Fruci, D.; Niedermann, G.; et al. Concerted peptide trimming by human ERAP1 and ERAP2 aminopeptidase complexes in the endoplasmic reticulum. Nat. Immunol. 2005, 6, 689–697.

5. Evnouchidou, I.; van Endert, P. Peptide trimming by endoplasmic reticulum aminopeptidases: Role of MHC class I binding and ERAP dimerization. Hum. Immunol. 2019, 80, 290–295.

6. Yao, Y.; Liu, N.; Zhou, Z.; Shi, L. Influence of ERAP1 and ERAP2 gene polymorphisms on disease susceptibility in different populations. Hum. Immunol. 2019, 80, 325–334.

7. Stamogiannos, A.; Koumantou, D.; Papakyriakou, A.; Stratikos, E. Effects of polymorphic variation on the mechanism of Endoplasmic Reticulum Aminopeptidase 1. Mol. Immunol. 2015, 67, 426–435.

8. López de Castro, J.A. How ERAP1 and ERAP2 Shape the Peptidomes of Disease-Associated MHC-I Proteins. Front. Immunol. 2018, 9.

9. Cagliani, R.; Riva, S.; Biasin, M.; Fumagalli, M.; Pozzoli, U.; Lo Caputo, S.; Mazzotta, F.; Piacentini, L.; Bresolin, N.; Clerici, M.; et al. Genetic diversity at endoplasmic reticulum aminopeptidases is maintained by balancing selection and is associated with natural resistance to HIV-1 infection. Hum. Mol. Genet. 2010, 19, 4705–4714.

10. Andrés, A.M.; Dennis, M.Y.; Kretzschmar, W.W.; Cannons, J.L.; Lee-Lin, S.-Q.; Hurle, B.; Schwartzberg, P.L.; Williamson, S.H.; Bustamante, C.D.; Nielsen, R.; et al. Balancing Selection Maintains a Form of ERAP2 that Undergoes Nonsense-Mediated Decay and Affects Antigen Presentation. PLoS Genet. 2010, 6.

11. Jostins, L.; Ripke, S.; Weersma, R.K.; Duerr, R.H.; McGovern, D.P.; Hui, K.Y.; Lee, J.C.; Philip Schumm, L.; Sharma, Y.; Anderson, C.A.; et al. Host–microbe interactions have shaped the genetic architecture of inflammatory bowel disease. Nature 2012, 491, 119–124.

12. Kuiper, J.J.W.; Van Setten, J.; Ripke, S.; Van 'T Slot, R.; Mulder, F.; Missotten, T.; Baarsma, G.S.; Francioli, L.C.; Pulit, S.L.; De Kovel, C.G.F.; et al. A genome-wide association study identifies a functional ERAP2 haplotype associated with birdshot chorioretinopathy. Hum. Mol. Genet. 2014, 23, 6081–6087.

13. Wiśniewski, A.; Kasprzyk, S.; Majorczyk, E.; Nowak, I.; Wilczyńska, K.; Chlebicki, A.; Zoń-Giebel, A.; Kuśnierczyk, P. ERAP1-ERAP2 haplotypes are associated with ankylosing spondylitis in Polish patients. Hum. Immunol. 2019, 80, 339–343.

14. Robinson, P.C.; Costello, M.E.; Leo, P.; Bradbury, L.A.; Hollis, K.; Cortes, A.; Lee, S.; Joo, K.B.; Shim, S.-C.; Weisman, M.; et al. ERAP2

is associated with ankylosing spondylitis in HLA-B27-positive and HLA-B27-negative patients. Ann. Rheum. Dis. 2015, 74, 1627–1629.

15. Chiaroni-Clarke, R.C.; Munro, J.E.; Chavez, R.A.; Pezic, A.; Allen, R.C.; Akikusa, J.D.; Piper, S.E.; Saffery, R.; Ponsonby, A.-L.; Ellis, J.A. Independent confirmation of juvenile idiopathic arthritis genetic risk loci previously identified by immunochip array analysis. Pediatr. Rheumatol. 2014, 12, 53.

16. Biasin, M.; Sironi, M.; Saulle, I.; de Luca, M.; la Rosa, F.; Cagliani, R.; Forni, D.; Agliardi, C.; lo Caputo, S.; Mazzotta, F.; et al. Endoplasmic reticulum aminopeptidase 2 haplotypes play a role in modulating susceptibility to HIV infection. AIDS Lond. Engl. 2013, 27, 1697–1706.

17. Saulle, I.; Ibba, S.V.; Torretta, E.; Vittori, C.; Fenizia, C.; Piancone, F.; Minisci, D.; Lori, E.M.; Trabattoni, D.; Gelfi, C.; et al. Endoplasmic Reticulum Associated Aminopeptidase 2 (ERAP2) Is Released in the Secretome of Activated MDMs and Reduces in vitro HIV-1 Infection. Front. Immunol. 2019, 10.

18. Evnouchidou, I.; Birtley, J.; Seregin, S.; Papakyriakou, A.; Zervoudi, E.; Samiotaki, M.; Panayotou, G.; Giastas, P.; Petrakis, O.; Georgiadis, D.; et al. A Common Single Nucleotide Polymorphism in Endoplasmic Reticulum Aminopeptidase 2 Induces a Specificity Switch That Leads to Altered Antigen Processing. J. Immunol. 2012.

19. Ye, C.J.; Chen, J.; Villani, A.-C.; Gate, R.E.; Subramaniam, M.; Bhangale, T.; Lee, M.N.; Raj, T.; Raychowdhury, R.; Li, W.; et al. Genetic analysis of isoform usage in the human anti-viral response reveals influenza-specific regulation of ERAP2 transcripts under balancing selection. Genome Res. 2018, 28, 1812–1825.

20. Saveanu, L.; Carroll, O.; Hassainya, Y.; Endert, P.V. Complexity, contradictions, and conundrums: Studying post-proteasomal proteolysis in HLA class I antigen presentation. Immunol. Rev. 2005, 207, 42–59.

21. Rossio, J.L.; Esser, M.T.; Suryanarayana, K.; Schneider, D.K.; Bess, J.W.; Vasquez, G.M.; Wiltrout, T.A.; Chertova, E.; Grimes, M.K.;

Sattentau, Q.; et al. Inactivation of Human Immunodeficiency Virus Type 1 Infectivity with Preservation of Conformational and Functional Integrity of Virion Surface Proteins. J. Virol. 1998, 72, 7992–8001.

22. Hui, K.P.Y.; Cheung, M.-C.; Perera, R.A.P.M.; Ng, K.-C.; Bui, C.H.T.; Ho, J.C.W.; Ng, M.M.T.; Kuok, D.I.T.; Shih, K.C.; Tsao, S.-W.; et al. Tropism, replication competence, and innate immune responses of the coronavirus SARS-CoV-2 in human respiratory tract and conjunctiva: An analysis in ex-vivo and in-vitro cultures. Lancet Respir. Med. 2020, 8, 687–695.
23. Batéjat, C.; Grassin, Q.; Manuguerra, J.-C.; Leclercq, I. Heat inactivation of the Severe Acute Respiratory Syndrome Coronavirus 2. bioRxiv 2020.
24. Merlini, E.; Tincati, C.; Biasin, M.; Saulle, I.; Cazzaniga, F.A.; d'Arminio Monforte, A.; Cappione, A.J.I.; Snyder-Cappione, J.; Clerici, M.; Marchetti, G.C. Stimulation of PBMC and Monocyte-Derived Macrophages via Toll-Like Receptor Activates Innate Immune Pathways in HIV-Infected Patients on Virally Suppressive Combination Antiretroviral Therapy. Front. Immunol. 2016, 7.
25. Saulle, I.; Ibba, S.V.; Vittori, C.; Fenizia, C.; Mercurio, V.; Vichi, F.; Caputo, S.L.; Trabattoni, D.; Clerici, M.; Biasin, M. Sterol metabolism modulates susceptibility to HIV-1 Infection. AIDS Lond. Engl. 2020.
26. Wu, T.G.; Rose, W.A.; Albrecht, T.B.; Knutson, E.P.; König, R.; Perdigão, J.R.; Nguyen, A.P.A.; Fleischmann, W.R. IFN-alpha-induced murine B16 melanoma cancer vaccine cells: Induction and accumulation of cell-associated IL-15. J. Interf. Cytokine Res. Off. J. Int. Soc. Interf. Cytokine Res. 2007, 27, 13–22.
27. Nordmark, G.; Ronnblom, M.-L.E.; Primary, L. Sjogren's Syndrome and the Type I Interferon System. Available online: https://www.eurekaselect.com/101237/article (accessed on 15 August 2020).
28. Kirou, K.A.; Gkrouzman, E. Anti-interferon alpha treatment in SLE. Clin. Immunol. 2013, 148, 303–312.

29. Raschi, E.; Chighizola, C.B.; Cesana, L.; Privitera, D.; Ingegnoli, F.; Mastaglio, C.; Meroni, P.L.; Borghi, M.O. Immune complexes containing scleroderma-specific autoantibodies induce a profibrotic and proinflammatory phenotype in skin fibroblasts. Arthritis Res. Ther. 2018, 20, 187.

30. Lappalainen, T.; Sammeth, M.; Friedländer, M.R.; Ac't Hoen, P.; Monlong, J.; Rivas, M.A.; Gonzàlez-Porta, M.; Kurbatova, N.; Griebel, T.; Ferreira, P.G.; et al. Transcriptome and genome sequencing uncovers functional variation in humans. Nature 2013, 501, 506–511.

31. The Crystal Structure of Human Endoplasmic Reticulum Aminopeptidase 2 Reveals the Atomic Basis for Distinct Roles in Antigen Processing | Biochemistry. Available online: https://pubs.acs.org/doi/10.1021/bi201230p (accessed on 17 August 2020).

32. de Castro, J.A.L.; Stratikos, E. Intracellular antigen processing by ERAP2: Molecular mechanism and roles in health and disease. Hum. Immunol. 2019, 80, 310–317.

33. Evnouchidou, I.; Weimershaus, M.; Saveanu, L.; Endert, P. van ERAP1–ERAP2 Dimerization Increases Peptide-Trimming Efficiency. J. Immunol. 2014.

34. Saulle, I.; Vicentini, C.; Clerici, M.; Biasin, M. An Overview on ERAP Roles in Infectious Diseases. Cells 2020, 9, 720.

35. Forni, D.; Cagliani, R.; Tresoldi, C.; Pozzoli, U.; Gioia, L.D.; Filippi, G.; Riva, S.; Menozzi, G.; Colleoni, M.; Biasin, M.; et al. An Evolutionary Analysis of Antigen Processing and Presentation across Different Timescales Reveals Pervasive Selection. PLoS Genet. 2014, 10, e1004189.

36. Lori, E.M.; Cozzi-Lepri, A.; Tavelli, A.; Mercurio, V.; Ibba, S.V.; Lo Caputo, S.; Castelli, F.; Castagna, A.; Gori, A.; Marchetti, G.; et al. Evaluation of the effect of protective genetic variants on cART success in HIV-1-infected patients. J. Biol. Regul. Homeost. Agents 2020, 34.

37. Genetic Risk Factors for Death with SARS-CoV-2 from the UK Biobank | medRxiv. Available online: https://www.medrxiv.org/content/10.1101/2020.07.01.20144592v1 (accessed on 15 August 2020).

38. Stamatakis, G.; Samiotaki, M.; Mpakali, A.; Panayotou, G.; Stratikos, E. Generation of SARS-CoV-2 S1 spike glycoprotein putative antigenic epitopes in vitro by intracellular aminopeptidases. bioRxiv 2020.

39. Hisatsune, C.; Ebisui, E.; Usui, M.; Ogawa, N.; Suzuki, A.; Mataga, N.; Takahashi-Iwanaga, H.; Mikoshiba, K. ERp44 Exerts Redox-Dependent Control of Blood Pressure at the ER. Mol. Cell 2015, 58, 1015–1027.

40. Cui, X.; Rouhani, F.N.; Hawari, F.; Levine, S.J. An Aminopeptidase, ARTS-1, Is Required for Interleukin-6 Receptor Shedding. J. Biol. Chem. 2003, 278, 28677–28685.

41. Michot, J.-M.; Albiges, L.; Chaput, N.; Saada, V.; Pommeret, F.; Griscelli, F.; Balleyguier, C.; Besse, B.; Marabelle, A.; Netzer, F.; et al. Tocilizumab, an anti-IL-6 receptor antibody, to treat COVID-19-related respiratory failure: A case report. Ann. Oncol. Off. J. Eur. Soc. Med. Oncol. 2020, 31, 961–964.

42. Ranjit, S.; Wong, J.Y.; Tan, J.W.; Sin Tay, C.; Lee, J.M.; Yin Han Wong, K.; Pojoga, L.H.; Brooks, D.L.; Garza, A.E.; Maris, S.A.; et al. Sex-specific differences in endoplasmic reticulum aminopeptidase 1 modulation influence blood pressure and renin-angiotensin system responses. JCI Insight 2019, 4.

43. Golovko, L.; Lyons, L.A.; Liu, H.; Sørensen, A.; Wehnert, S.; Pedersen, N.C. Genetic susceptibility to feline infectious peritonitis in Birman cats. Virus Res. 2013, 175, 58–63.

44. Cong, F.; Liu, X.; Han, Z.; Shao, Y.; Kong, X.; Liu, S. Transcriptome analysis of chicken kidney tissues following coronavirus avian infectious bronchitis virus infection. BMC Genom. 2013, 14, 743.

www.ingramcontent.com/pod-product-compliance
Lightning Source LLC
LaVergne TN
LVHW011907200926
848484LV00002B/555